T0310297

STATISTICAL INFERENCE

STATISTICAL INFERENCE

A Short Course

MICHAEL J. PANIK
University of Hartford
West Hartford, CT

WILEY
A JOHN WILEY & SONS, INC., PUBLICATION

Published by John Wiley & Sons, Inc., Hoboken, New Jersey
Published simultaneously in Canada

Library of Congress Cataloging-in-Publication Data:

Panik, Michael J.
Statistical inference : a short course / Michael J. Panik.
p. cm.
Includes index.
ISBN 978-1-118-22940-8 (cloth)
1. Mathematical statistics–Testbooks. I. Title.
QA276.12.P36 2011
519.5–dc23
2011047632

Printed in the United States of America

ISBN: 9781118229408

10 9 8 7 6 5 4 3 2 1

To the memory of
Richard S. Martin

CONTENTS

PREFACE

Statistical Inference: A Short Course is a condensed and to-the-point presentation of the essentials of basic statistics for those seeking to acquire a working knowledge of statistical concepts, measures, and procedures. While most individuals will not be performing high-powered statistical analyses in their work or professional environments, they will be, on numerous occasions, reading technical reports, reviewing a consultant's findings, perusing through academic, trade, and professional publications in their field, and digesting the contents of diverse magazine/newspaper articles (online or otherwise) wherein facts and figures are offered for appraisal. Let us face it—there is no escape. We are a society that generates a virtual avalanche of information on a daily basis.

That said, correctly understanding notions such as: a *research hypothesis*, *statistical significance*, *randomness*, *central tendency*, *variability*, *reliability*, *cause and effect*, and so on are of paramount importance when it comes to being an informed consumer of statistical results. Answers to questions such as:

"How precisely has some population value (e.g., the mean) been estimated?"
"What level of reliability is associated with any such estimate?"
"How are probabilities determined?"
"Is probability the same thing as odds?"
"How can I predict the level of one variable from the level of another variable?"
"What is the strength of the relationship between two variables?"

and so on, will be offered and explained.

Statistical Inference: A Short Course is general in nature and is appropriate for undergraduates majoring in the natural sciences, the social sciences, or in business. It can also be used in first-year graduate courses in these areas. This text offers what can be considered as "just enough" material for a one-semester course without overwhelming the student with "too fast a pace" or "too many" topics. The essentials of the course appear in the main body of the chapters and interesting "extras" (some might call them "essentials") are found in the chapter appendices and chapter exercises. While Chapters 1–10 are fundamental to any basic statistics course, the instructor can "pick and choose" items from Chapters 11–13. This latter set of chapters is optional and the topics therein can be selected with an eye toward student interest and need.

This text is highly readable, presumes only a knowledge of high school algebra, and maintains a high degree of rigor and statistical as well as mathematical integrity in the presentation. Precise and complete definitions of key concepts are offered throughout and numerous example problems appear in each chapter. Solutions to all the exercises are provided, with the exercises themselves designed to test the student's mastery of the material rather than to entertain the instructor.

While all beginning statistics texts discuss the concepts of simple random sampling and normality, this book takes such discussions a bit further. Specifically, a couple of the key assumptions typically made in the areas of estimation and testing are that we have a "random sample" of observations drawn from a "normal population." However, given a particular data set, how can we determine if it actually constitutes a random sample and, secondly, how can we determine if the parent population can be taken to be normal? That is, can we proceed "as if" the sample is random? And can we operate "as if" the population is normal? Answers to these questions will be provided by a couple of formal test procedures for randomness and for the assessment of normality. Other topics not usually found in introductory texts include determining a confidence interval for a population median, ratio estimation (a technique akin to estimating a population proportion), general discussions of randomness and causality, and some nonparametric methods that serve as an alternative to parametric routines when the latter are not strictly applicable. As stated earlier, the instructor can pick and choose from among them or decide to bypass them altogether.

Looking to specifics:

Chapter 1 (The Nature of Statistics): Defines the subject matter, introduces the concepts of population and sample, and discusses variables, sampling error, and measurement scales.

Chapter 2 (Analyzing Quantitative Data): Introduces tabular and graphical techniques for ungrouped as well as grouped data (frequency distributions and histograms).

Chapter 3 (Descriptive Characteristics of Quantitative Data): Covers basic summary characteristics (mean, median, and so on) along with the weighted mean, the empirical rule, Chebysheff's theorem, *Z*-scores, the coefficient of variation, skewness, quantiles, kurtosis, detection of outliers, the trimmed mean, and boxplots. The appendix introduces descriptive measures for the grouped data case.

Chapter 4 (Essentials of Probability): Reviews set notation and introduces events within the sample place, random variables, probability axioms and corollaries, rules for calculating probabilities, types of probabilities, independent events, sources of probabilities, and the law of large numbers.

Chapter 5 (Discrete Probability Distributions and their properties): Covers discrete random variables, the probability mass and cumulative distribution functions, expectation and variance, permutations and combinations, and the Bernoulli and binomial distributions.

Chapter 6 (The Normal Distribution): Introduces continuous random variables, probability density functions, empirical rule, standard normal variables, and percentiles. The appendix covers the normal approximation to binomial probabilities.

Chapter 7 (Simple Random Sampling and the Sampling Distribution of the Mean): Covers simple random sampling, the concept of a point estimator, sampling error, the sampling distribution of the mean, standard error of the mean, standardized sample mean, and a central limit theorem. Appendices house the use of a table of random numbers, systematic random sampling, assessing normality via a normal probability plot, and provide an extended discussion on the concepts of randomness, risk, and uncertainty.

Chapter 8 (Confidence Interval Estimation of μ): Presents the error bound concept, degree of precision, confidence probability, confidence statements and confidence coefficients, reliability, the t distribution, confidence limits for the population mean using the standard normal and t distributions, and sample size requirements. Order statistics, and a confidence interval for the median are treated in the appendix.

Chapter 9 (The Sampling Distribution of a Proportion and Its Confidence Interval Estimation): Looks at the sampling distribution of a sample proportion and its standard error, the standardized observed relative frequency of a success, error bound, a confidence interval for the population proportion, degree of precision, reliability, and sample size requirements. The appendix introduces ratio estimation.

Chapter 10 (Testing Statistical Hypotheses): Covers the notion of a statistical hypothesis, null and alternative hypotheses, types of errors, test statistics, critical region, level of significance, types of tests, decision rules, hypothesis tests for the population mean, statistical significance, research hypothesis, p-values, and hypothesis tests for the population proportion. Assessing randomness, a runs test, parametric versus nonparametric tests, the Wilcoxon signed rank test, and the Lilliefors' test for normality appear in appendices.

Chapter 11 (Comparing Two Population Means and Two Population Proportions): Considers confidence intervals and hypothesis tests for the difference of means when sampling from two independent normal populations, confidence intervals, and hypothesis tests for the difference of means when sampling from dependent populations, and confidence intervals and hypothesis tests for the difference of proportions when sampling from two independent binomial populations. Appendices introduce a runs test for two independent populations, and the Wilcoxon signed rank test when sampling from two dependent populations.

Chapter 12 (Bivariate Regression and Correlation): Covers scatter diagrams, linear relationships, a statistical equation versus a strict mathematical equation, population and sample regression equations, random error term, the principle of least squares, least squares normal equations, the Gauss–Markov theorem, the partitioned sum of squares table, the coefficient of determination, confidence intervals and hypothesis tests for the population regression intercept and slope, predicting the average value of Y given X and the confidence band, predicting a particular value of Y given X and prediction limits, correlation, and inferences about the population correlation coefficient. Assessing normality via regression analysis and a discussion of the notion of cause and effect are treated in appendices.

Chapter 13 (An Assortment of Additional Statistical Tests): Introduces the concept of a distributional hypothesis, the multinomial distribution, Pearson's goodness-of-fit test, the chi-square distribution, testing independence, contingency tables, testing k proportions, Cramer's measure of strength of association, a confidence interval for a population variance, the F distribution, and the application of the F statistic to regression analysis.

While the bulk of this text was developed from class notes used in courses offered at the University of Hartford, West Hartford, CT, the final draft of the manuscript was written while the author was Visiting Professor of Mathematics at Trinity College, Hartford, CT. Sincere thanks go to my colleagues Bharat Kolluri, Rao Singamsetti, Frank DelloIacono, and Jim Peta at the University of Hartford for their support and encouragement and to David Cruz-Uribe and Mary Sandoval of Trinity College for the opportunity to teach and to participate in the activities of the Mathematics Department.

A special note of thanks goes to Alice Schoenrock for her steadfast typing of the various iterations of the manuscript and for monitoring the activities involved in obtaining a complete draft of the same. I am also grateful to Mustafa Atalay for drawing most of the illustrations and for sharing his technical expertise in graphical design.

An additional offering of appreciation goes to Susanne Steitz-Filler, Editor, Mathematics and Statistics, at John Wiley & Sons for her professionalism and vision concerning this project.

Michael J. Panik

Windsor, CT

1

THE NATURE OF STATISTICS

1.1 STATISTICS DEFINED

Broadly defined, *statistics* involves the theory and methods of

collecting,
organizing,
presenting,
analyzing, and
interpreting

data so as to determine their essential characteristics. While some discussion will be devoted to the collection, organization, and presentation of data, we shall, for the most part, concentrate on the analysis of data and the interpretation of the results of our analysis.

How should the notion of *data* be viewed? It can be thought of as simply consisting of "information" that can take a variety of forms. For example, data can be *numerical* (test scores, weights, lengths, elapsed time in minutes, etc.) or *non-numerical* (such as an attribute involving color or texture or a category depicting the sex of an individual or their political affiliation, if any, etc.) (See Section 1.4 of this chapter for a more detailed discussion of data forms or varieties.)

Statistical Inference: A Short Course, First Edition. Michael J. Panik.

Two major types of statistics will be recognized: (1) *descriptive*; and (2) *inductive*[1] or *inferential.*

Descriptive Statistics: Deals with summarizing data. Our goal here is to arrange data in a readable form. To this end, we can construct tables, charts, and graphs; we can also calculate percentages, rates of change, and so on. We simply offer a picture of "what is" or "what has transpired."

Inductive Statistics: Employs the notion of *statistical inference*, that is, inferring something about the entire data set from an examination of only a portion of the data set. How is this inferential process carried out? Through *sampling*—a representative group of items is subject to study and the conclusions derived therefrom are assumed to characterize the entire data set. Keep in mind, however, that since we are only sampling and not undertaking an exhaustive census of the entire data set, some "margin of error" associated with our inference will most assuredly emerge. Hence, our sample result must be accompanied by a measure of the uncertainty of the inference made. Questions such as "How reliable is our result?" or, "What is the level of confidence associated with our result?" must be addressed before presenting our findings. This is why inferential statistics is often referred to as "decision making under uncertainty." Clearly inferential statistics enables us to go beyond a purely descriptive treatment of data—it enables us to make estimates, forecasts or predictions, and generalizations.

In sum, if we want to only summarize or present data or just catalog facts then descriptive techniques are called for. But if we want to make inferences about the entire data set on the basis of sample information or, more generally, make decisions in the face of uncertainty then the use of inductive or inferential techniques is warranted.

1.2 THE POPULATION AND THE SAMPLE

The concept of the "entire data set" alluded to above will be called the *population*; it is the group to be studied. (Remember that "population" does not refer exclusively to "people;" it can be a group of states, countries, cities, registered democrats, cars in a parking lot, students at a particular academic institution, and so on.) We shall let N denote the population size or the number of elements in the population.

Each separate characteristic of an element in the population will be represented by a variable (usually denoted as X). We may think of a *variable* as describing any qualitative or quantitative aspect of a member of the population. A *qualitative* variable has values that are only "observed." Here a characteristic pertains to some

[1] *Induction* is a process of reasoning from the specific to the general.

attribute (such as color) or category (male or female). A *quantitative* variable will be classified as either *discrete* (it takes on a finite or countable number of values) or *continuous* (it assumes an infinite or uncountable number of values). Hence, discrete values are "counted;" continuous values are "measured." For instance, a discrete variable might be the number of blue cars in a parking lot, the number of shoppers passing through a supermarket check-out counter over a 15 min time interval, or the number of sophomores in a college-level statistics course. A continuous variable can describe weight, length, the amount of water passing through a culvert during a thunderstorm, elapsed time in a race, and so on.

While a population can consist of all conceivable observations on some variable X, we may view a *sample* as a subset of the population. The sample size will be denoted as n, with $n < N$. It was mentioned above that, in order to make a legitimate inference about a population, a *representative sample* was needed. Think of a representative sample as a "typical" sample—it should adequately reflect the attributes or characteristics of the population.

1.3 SELECTING A SAMPLE FROM A POPULATION

While there are many different ways of constructing a sampling plan, our attention will be focused on the notion of simple random sampling. Specifically, a sample of size n drawn from a population of size N is obtained via *simple random sampling* if every possible sample of size n has an equal chance of being selected. A sample obtained in this fashion is then termed a *simple random sample*; each element in the population has the same chance of being included in a simple random sample.

Before any sampling is actually undertaken, a list of items (called the *sampling frame*) in the population is formed and thus serves as the formal source of the sample, with the individual items listed on the frame termed *elementary sampling units*. So, given the sampling frame, the actual process of random sample selection will be accomplished *without replacement*, that is, once an item from the population has been selected for inclusion in the sample, it is not eligible for selection again—it is not returned to the population pool (it is, so to speak, "crossed off" the frame) and consequently cannot be chosen, say, a second time as the simple random sampling process commences. (Under *sampling with replacement*, the item chosen is returned to the population before the next selection is made.)

Will the process of random sampling guarantee that a representative sample will be acquired? The answer is, "probably." That is, while randomization does not absolutely guarantee representativeness (since random sampling gives the same chance of selection to every sample—representative ones as well as nonrepresentative ones), we are highly likely but not certain to get a representative sample. Then why all the fuss about random sampling? The answer to this question hinges upon the fact that it is possible to make erroneous inferences from sample data. (After all, we are not examining the entire population.) Under simple random sampling, we can validly apply the rules of probability theory to calculate the chances or magnitudes of such errors; and their rates enable us to assess the reliability of, or form a degree of confidence in, our inferences about the population.

Let us recognize two basic types of errors that can creep into our data analysis. The first is *sampling error*, which is reflective of the inherent natural variation between samples (since different samples possess different sample values); it arises because sampling gives incomplete information about a population. This type of error is inescapable—it is always present. If one engages in sampling then sampling error is a fact of life. The other variety of error is *nonsampling* error—human or mechanical factors tend to distort the observed values. Nonsampling error can be controlled since it arises essentially from unsound experimental techniques or from obtaining and recording information. Examples of nonsampling error can range from using poorly calibrated or inadequate measuring devices to inaccurate responses (or nonresponses) to questions on a survey form. In fact, even poorly worded questions can lead to such errors. And if preference is given to selecting some observations over others so that, for example, the underrepresentation of some group of individuals or items occurs, then a *biased* sample results.

1.4 MEASUREMENT SCALES

We previously referred to *data*[2] as "information," that is, as a collection of facts, values, or observations. Suppose then that our data set consists of observations that can be "measured" (e.g., classified, ordered, or quantified). At what level does the measurement take place? In particular, what are the "forms" in which data are found or the "scales" on which data are measured? These scales, offered in terms of increasing information content, are classified as nominal, ordinal, interval, and ratio.

1. *Nominal Scale*: Nominal should be associated with the word "name" since this scale identifies categories. Observations on a nominal scale possess neither numerical value nor order. A variable whose values appear on a nominal scale is termed qualitative or categorical. For example, a variable X depicting the sex of an individual (male or female) is nominal in nature as are variables depicting religion, political affiliation, occupation, marital status, color, and so on. Clearly, nominal values cannot be ranked or ordered—all items are treated equally. The only valid operations for variables treated on a nominal scale are the determination of "$=$" or "$\neq$." For nominal data, any statistical analysis is limited and usually relegated to the calculation of percentages.
2. *Ordinal Scale*: (think of the word "order") Includes all properties of the nominal scale with the additional property that the observations can be ranked from the "least important" to the "most important." For instance, hierarchical

[2] "Data" is a plural noun; "datum" is the singular of data.

organizations within which some members are more important or ranked higher than others have observations that are considered to be ordinal since a "pecking order" can be established. For example, military organizations exhibit a well-defined hierarchy (although it is "better" to be a colonel than a private, the ranking does not indicate "how much better"). Other examples are as follows:

Performance Ratings	Corporate Hierarchy
Excellent	President
Very good	Senior vice president
Good	Vice president
Fair	Assistant vice president
Poor	Senior manager
	Manager

The only valid operations for ordinally scaled variables are "$=, \neq, <, >$." Both nominal and ordinal scales are *nonmetric scales* since differences among their values are meaningless.

3. *Interval Scale*: Includes all the properties of the ordinal scale with the additional property that the distance between observations is meaningful; the numbers assigned to the observations indicate order and possess the property that the difference between any two consecutive values is the same as the difference between any other two consecutive values. Hence, the difference $3 - 2 = 1$ has the same meaning as $5 - 4 = 1$. While an interval scale has a zero point, its location may be arbitrary so that ratios of interval scale values have no meaning. For instance, 0°C does not imply the absence of heat (it is simply the temperature at which water freezes); or 60°C is not twice as hot as 30°C. Also, a score of zero on a standardized test does not imply a lack of knowledge; and a student with a score of 400 is not four times as smart as a student who scored 100. The operations for handling variables measured on an interval scale are "$=, \neq, <, >, +, -$."
4. *Ratio Scale*: Includes all the properties of the interval scale with the added property that ratios of observations are meaningful. This is because "absolute zero is uniquely defined." In this regard, if a variable X is measured in dollars ($), then $0 represents the "absence of monetary value;" and a price of $20 is twice as costly as a price of $10 (the ratio is $2/1 = 2$). Other examples of ratio scale measurements are as follows: weight, height, age, GPA, income, and so on. Valid operations for variables measured on a ratio scale are "$=, \neq, <, >, +, -, \times, \div$."

Both interval and ratio scales are said to be *metric scales* since differences between values measured on these scales are meaningful; and variables measured on these scales are said to be *quantitative variables*.

It should be evident from the preceding discussion that any variable measured on one scale automatically satisfies all the properties of a less informative scale.

TABLE 1.1 Characteristics of the Residence at 401 Elm Street

Characteristic	Variable	Observation Values
Number of families	X_1	One family (1) or two family (2)
Attached garage	X_2	Yes (1) or no (0)
Number of rooms	X_3	6
Number of bathrooms	X_4	2
Square feet of living space	X_5	2100
Assessed value	X_6	\$230,500
Year constructed	X_7	1987
Lot size (square feet)	X_8	2400

Example 1.1

Suppose our objective is to study the residential housing stock of a particular city. Suppose further that our inquiry is to be limited to one- and two-family dwellings. These categories of dwellings make up the *target population*—the population about which information is desired. How do we obtain data on these housing units? Should we simply stroll around the city looking for one- and two-family housing units? Obviously this would be grossly inefficient. Instead, we will consult the City Directory. This directory is the *sampled population* (or sampling frame)—the population from which the sample is actually obtained. Now, if the City Directory is kept up to date then we have a *valid sample*—the target and sampled populations have similar characteristics.

Let the individual residences constitute the elementary sampling units, with an *observation* taken to be a particular data point or value of some characteristic of interest. We shall let each such characteristic be represented by a separate variable, and the value of the variable is an observation of the characteristic.

Assuming that we have settled on a way of actually extracting a random sample from the directory, suppose that one of the elementary sampling units is the residence located at 401 Elm Street. Let us consider some of its important characteristics (Table 1.1).

Note that X_1 and X_2 are qualitative or nonmetric variables measured on a nominal scale while variables $X_3, \ldots, X_8$ are quantitative or metric variables measured on a ratio scale.

1.5 LET US ADD

Quite often throughout this text the reader will be asked to total or form the sum of the values appearing in a variety of data sets. We have a special notation for the operation of addition. We will let the Greek capital sigma or Σ serve as our "summation sign." Specifically, for a variable X with values $X_1, X_2, \ldots, X_n$,

$$X_1 + X_2 + \cdots + X_n = \sum_{i=1}^{n} X_i. \tag{1.1}$$

Here the right-hand side of this expression reads "the sum of all observations X_i as i goes from 1 to n." In this regard, Σ is termed an *operator*—it operates only on those items having an i index, and the operation is addition. When it is to be understood that we are to add over all i values, then Equation (1.1) can be rewritten simply as $X_1 + X_2 + \cdots + X_n = \Sigma X_i$.

EXERCISES

1 Are the following variables qualitative or quantitative?

a. Color
b. Gender
c. Zip code
d. Temperature
e. Your cell phone number
f. Posted speed limit
g. Life expectancy
h. Size of your family
i. Number on a uniform
j. Marital status
k. Drivers license number
l. Cost to fill up your car's gas tank

2 Are the following variables discrete or continuous?

a. Number of trucks parked at a truck stop
b. Number of tails obtained in three flips of a fair coin
c. Time taken to walk up a flight of stairs
d. The height attained by a high jumper
e. Life expectancy
f. Number of runs scored in a baseball game
g. Length of your favorite song
h. Weight loss after dieting for a month

3 Are the following variables nominal or ordinal?

a. Gender
b. Brand of a loaf of bread
c. Response to a customer satisfaction survey: poor, fair, good, or excellent
d. Letter grade in a college course
e. Faculty rank at a local college

4 Are the following variables all measured on a ratio scale?

a. Cost of a new pair of shoes

b. A day's wages for a laborer

c. Your house number

d. Porridge in the pot 9 days old

e. Your shoe size

f. An IQ score of 130

2

ANALYZING QUANTITATIVE DATA

2.1 IMPOSING ORDER

In this and the next chapter, we shall work in the area of descriptive statistics. As indicated above, descriptive techniques serve to summarize data; we want to put the data into a "readable form" or to "create order." In doing so we can determine, for instance, if a "pattern of behavior" emerges.

2.2 TABULAR AND GRAPHICAL TECHNIQUES: UNGROUPED DATA

Suppose we have a sample of n observations on some variable X. This can be written as

$$X : X_1, X_2, \ldots, X_n \text{ or } X : X_i,\ i = 1, \ldots, n$$

Our first construct is what is called an *absolute frequency distribution*—it shows the absolute frequencies with which the "different" values of a variable X occur in a set of data, where *absolute frequency* (f_j) is the number of times a particular value of X is recorded.

Example 2.1

Let the $n = 15$ values of a variable X appear as

$$X : 1, 3, 1, 2, 4, 1, 3, 5, 6, 6, 4, 2, 3, 4, 3.$$

Statistical Inference: A Short Course, First Edition. Michael J. Panik.

TABLE 2.1 Absolute Frequency Distribution for X

X	Absolute Frequency (f_j)
1	3
2	2
3	4
4	3
5	1
6	2
	$15 = \Sigma f_j = n$

Table 2.1 describes the absolute frequency distribution for this data set. Note that while there are 15 individual values for X (we have X: X_i, $i = 1, \ldots, 15$), there are only six "different" values of X since most of the X values occur more than once. Hence, the j subscript indexes the different X's (there are only six of them so $j = 1, \ldots, 6$). So given this absolute frequency distribution, what sort of pattern emerges? Is there a "most frequent" value or a "least frequent" value?

Our next descriptive tool is the *relative frequency distribution*—it expresses each absolute frequency (f_j) as a fraction of the total number of observations n. Hence, a relative frequency is calculated as f_j/n.

Example 2.2

Given the Example 2.1 data set, determine the associated relative frequency distribution. To form Table 2.2, we use the same X heading as in Table 2.1 but replace f_j by f_j/n. So while the absolute frequencies sum to 15, the relative frequencies must sum to 1.

Finally, let us form a *percent distribution*—we simply multiply the relative frequencies by 100 to convert them to percentages.

Example 2.3

Using the Example 2.2 relative frequency table, determine the corresponding percent distribution. A glance at Table 2.3 reveals the desired set of percentages. Clearly the percent column must total 100 (%).

TABLE 2.2 Relative Frequency Distribution for X

X	Relative Frequency (f_j/n)
1	$3/15 = 0.2000$
2	$2/15 = 0.1333$
3	$4/15 = 0.2666$
4	$3/15 = 0.2000$
5	$1/15 = 0.0666$
6	$2/15 = 0.1333$
	$\approx 1.000 = \Sigma f_j/n$

TABLE 2.3 Percent Distribution for *X*

X	% ($=100 \times f_j/n$)
1	20.00
2	13.33
3	26.66
4	20.00
5	6.66
6	13.33
	$\approx$100

It is important to remember that these three distributions are all equivalent ways of conveying basically the same information. Which one you choose to employ depends upon the type of information you wish to present. For instance, from Tables 2.1 to 2.3, it is apparent that $X = 3$ occurs four times (Table 2.1) or accounts for 4/15 of all data values (Table 2.2) or appears 26.66% of the time (Table 2.3).

We can easily graphically illustrate the distributions appearing in Tables 2.1–2.3 by constructing *absolute* or *relative frequency functions* as well as a *percent function.* These graphs appear in Fig. 2.1a–c, respectively.

2.3 TABULAR AND GRAPHICAL TECHNIQUES: GROUPED DATA

The preceding set of descriptive techniques (tabular and graphical) was easily applied to discrete sets of data with a relatively small number of observations. (A small sample in statistics typically has about 30 observations (give or take a few data points) whereas a medium-sized sample has around 50 observations.) Much larger data sets, involving either discrete or continuous observations, can be handled in a much more efficient manner. That is, instead of listing each and every different value of some variable X (there may be hundreds of them) and then finding its associated absolute or relative frequency or percent, we can opt to group the values of X into specified classes.

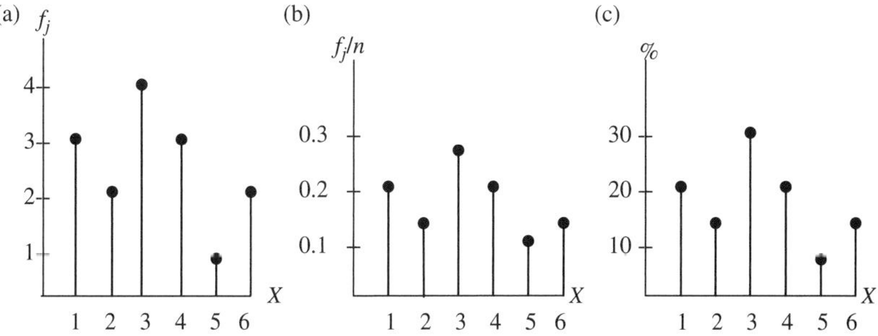

FIGURE 2.1 (a) Absolute frequency function. (b) Relative frequency function. (c) Percent function.

Paralleling the above development of tabular and graphical methods, let us first determine an *absolute frequency distribution*—it shows the absolute frequencies with which the various values of a variable X are distributed among chosen classes. Here an absolute frequency is now a *class frequency*—it is the number of items falling into a given class.

The process of constructing this distribution consists of the following steps:

1. Choose the classes into which the data are to be grouped. Here, we must do the following:
   (a) Determine the number of classes.
   (b) Specify the range of values each class is to cover.
2. Sort the observations into appropriate classes.
3. Count the number of items in each class so as to determine absolute class frequencies.

While there are no hard and fast rules for carrying out step 1, certain guidelines or "rules of thumb" can be offered. For instance, as follows:

1. Usually the number of classes ranges between 5 and 20 inclusive.
2. Obviously one must choose the classes so that all of the data can be accommodated.
3. Each X_i belongs to one and only one class (avoid overlapping classes).
4. The class intervals should be equal (the *class interval* is the length of a class).[1]
5. Try to avoid open-ended classes (e.g., the class "65 and over" is open ended).
6. The resulting distribution should have a single peak (i.e., we would like to be able to identify a major point of concentration of data).

Example 2.4

Let us examine the array of 50 observations on a variable X offered in Table 2.4.

Given that the smallest observation or min $X_i = 1$ and the largest observation or max $X_i = 89$, let us determine the *range* of our data set (a measure of the total spread of the data) as

$$\text{Range} = \text{Max } X_i - \text{Min } X_i = 89 - 1 = 88. \tag{2.1}$$

TABLE 2.4 Observations on X

15	58	35	49	78	31	45	13	33	41
9	24	12	25	63	27	76	31	24	21
35	16	52	19	1	16	27	48	33	35
17	89	24	37	43	28	58	72	28	42
34	30	31	46	60	44	36	52	18	49

[1] See Appendix 2.A for the case of unequal class intervals.

Since the range of our data set can be viewed as being made up of classes and class intervals, we can write

$$\text{Range} \approx (\text{number of classes}) \times (\text{class interval}). \tag{2.2}$$

So given that we know the range, we can choose the number of classes and then solve for the common class interval or class length; or we can pick a class interval and then solve for the required number of classes. Usually the class interval is chosen as something easy to work with, such as 5 or 10, or 20, and so on. If we decide to have the length of each class equal to 10, then from Equation (2.2) we get,

$$\text{Number of classes} \approx \frac{\text{range}}{\text{class interval}} = \frac{88}{10} = 8.8.$$

Since we cannot have a fractional part of a class, we round to nine classes. At the outset, had we decided to use nine classes, then, from Equation (2.2), we would solve for

$$\text{Class interval} \approx \frac{\text{Range}}{\text{Number of classes}} = \frac{88}{9} = 9.8.$$

Then rounding to 10 gives us a convenient class length to work with.

In view of the preceding set of guidelines, Table 2.5 (columns one and two) displays our resulting absolute frequency distribution.

A few points to ponder:

1. The j subscript is now used to index classes, that is, since we have nine classes, $j = 1, \ldots, 9$.
2. We may define a set of *class limits* as indicating the smallest and largest values that can go into any class. Obviously the values 0, 10, 20, and so on are *lower class limits* while the values such as 9, 19, and 29 constitute *upper class limits*.
3. Given Table 2.5, how can we find the class interval or the common class length? We need only take the difference between successive lower class limits, for example, $10 - 0 = 10$, $20 - 10 = 10$, and so on. (Note: the class interval is "not"

TABLE 2.5 Absolute Frequency Distribution for X

Classes of X	Absolute Class Frequency (f_j)	Class Mark (m_j)
0–9	2	4.5
10–19	8	14.5
20–29	10	24.5
30–39	12	34.5
40–49	9	44.5
50–59	4	54.5
60–69	2	64.5
70–79	2	74.5
80–89	1	84.5
	$50 = \sum f_j$	

the difference between the upper and lower class limits of a given class, that is, for the third class, $29 - 20 = 9 \neq$ class length $= 10$.)

4. The *class mark* (m_j) of a given class is the midpoint of a class and is obtained by averaging the class limits, for example, for the second class $m_2 = \frac{10+19}{2} = 14.5$. What is the role of a class mark? To answer this, we need only note that in the grouping process the individual observations on X "lose their identity," so to speak. For instance, when faced with a table such as Table 2.5, in the class 40–49, we have nine data points. However, we do not know their individual values. Hence, we can only identify a typical or representative value for this class—it is $m_5 = 44.5$, the class midpoint. (Note that the class interval can be computed as the difference between two successive class marks, for example, $m_4 - m_3 = 34.5 - 24.5 = 10$ as required.)
5. Finally, it is obvious that each successive class is separated from its predecessor by a value of one unit, that is, we have a one unit gap between successive classes. To avoid these gaps between classes, sometimes the notion of a *class boundary* is used—the boundary between the first and second classes is 9.5 (9.5 is halfway between 9 and 10) and so on. (Note that the class interval can be calculated as the difference between the upper and lower boundaries of a given class (e.g., for the class 20–29, its length is $29.5 - 19.5 = 10$.)) Query: Why are class boundaries always impossible values? (The answer is obvious so do not over think it.)

Example 2.5

Given the classes and absolute frequencies appearing in Table 2.5, the reader should now be able to construct a relative frequency distribution (relative frequency is f_j/n) and a percent distribution (multiply the relative frequencies by 100 to convert then to percents). These distributions appear in Tables 2.6 and 2.7, respectively.

A convenient graphical device for illustrating the distributions appearing in Tables 2.5–2.7 is the *histogram*—a set of vertical bars or rectangles whose heights correspond to the absolute frequencies (or relative frequencies or percents) of the

TABLE 2.6 Relative Frequency Distribution for X

Classes of X	Relative Class Frequency (f_j/n)
0–9	2/50 (=0.0400)
10–19	8/50 (=0.1600)
20–29	10/50 (=0.2000)
30–39	12/50 (=0.2400)
40–49	9/50 (=0.1800)
50–59	4/50 (=0.0800)
60–69	2/50 (=0.0400)
70–79	2/50 (=0.0400)
80–89	1/50 (=0.0200)
	=1.000

TABLE 2.7 Percent Distribution for X

Classes of X	% $(=100f_i/n)$
0–9	4.00
10–19	16.00
20–29	20.00
30–39	24.00
40–49	18.00
50–59	8.00
60–69	4.00
70–79	4.00
80–89	2.00
	= 100

classes of X and whose bases are the class intervals. To avoid gaps between the rectangles, we need only plot the various lower class limits on the X-axis.

Example 2.6

Figure 2.2 houses an absolute frequency histogram, a relative frequency histogram, and a percent histogram.

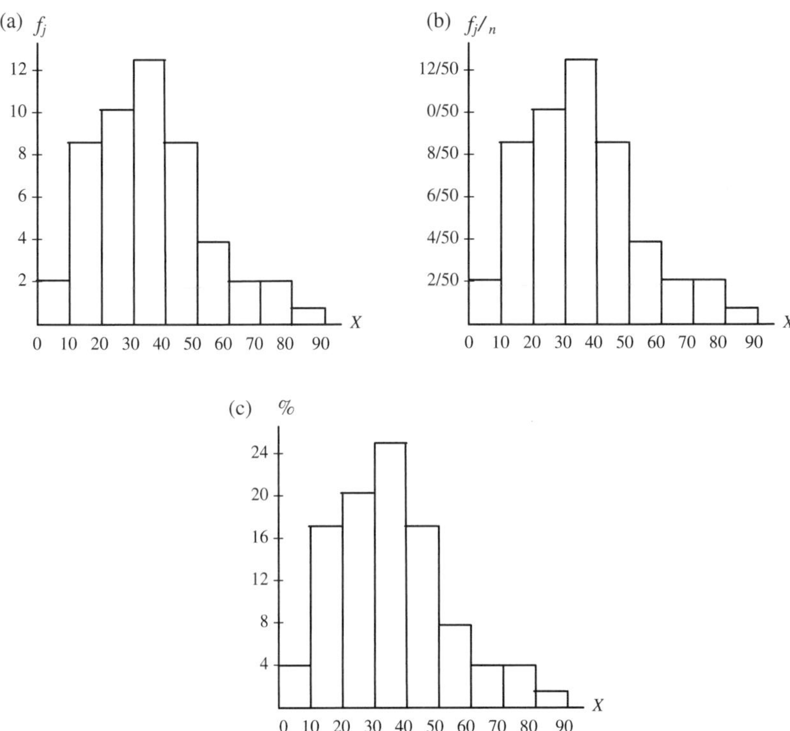

FIGURE 2.2 (a) Absolute frequency histogram. (b) Relative frequency histogram. (c) Percent histogram.

EXERCISES

1 Given the following $n = 25$ observations on a variable X, find:

a. Absolution frequency distribution for X.

b. Relative frequency distribution for X.

c. Percent distribution for X.

d. Graph the absolute and relative frequency functions for X.

e. Graph the percent function for X.

$$X : 5,\ 5,\ 10,\ 11,\ 7,\ 7,\ 6,\ 10,\ 6,\ 12,\ 6,\ 8,\ 9,\ 10,\ 7,\ 7,\ 8,\ 9,$$
$$10,\ 9,\ 9,\ 11,\ 10,\ 13,\ 7$$

2 Table E.2.2 provides observations on a variable X (the weight, in grams, of $n = 40$ samples of wheat).

TABLE E.2.2 Observations on a Variable *X*

90	48	80	81	87	76	93	67
97	90	84	91	82	61	91	88
80	92	71	56	74	99	74	71
60	85	72	65	83	83	60	89
66	83	40	74	86	88	63	79

a. What is the range of this data set?

b. Suppose we decide to use six classes. What is the class interval?

c. Suppose the first class has limits 40–49. Determine the limits of the remaining five classes.

d. Construct absolute and relative frequency distributions and the percent distribution for X.

e. Find the class marks and class boundaries.

f. Graph the absolute and relative frequency histograms along with the percent histogram for X.

3 If the class limits for a frequency distribution are 25–28, 29–32, 33–36, 37–40, and 41–44, determine the class marks and the class boundaries. Are these classes of equal length?

4 Table E.2.4 houses the gains in weight (in pounds) of $n = 48$ steers taking a dietary supplement.

TABLE E.2.4 Gains in Weight (in Pounds)

182	169	116	113	81	124	85	112
183	173	115	110	65	121	101	129
185	162	136	105	129	76	107	85
194	150	137	109	116	132	72	115
126	78	116	185	133	119	57	113
71	124	98	118	93	162	149	136

a. Find the range of this data set.

b. Suppose we decide on a class interval of 16. How many classes should be used?

c. Let the first class have limits 55–70. Find the limits of the remaining classes.

d. Construct absolute and relative frequency distributions and the percent distribution for X.

e. What are the class marks and class boundaries?

f. Graph the absolute and relative frequency histograms along with the percent histogram for X.

5 Selecting the number of classes when working with grouped data is rather arbitrary and, for the most part, is done by trial and error. An empirical relationship often used to determine the number of classes is *Sturge's Rule*

$$K = 1 + 3.3\log(n),$$

where K is the optimal number of classes, n is the total number of observations, and log (n) is the common (base 10) logarithm of n. For $n = 50$, what is the optimal number of classes? When $n = 100$, how many classes does Sturge's Rule say we should use?

6 Given $n = 200$ data values, with 10.6 the smallest and 75.2 the largest value of a variable X, use Sturge's Rule to choose the classes for grouping the data values.

7 Given the following absolute frequency distribution (Table E.2.7):

TABLE E.2.7 Absolute Frequency Distribution

Classes of X	Absolute Frequency
25–49	16
50–74	25
75–99	30
100–124	20
125–149	9
	100

a. Determine the class marks and class boundaries. What is the class interval?
b. Determine the relative frequency and percent distributions.

8 Given the following relative frequency distribution for $n = 50$ data points, find the absolute frequency distribution (Table E.2.8).

TABLE E.2.8 Absolute Frequency Distribution

Classes of X	Relative Frequency
20–39	0.10
40–59	0.40
60–79	0.30
80–99	0.20
	1.00

9 Table E.2.9 houses the distribution for the number of defective items found in 50 lots of manufactured items. Find the following:

TABLE E.2.9 Distribution for the Number of Defective Items

Number of Defects	Absolute Frequency
0	9
1	10
2	9
3	8
4	7
5	5
6	1
7	1
	50

a. The percentage of lots containing at least four defective items.
b. The percentage of lots containing no more than four defective items.
c. The percentage of lots that contain no more than two defective items.
d. The percentage of lots that contain fewer than two defective items.

APPENDIX 2.A HISTOGRAMS WITH CLASSES OF DIFFERENT LENGTHS

We previously defined a histogram as a set of vertical bars whose heights correspond to absolute class frequencies and whose bases are the class intervals. This definition implicitly assumes that all of the classes are of equal length. However, for some data sets, this might not be the case.

A more general definition of a *histogram* is as follows: a set of vertical bars whose areas are directly proportional to the absolute class frequencies presented, where

$$\begin{aligned}\text{Area} &= \text{Class interval} \times \text{Height} \\ &= \text{Class interval} \times \text{Frequency density}\end{aligned} \tag{2.A.1}$$

and

$$\text{Frequency density} = \frac{\text{Absolute class frequency}}{\text{Class interval/Standard width}}. \tag{2.A.2}$$

Hence, *frequency density* amounts to absolute class frequency on a per standard width basis—it is the absolute class frequency per interval length. So for classes of equal length,

$$\begin{aligned}\text{Height} &= \text{Frequency density} \\ &= \text{Absolute class frequency}\end{aligned}$$

and for classes of different lengths,

$$\text{Height} = \text{Frequency density} \neq \text{Absolute class frequency}.$$

Given Equation (2.A.2), we may now express Equation (2.A.1) as

$$\text{Area} = \left[\text{Class interval}\left(\frac{\text{Standard width}}{\text{Class interval}}\right)\right] \times \text{Absolute class frequency}, \tag{2.A.3}$$

where the constant of proportionality mentioned in the preceding definition of a histogram is

$$\left[\text{Class interval}\left(\frac{\text{Standard width}}{\text{Class interval}}\right)\right].$$

Example 2.A.1

Table 2.A.1 involves an absolute frequency distribution and its accompanying histogram (Fig. 2.A.1). Here the standard width of each class is three and the height of each class is its frequency density = absolute class frequency (Eq. (2.A.2)). Since the classes are of equal length, the areas are determined as standard width (three) × absolute class frequency.

Table 2.A.2 involves an absolute frequency distribution with a slightly more complicated class structure. Its associated histogram appears as Fig. 2.A.2. Here the

TABLE 2.A.1 Absolute Frequency Distribution for *X*

Classes of *X*	Absolute Frequency	Height = Frequency Density = Absolute Class Frequency	Area of Rectangle
5–7	2	2/(3/3) = 2	3 × 2 = 6
8–10	5	5/(3/3) = 5	3 × 5 = 15
11–13	22	22/(3/3) = 22	3 × 22 = 66
14–16	13	13/(3/3) = 13	3 × 13 = 39
17–19	7	7/(3/3) = 7	3 × 7 = 21
20–22	1	1/(3/3) = 1	3 × 1 = 3
	50		

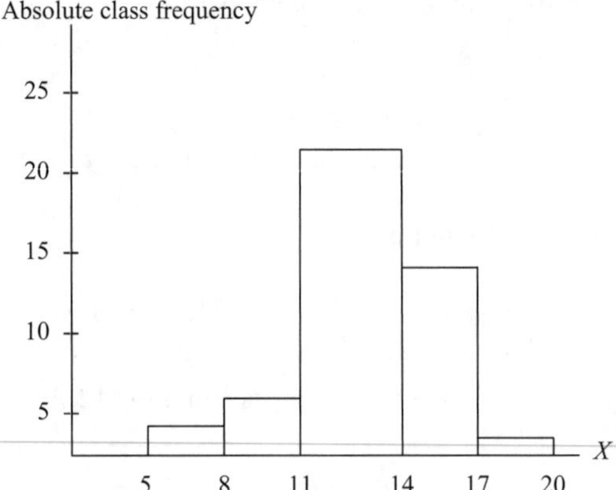

FIGURE 2.A.1 Absolute frequency histogram.

TABLE 2.A.2 Absolute Frequency Distribution for *X*

Classes of *X*	Absolute Frequency	Height = Frequency Density	Area of Rectangle = Height × Class Interval
5–9	6	6/(5/10) = 12	12 × 5 = 60
10–19	8	8/(10/10) = 8	8 × 10 = 80
20–29	11	11/(10/10) = 11	11 × 10 = 110
30–39	13	13/(10/10) = 13	13 × 10 = 130
40–49	20	20/(10/10) = 20	20 × 10 = 200
50–59	7	7/(10/10) = 7	7 × 10 = 70
60–69	5	5/(10/10) = 5	5 × 10 = 50
70–79	2	2/(10/10) = 2	2 × 10 = 20
80–109	2	2/(30/10) = 2/3	2/3 × 30 = 20
	74		

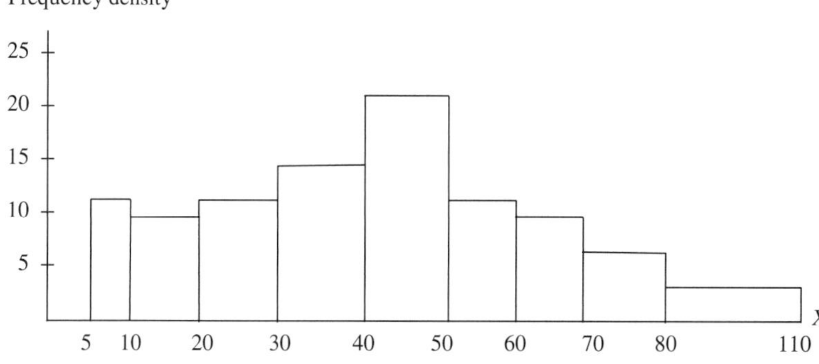

FIGURE 2.A.2 Frequency density histogram.

width of the first class is 5, the width of the last class is 30, and the classes 10–19 through 70–79 each have a "standard width" of 10. The height of each class is again determined by its frequency density (Table 2.A.2).

3

DESCRIPTIVE CHARACTERISTICS OF QUANTITATIVE DATA

3.1 THE SEARCH FOR SUMMARY CHARACTERISTICS

Up to this point in our data analysis, we have been working with the entire set of observations, either in tabular form (e.g., an absolute frequency distribution) or in graphical form (e.g., an absolute frequency function or an absolute frequency histogram). Is there a more efficient way of squeezing information out of a set of quantitative data? As one might have anticipated, the answer is yes. Instead of working with the entire mass of data in the form of a table or graph, let us determine concise summary characteristics of our variable X, which are indicative of the properties of the data set itself.

If we are working with a population of X values, then these descriptive measures will be termed "parameters." Hence a *parameter* is any descriptive measure of a population. But if we are dealing with a sample of X values drawn from the X population, then these descriptive measures will be called "statistics." Thus a *statistic* is any descriptive measure of a sample. (You now know what this book is ultimately all about—you are going to study descriptive measures of samples.) As we shall see later on, statistics are used to measure or estimate (unknown) parameters. For any population, parameters are constants. However, the value of a statistic is variable; its value depends upon the particular sample chosen, that is, it is sample sensitive. (This is why sampling error is an important factor in estimating a parameter.)

Statistical Inference: A Short Course, First Edition. Michael J. Panik.

TABLE 3.1 Concise Summary Characteristics of a Variable X

1. Measures of central location (used to describe a typical data value):
   a. Mean (unweighted or weighted)
   b. Quantiles (the median and other measures of position)
   c. Mode
2. Measures of dispersion (used to describe variation in a data set):
   a. Range and interquartile range
   b. Standard deviation
   c. Coefficient of variation
   d. Z-scores
3. Measures of shape
   a. Coefficient of skewness
   b. Coefficient of kurtosis
4. Measures which detect outliers

This said, what are some of the important characteristics of interest of a data set (being either a population or sample)? Table 3.1 lists the various descriptive measures to be introduced in this chapter. (In what follows, these descriptive measures are determined exclusively for sets of "ungrouped observations." A parallel presentation of these measures involving "grouped data" is offered in the appendix.)

3.2 THE ARITHMETIC MEAN

First, some notation:

a. We shall denote the *population mean* as follows:

$$\mu = \frac{\sum X_i}{N} \tag{3.1}$$

(the Greek letter μ is pronounced "mu");

b. We shall denote the *sample mean* as follows:

$$\bar{X} = \frac{\sum X_i}{n}. \tag{3.2}$$

Clearly, Equations (3.1) and (3.2) are "simple averages."

Example 3.1

Suppose we have a sample of $n = 5$ observations on a variable X: 1, 4, 10, 8, 10. What is the arithmetic mean of X or the value of $\bar{X}$? It is readily seen that

$$\bar{X} = \frac{1 + 4 + 10 + 8 + 10}{5} = \frac{33}{5} = 6.6.$$

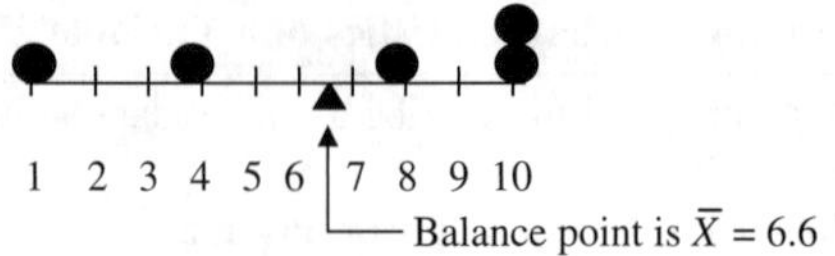

FIGURE 3.1 $\bar{X}$ is the center of gravity of X.

How should we interpret 6.6? A "physical" interpretation of the mean is that it represents X's center of gravity, that is, X's absolute frequency distribution will "balance" at the mean (Fig. 3.1).

So much for physics. How does the statistician interpret the word "balance?" To answer this question, let us introduce the concept of the ith deviation from the mean. Specifically, given a variable X:$X_1, X_2, \ldots, X_n$, *the ith deviation from the mean of* X is written as $x_i = X_i - \bar{X}$, $i = 1, \ldots, n$. In this regard:

if $X_i > \bar{X}$, then $x_i > 0$
if $X_i < \bar{X}$, then $x_i < 0$
if $X_i = \bar{X}$, then $x_i = 0$

Let us determine the set of deviations from the mean for the above five values of X (Table 3.2). Note that if we total the x_i's, $i = 1, \ldots, n$, we get zero. Thus $\Sigma\, x_i = \Sigma\,(X_i - \bar{X}) = 0$—the sum of all the deviations from the mean is always zero (whether one deals with a sample or a population.) This, then, is how we interpret the word "balance" in statistics.

Next, let us look to the properties of the mean (Equations (3.1) and (3.2)):

1. The mean always exists.
2. The mean is unique.
3. The mean is affected by outliers.
   (If in the above data set X: 1, 4, 10, 8, 10, we replace one of the terms by 10,000, it is obvious that the mean will essentially "explode" in value.[1])
4. The mean is said to be "relatively reliable," that is, it does not vary considerably under repeated sampling from the same population (more on the notion of reliability later on.)
5. Each observation used in the calculation of $\bar{X}$ has the same weight or the same relative importance.

[1] The values 1 and 10 are the *extreme values* in this data set. They may or may not be outliers—further analysis is needed to deem them so. However, an *outlier* is a very large or very small extreme value in a data set. The point being made here is that the terms outlier and extreme value are not interchangeable; every data set has its extreme values (max X_i and min X_i), which may or may not be outliers.

TABLE 3.2 Deviations from the Mean of X

$$\left.\begin{array}{l} 1-\bar{X}=1-6.6=-5.6 \\ 4-\bar{X}=4-6.6=-2.6 \end{array}\right\}=-8.2$$

$$\left.\begin{array}{l} 10-\bar{X}=10-6.6=3.4 \\ 8-\bar{X}=8-6.6=1.4 \\ 10-\bar{X}=10-6.6=3.4 \end{array}\right\}=8.2 \qquad 0.0\left(=\sum(X_i-\bar{X})\right)$$

What if property No. 5 is deemed inappropriate in certain circumstances? That is, what if some observations are considered to be more important than others? In the event that the data values display a differing relative importance, we must abandon, say, Equation (3.2) and look to the calculation of a *weighted mean*

$$\bar{X}_w = \frac{\Sigma X_i w_i}{\Sigma w_i}, \tag{3.3}$$

where w_i is the weight attached to X_i, $i = 1, \ldots, n$.

Example 3.2

Suppose a student takes three exams and obtains the scores X: 40, 90, and 85. Clearly, the student's average is $\bar{X} = 215/3 = 72$. However, suppose we are given the additional information that the first exam took half an hour, the second exam took 1 hr, and the third exam took 2 hr to complete. This added time dimension readily prompts us to realize that the tests should not be treated as equally important; those exams that took longer to complete should be more heavily weighted in the calculation of the average. To take account of their different relative importance, let us use Equation (3.3) to calculate the mean score (Table 3.3). How were the weights determined? We give the least important data value (40) a weight of $w_1 = 1$—then all other observations are compared back to the least important data point. Thus, since exam 2 took twice as long to complete as exam 1, its weight is $w_2 = 2$; and since exam 3 took four times longer to finish than exam 1, its weight is $w_3 = 4$. Then, via Equation (3.3), $\bar{X}_w = 560/7 = 80$.

TABLE 3.3 Determining the Weighted Mean of X

X	w_i	$X_i w_i$
40	1	40
90	2	180
85	4	340
	7 $(=\Sigma w_i)$	560 $(=\Sigma X_i w_i)$

Why is the weighted average higher than the unweighted average? Because the person did better on those exams that carried the larger weights. (What if the second exam took 1 hr and 15 min to complete and the third exam was finished in 2 hr and 30 min? The reader should be able to determine that the revised set of weights is now: $w_1 = 1, w_2 = 2.5,$ and $w_3 = 5.$)

3.3 THE MEDIAN

For a variable X (representing a sample or a population), let us define the *median* of X as the value that divides the observations on X into two equal parts; it is a positional value—half of the observations on X lie below the median, and half lie above the median. Since there is no standard notation in statistics for the median, we shall invent our own: the population median will be denoted as "*Med*;" the sample median will appear as "*med*."

Since the median is not a calculated value (as was the mean) but only a positional value that locates the middle of our data set, we need some rules for determining the median. To this end, we have rules for finding the median:

1. Arrange the observations on a variable X in an increasing sequence.
2. (a) For an odd number of observations, there is always a middle term whose value is the median.
   (b) For an even number of observations, there is no specific middle term. Hence take as the median the average of the two middle terms.

Note that for either case 2a or 2b, the median is the term that occupies the position $(n + 1)/2$ (or $(N + 1)/2$) in the increasing sequence of data values.

Example 3.3

Given the variable X: 8, 7, 12, 8, 6, 2, 4, 3, 5, 11, 10, locate the median. Arranging the observations in an increasing sequence yields 2, 3, 4, 5, 6, 7, 8, 8, 10, 11, 12. Given that we are looking for the middle term and we have an odd number of observations ($n = 11$ data points), it is easily determined that the median is 7—half of the data values are below 7 and half are above 7. Alternatively, the median is the term with position $(n + 1)/2 = 12/2 = 6$. Clearly, this is again 7.

For X: 1, 8, 5, 10, 15, 2, the arrangement of these data points in an increasing sequence gives 1, 2, 5, 8, 10, 15. With an even number of data points (there are $n = 6$ of them), the median is the average of the two middle terms or the median is $(5 + 8)/2 = 6.5$—half of the observations are below 6.5 and half are above 6.5. Note that the position of the median is again given by $(n + 1)/2 = 7/2 = 3.5$, that is, the median lies half the distance between the third and fourth data points. That distance is $8 - 5 = 3$. Then $0.5\ (3) = 1.5$. Thus the median equals $5 + 1.5 = 6.5$ as expected.

Looking to the properties of the median:

1. The median may or may not equal the mean.
2. The median always exists.
3. The median is unique.
4. The median is not affected by outliers (for X: 1, 8, 5, 10, 15, 2, if 15 is replaced by 1000, the median is still 6.5).

3.4 THE MODE

The *mode* of X is the value that occurs with the highest frequency; it is the most common or most probable value of X. Let us denote the population mode as "*Mo*;" the sample mode will appear as "*mo*."

Example 3.4

For the sample data set X: 1, 2, 4, 4, 5, the modal value is $mo = 4$ (its frequency is two); for the variable Y: 1, 2, 4, 4, 5, 6, 6, 9, there are two modes—4 and 6 (Y is thus a *bimodal* variable); and for Z: 1, 2, 3, 7, 9, 11, there is no mode (no one value appears more often than any other).

For the properties of the mode:

1. The mode may or may not equal the mean and median.
2. The mode may not exist.
3. If the mode exists, it may not be unique.
4. The mode is not affected by outliers (for the data set X: 1, 2, 4, 4, 5, if 5 is replaced by 1000, the mode is still four).
5. The mode always corresponds to one of the actual values of a variable X

We next turn to measures of dispersion or variation within a data set. Here we are attempting to get a numerical measure of the spread or scatter of the values of a variable X along the horizontal axis.

3.5 THE RANGE

We previously defined the *range* of a variable X as range $= \max X_i - \min X_i$. Clearly, the range is a measure of the total spread of the data. However, knowing the total spread of the data values tells us nothing about what is going on between the extremes of $\max X_i$ and $\min X_i$. To specify a more appealing measure of dispersion, let us consider the spread or scatter of the data points about the mean. In particular, let us consider the "average variability about the mean." This then takes us to the concept of the standard deviation of a variable X.

3.6 THE STANDARD DEVIATION

We previously denoted the ith deviation from the (sample) mean as $X_i - \bar{X}$. Since the sum of all the deviations from the mean is zero, it is evident that one way to circumvent the issue of the sign of the difference $X_i - \bar{X}$ is to square it. Then all of these squared deviations from the mean are nonnegative.

Suppose we have a population of X values or X: $X_1, X_2, \ldots, X_N$. Then we can define the variance of X as follows:

$$\sigma^2 = \frac{\Sigma(X_i - \mu)^2}{N}, \tag{3.4}$$

that is, the *population variance* of X is the average of the squared deviations from the mean of X or from μ. Since the variance of any variable is expressed in terms of (units of X)^2 (e.g., if X is measured in "inches," then σ^2 is expressed in terms of "inches^2"), a more convenient way of assessing variability about μ is to work with the *standard deviation* of X, which is defined as the positive square root of the variance of X or

$$\sigma = \sqrt{\frac{\Sigma(X_i - \mu)^2}{N}} = \sqrt{\frac{\Sigma X_i^2}{N} - \mu^2}. \tag{3.5}$$

Here σ is expressed in the original units of X.

If we have a sample of observations on X or X: $X_1, X_2, \ldots, X_n$, then the *sample variance* of X is denoted as

$$s^2 = \frac{\Sigma(X_i - \bar{X})^2}{n-1} \tag{3.6}$$

and the (sample) *standard deviation* of X is

$$s = \sqrt{\frac{\Sigma(X_i - \bar{X})^2}{n-1}} = \sqrt{\frac{\Sigma X_i^2}{n-1} - \frac{(\Sigma X_i)^2}{n(n-1)}}. \tag{3.7}$$

Here too s is measured in the same units as X. (It is important to remember that, in Equation (3.7), ΣX_i^2 is a "sum of squares" while $(\Sigma X_i)^2$ is the "square of a sum." These quantities are different entities and thus are not equal or interchangeable expressions.)

Note that in Equations (3.6) and (3.7), we have $n-1$ in the denominator instead of n itself. Here dividing by $n-1$ is termed a correction for *degrees of freedom* (denoted d.f.). We may view degrees of freedom as the number of independent observations remaining in the sample, that is, it is the sample size n less the number of prior

TABLE 3.4 Values of X and X^2

X	X^2
2	4
4	16
6	36
10	100
12	144
14	196
48	496

estimates made [$\bar{X}$is our estimate of μ in Equation (3.6)].[2] (Also, if we had divided by n instead of $n-1$, then Equation (3.6) would provide us with what is called a "biased" estimator for σ^2. More on this point later.)

Example 3.5

Given the sample values X: 2, 4, 6, 10, 12, 14, find and interpret s. As a computational device, let us construct the following table (Table 3.4). Then, from Equation (3.7), we have the following:

$$s = \sqrt{\frac{496}{5} - \frac{(48)^2}{6 \times 5}} = \sqrt{22.4} = 4.7328.$$

Thus, on an average, the individual X values are approximately 4.7 units away from the mean $\bar{X}$, being either above or below $\bar{X}$. So while some of the observations on X are above the mean and others are below the mean, on an average, they are approximately 4.7 units away from the mean.

We now turn to a set of comments on the uses of the standard deviation. In what follows, we shall assume that we have a sample of size n (with parallel arguments holding for a population of size N):

1. s is a measure of the scatter or dispersion of the values of a variable X about its mean $\bar{X}$. If s is "relatively small," the X_i's are clustered closely about $\bar{X}$; and if s is "relatively large," then the X_i's are scattered widely about $\bar{X}$.

[2] Technically speaking, d.f. is a property of the sum of squares $\Sigma(X_i - \bar{X})^2$ under the linear restriction $\Sigma(X_i - \bar{X}) = 0$. For instance, suppose we have $n = 3$ values of X:X_1, X_2, X_3 and we are told that $\bar{X} = 5$. Then under the requirement that $\Sigma(X_i - \bar{X}) = 0$, we are only "free" to arbitrarily choose the values of any two of the X_is. Once any two X_is are specified given that $\bar{X} = 5$, the remaining value of X is uniquely determined. For example, if $X_1 = 2$, $X_2 = 4$, and $\bar{X} = 5$, then

$$(X_1 - 5) + (X_2 - 5) + (X_3 - 5) = 0$$

or

$$(2 - 1) + (4 - 5) + (X_3 - 5) = 0$$

and thus $X_3 = 9$. So when $\bar{X}$is known, only two of the three X's are arbitrary—the remaining X value is completely specified. Hence only two observations are "free" to vary or d.f. $= n - 1 = 2$.

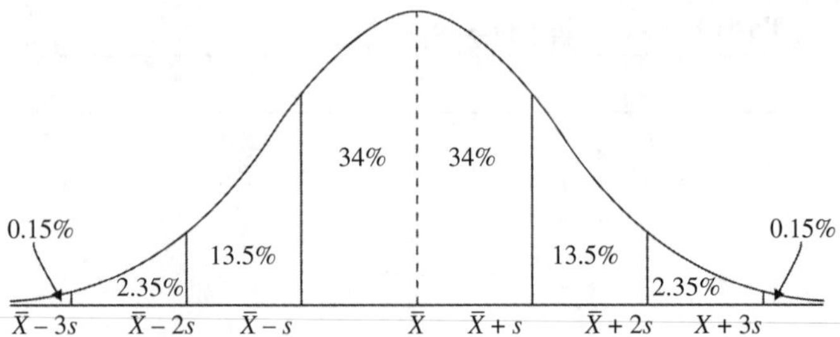

FIGURE 3.2 The empirical rule (X normally distributed).

2. Given a *normal distribution* (one which is continuous, bell-shaped, and symmetrical), the following *empirical rule* holds:

   (a) 68% of all observations on X will fall within 1 standard deviation of the mean or within the interval $\bar{X}\pm s$ (Fig. 3.2).

   (b) 95% of all observations on X will fall within 2 standard deviations of the mean or within $\bar{X}\pm 2s$.

   (c) 99.7% of all observations on X will fall within 3 standard deviations of the mean or within $\bar{X}\pm 3s$.

Suppose X is not normally distributed. Can we still reach a conclusion that is similar to that offered by the empirical rule? Interestingly enough, the answer is yes. To this end, we have *Chebysheff's Theorem*: given any set of observations (sample or population) on a variable X and any constant $k > 1$ (typically $k=2, 3$), "at least" $1-1/k^2$ of all X values will lie within k standard deviations of the mean or within the interval $\bar{X}\pm ks$. (See Fig. 3.3 for $k=2, 3$.)

a. For $k=2$, at least $1-1/k^2=3/4$ or 75% of all observations on X will lie within the interval $\bar{X}\pm 2s$ or within $(\bar{X}-2s, \bar{X}+2s)$.

b. For $k=3$, at least $1-1/k^2=8/9$ or 88.9% of all observations on X will fall within $\bar{X}\pm 3s$ or within $(\bar{X}-3s, \bar{X}+3s)$.

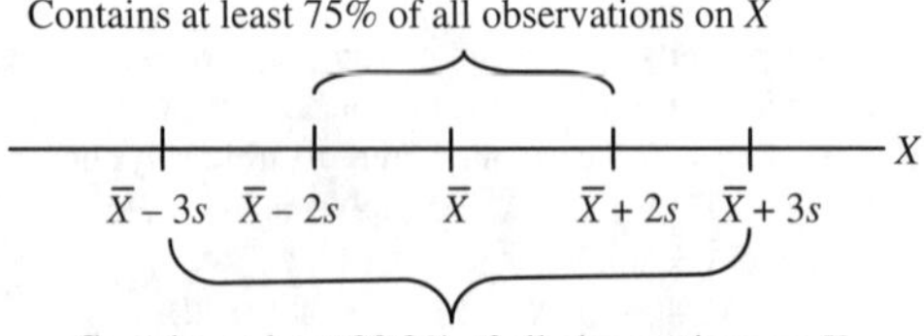

FIGURE 3.3 Contains at least 75% of all observations on X; contains at least 88.9% of all observations on X; Chebysheff's Theorem for $k=2, 3$.

Clearly, Chebysheff's Theorem provides us with a "lower bound" on the percentage of observations that lie within k standard deviations of the mean.

Example 3.6

Suppose a variable X is normally distributed with $\bar{X}=10$ and $s=5$.

a. What percentage of X values lie between 5 and 15? Since 5 and 15 correspond to the end points of the interval $\bar{X} \pm s$, it follows that 68% of all X values lie within this interval.
b. What percentage of X values lie below 0 and above 20? Given that 0 and 20 correspond to the end points of the interval $\bar{X} \pm 2s$ and this interval contains 95% of all observations on X, it follows that 5% of the data points lie below 0 and above 20.
c. What percentage of X values lie below -5? The value -5 is located 3 standard deviations below $\bar{X}$ so that only 0.15% of all Xs lie below -5.

Example 3.7

Suppose that a variable X has $\bar{X}=25$ and $s=8$.

a. What percentage of X values lie within 2 standard deviations of the mean? What is the implied interval? Since it is not specified that X is normally distributed, we must rely upon Chebysheff's Theorem. That is, for $k=2$, at least $1-1/k^2=1-1/4=3/4$ of all X values lie within the interval $\bar{X} \pm 2s \rightarrow 25 \pm 16$ or within (9, 41).
b. What is the minimum percentage of X values found between the limits 13 and 37? Given the interval (13, 37), let us work backward to find k, that is, $\bar{X} \pm ks \rightarrow 25 \pm k8$. Clearly, we must have $13=25-k8$ and $37=25+k8$. Then solving either of these latter two equalities for k renders $k=1.5$ so that at least $1-1/(1.5)^2=0.55556$ or 55.6% of all observations lie within (13, 37).

3. The distance of any observation X_i from the mean can be put in terms of standard deviations. For instance, if X has a mean of $\bar{X}=100$ and a standard deviation of $s=5$, then the X value 110 lies 10 units above $\bar{X}$ or it is found to be 2 standard deviations above $\bar{X}$; and the X value of 85 lies 15 units below $\bar{X}$ or it is located 3 standard deviations below $\bar{X}$. Hence the standard deviation serves as an "index of distance from the mean."
4. When all the values of a variable X have been divided by X's standard deviation, we say that X has been *standardized*. How is this standardization carried out? We start with X: $X_1, X_2, \ldots, X_n$ and compute s. We then form the new variable

$$\frac{X}{s}: \frac{X_1}{s}, \frac{X_2}{s}, \ldots, \frac{X_n}{s}, \quad s \neq 0.$$

What are the characteristics of this standardized variable X/s? It can be demonstrated that its mean is $\bar{X}/s$ and its standard deviation is 1. In fact, the standard deviation of "any" standardized data set is always 1 (provided $s \neq 0$).

5. Let us take the standardization process just outlined in comment No. 4 a step further. Specifically, let us also translate the mean to the origin of X and then standardize the adjusted values. This can be accomplished by the following sequence of steps:
   (a) Given X: $X_1, X_2, \ldots, X_n$, find $\bar{X}$ and s.
   (b) Shift the origin to the mean, that is, form $X - \bar{X}: X_1 - \bar{X}, X_2 - \bar{X}, \ldots, X_n - \bar{X}$.
   (c) Standardize the adjusted variable to form the new variable.

$$Z = \frac{X - \bar{X}}{s} : \underset{(Z_1)}{\frac{X_1 - \bar{X}}{s}}, \underset{(Z_2)}{\frac{X_2 - \bar{X}}{s}}, \ldots, \underset{(Z_n)}{\frac{X_n - \bar{X}}{s}}. \tag{3.8}$$

Then, it can be shown that the mean of Z is zero ($\bar{Z} = 0$) and the standard deviation of Z is 1 ($s_z = 1$). In fact, any distribution with a mean of zero and a standard deviation of 1 is termed a *unit distribution*.

The values of Z specified by Equation (3.8) are appropriately called *Z-scores*. Note that by virtue of their structure, Z-scores are dimensionless or independent of units. The role of the Z-score is that it specifies the relative position of any X_i value in the X data set—it expresses distance from the mean in standard deviation units, that is, the size of the Z-score Z_i indicates how many standard deviations separate X_i and $\bar{X}$. To see this, let us first note that if $X_i > \bar{X}$, then $Z_i > 0$; if $X_i < \bar{X}$, then $Z_i < 0$; and if $X_i = \bar{X}$, then $Z_i = 0$. So if, for instance, $Z_i = 2$, this tells us that X_i lies 2 standard deviations above $\bar{X}$; and if $Z_i = -0.6$, then X_i lies 6/10 of a standard deviation below $\bar{X}$.

In addition, Z-scores offer a basis for comparing two (or more) variables having different means and standard deviations. To achieve comparability, we convert the values of each of these variables to a set of Z-scores. They now have the same mean (of zero) and same standard deviation (of 1). Thus, we need only work with each variables set of Z-scores. (More on this approach later.)

Example 3.8

Given the sample values X: 2, 3, 4, 7, construct a unit distribution (or find the set of Z-scores for these data points). From the first three columns of Table 3.5 we can obtain the following:

$$\bar{X} = \frac{\Sigma X_i}{n} = \frac{16}{4} = 4, \quad s = \sqrt{\frac{\Sigma(X_i - \bar{X})^2}{n-1}} = \sqrt{\frac{14}{3}} \approx 2.16.$$

TABLE 3.5 Derivation of Z-Scores

X	$X_i - \bar{X}$	$(X_i - \bar{X})^2$	$Z_i = (X_i - \bar{X})/s$
2	−2	4	−2/2.16 = −0.93
3	−1	1	−1/2.16 = −0.46
4	0	0	0/2.16 = 0
7	3	9	3/2.16 = 1.39
16	0	14	

Then the set of Z-scores appears in the last column of Table 3.5. It is now readily seen that: $X_1 = 2$ lies 0.93 standard deviations below the mean (its Z-score Z_1 is negative); $X_2 = 3$ lies 0.46 standard deviations below the mean; $X_3 = 4$ equals the mean $\bar{X}$; and $X_4 = 7$ lies 1.39 standard deviations above the mean (since its Z-score Z_4 is positive). (As an exercise, the reader should demonstrate that the average of these Z-scores is zero and the standard deviation of the same is one.) Figure 3.4 compares the X and Z values.

3.7 RELATIVE VARIATION

How may we compare the dispersions of two (or more) distributions? Suppose we have distributions A and B and they have the same (or approximately the same) means and are expressed in the same units. Then the distribution having the larger standard deviation (a measure of *absolute variation*) has more variability or heterogeneity among its values. (The one with the smaller standard deviation has more homogeneity or uniformity among its values.)

It should be evident that if distributions A and B are expressed in different units (one might be expressed in cubic feet and the other in lumens), then there is no realistic basis for comparing their dispersions. Equally troublesome can be the instance in which their means display a marked difference, for example, we may ask: "is there more variability among the weights of elephants in a herd of elephants or among the weights of ants in an anthill?" While these two data sets can be converted to the same units, won't the standard deviation of the weights of the elephants "always" be a bigger number than the standard deviation of the weights of the ants?

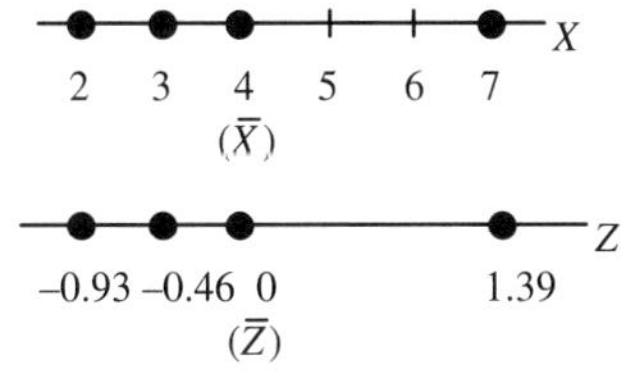

FIGURE 3.4 Distance from the mean expressed in terms of standard deviations.

This said, to induce comparability, let us convert the measures of absolute variation to "relative forms"—let us express each standard deviation as a percentage of the average about which the deviations are taken. This then leads us to define the *coefficient of variation* as:

$$\text{For a population, } V = \frac{\sigma}{\mu} \times 100. \tag{3.9a}$$

$$\text{For a sample, } v = \frac{s}{\bar{X}} \times 100. \tag{3.9b}$$

Here v, say, is a pure number that is independent of units. Once v is determined, we are able to conclude that "the standard deviation is $v\%$ of the mean." Hence the distribution with the smaller coefficient of variation has more uniformity among its values (or the distribution with the larger coefficient of variation has more heterogeneity among its values).

Example 3.9

The incomes received by individuals in group A have a mean of $\bar{X}_A = \$40{,}000$ and a standard deviation of $s_A = \$3{,}000$ while the incomes of individuals in group B have $\bar{X}_B = \$24{,}000$, $s_B = \$2{,}000$. Which group displays a greater variability regarding incomes received by its members? While there is no units problem, we see that the means of the two groups differ substantially. Hence we must apply Equation (3.9b):

$$v_A = \frac{3000}{40000} \times 100 = 7.5\%,$$

$$v_B = \frac{2000}{24000} \times 100 = 8.3\%.$$

So for group A, the standard deviation is 7.5% of the mean; for group B, the standard deviation is 8.3% of the mean. Thus group A has more uniformity (less variability) among the incomes earned by its members.

3.8 SKEWNESS

We may view the notion of *skewness* as a departure from symmetry.[3] Suppose we have a sample of X values. Then if the distribution of X is symmetrical, $\bar{X} = med = mo$ (Fig. 3.5). If the distribution is *skewed to the right* or *positively skewed* (its right-hand tail is elongated) the $\bar{X} > med > mo$. And if the distribution is *skewed to the left* or *negatively skewed* (its left-hand tail is elongated), then $\bar{X} < med < mo$.

[3] Think of a distribution as being cut out from a sheet of paper. Now, crease the sheet (distribution) in the middle or at its median and then fold the right-hand side onto the left-hand side. If the distribution is *symmetrical*, then the right-hand side will be the mirror image of the left-hand side. The overlapping fit will be exact.

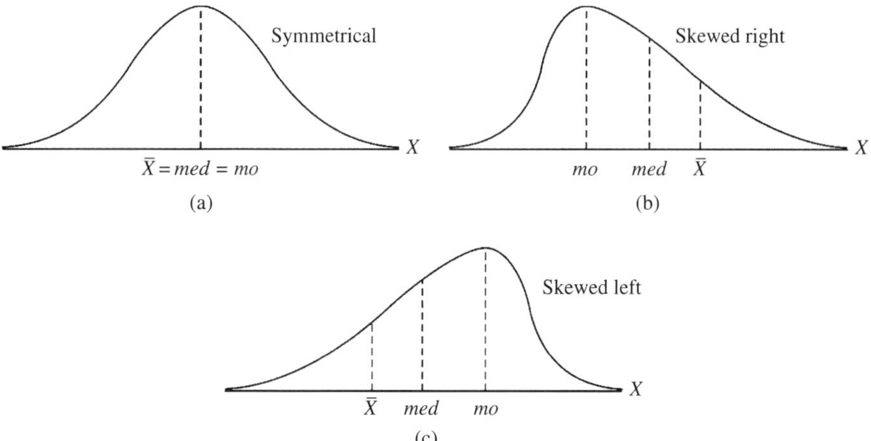

FIGURE 3.5 Comparing the relative positions of the mean, median, and mode.

Can we measure skewness? While there are a few different methods for detecting skewness in a data set, one measure, which is particularly easy to apply, is the *Pearson Coefficient of Skewness*:

$$\text{For a population, } Sk = \frac{\mu - Mo}{\sigma} \text{ and} \tag{3.10a}$$

$$\text{For a sample, } sk = \frac{\bar{X} - mo}{s}. \tag{3.10b}$$

Here each of these coefficients of skewness is a pure number (independent of units) and serves as an index of the degree of skewness. So if, for instance, $sk = 0$, the distribution is symmetrical; if $sk > 0$, the distribution is skewed to the right; and if $sk < 0$, the distribution is skewed to the left. By convention, if $|sk| > 1$,[4] the distribution is said to exhibit a substantial amount of skewness.

Example 3.10

Is the data set in Example 3.8 skewed? If so, which way? We previously found that $\bar{X} = 4$ and $s = 2.16$. Since this data set has no mode, Equation (3.10a) cannot be used. However, in those instances when the mode cannot be uniquely determined, an alternative to Equation (3.10a) is

$$sk = \frac{3(\bar{X} \quad \text{med})}{s}. \tag{3.10.1}$$

[4] absolute value function is defined as $|a| = \begin{cases} a, & a \geqslant 0 \\ -a, & a < 0 \end{cases}$

Since the median for the variable X is $med = 3.5$, we can readily find that

$$sk = \frac{3(4 - 3.5)}{2.16} = 0.69.$$

Hence this data set is moderately skewed to the right.

Another option is to use as our measure of skewness the average of the cubed Z-scores (called *standard skewness*) or

$$k_3 = \frac{\Sigma_{i=1}^{n} Z_i^3}{n-1} = \frac{1}{S^3} \frac{\Sigma_{i=1}^{n} (X_i - \bar{X})^3}{n-1}. \tag{3.11}$$

When $k_3 \approx 0$, the data set is taken to be symmetrical; for $k_3 > 0$, the data set is skewed right; and for $k_3 < 0$, the data set is skewed left. If $|k_3| > 0.5$, the data set is said to display a substantial amount of skewness. For instance, given the Example 3.8 data set, the reader can easily verify that $\Sigma(X_i - \bar{X})^3 = 18$, $s^3 = 10.08$, and thus $k_3 = \frac{1}{10.08}\left(\frac{18}{3}\right) = 0.5952$.

Hence, according to Equation (3.11), this data set is highly skewed to the right.

3.9 QUANTILES

In general, we can view *quantiles* as positional measures that divide the observations on a variable X into a number of equal portions (given that the data values are in an increasing sequence). We have already encountered the median of a data set; it is a positional value that splits the data into two equal parts. Others are:

Quartiles: Split the data into four equal parts
Deciles: Split the data into 10 equal parts
Percentiles: Split the data into 100 equal parts

There are three computed quartiles that will divide the observations on a variable X into four equal parts: Q_1, Q_2, and Q_3. In this regard, 25% of all observations on X lie below Q_1; 50% of all observations on X lie below Q_2 (= median); and 75% of all observations on X lie below Q_3. Remember that quartiles are "positional values." Hence the following:

$$Q_1 \text{ is located at } \frac{n+1}{4} \tag{3.12a}$$

$$Q_2 \text{ is located at } \frac{n+1}{2} \tag{3.12b}$$

$$Q_3 \text{ is located at } \frac{3(n+1)}{4} \tag{3.12c}$$

(provided, of course, that our data have been arranged in an increasing sequence). Given Equation (3.12), we can easily calculate the *interquartile range* (*IQR*) as $\text{IQR} = Q_3 - Q_1$—it is the range between the first and third quartiles and serves to locate the middle 50% of the observations on a variable X. Then the *quartile deviation* is readily formed as

$$q.d. = \text{IQR}/2 = (Q_3 - Q_1)/2 \tag{3.13}$$

and can be used as a measure of dispersion in the middle half of a distribution—a device that is particularly useful when outliers are present.

There are nine computed deciles that divide all the observations on X into 10 equal parts: $D_1, D_2, \ldots, D_9$. Thus 10% of all observations on X are below D_1, 20% of all data points for X lie below D_2, and so on. Here too deciles are "positional values" so that:

$$D_j \text{ is located at } \frac{j(n+1)}{10}, \ j = 1, \ldots, 9. \tag{3.14}$$

For instance, once all the observations on X are arranged in an increasing sequence, D_7 occupies the position $(7/10)(n+1)$.

Finally, 99 computed percentiles divide the observations on X into 100 equal parts: $P_1, P_2, \ldots, P_{99}$. Here 1% of all observations on X lie below P_1, 2% of the X values lie below P_2, and so on. Looking to the positions of these percentiles, we find that:

$$P_j \text{ is located at } \frac{j(n+1)}{100}, \ j = 1, \ldots, 99, \tag{3.15}$$

with, say, P_{40} found at $(40/100)(n + 1)$.

It is important to note that we can actually move in the reverse direction and find the percentile that corresponds to a particular value of X. That is, to determine the percentile of the score X', find

$$\text{Percentile of } X' = \frac{\text{Number of data points} < X'}{n} \times 100, \tag{3.16}$$

provided, of course, that the observations on X are in an increasing sequence. The value obtained in Equation (3.16) is rounded to the nearest integer.

Example 3.11

Given the following set of observations on X: 3, 8, 8, 6, 7, 9, 5, 10, 11, 20, 19, 15, 11, 16, find Q_1, D_2 and P_{30}. Upon arranging these $n = 14$ X_i's in increasing order,

3, 5, 6, 7, 8, 8, 9, 10, 11, 11, 15, 16, 19, 20,

we can easily determine that:

a. From Equation (3.12a), Q_1 is located at the position $(14+1)/4 = 15/4 = 3.75$, that is, Q_1 is $^3\!/_4$ of the distance between the third and fourth data points. Hence $Q_1 = 6.75$.
b. Using Equation (3.14), D_2 is located at the position $(2(14+1))/10 = 3$ or D_2 is located at the third observation. Hence $D_2 = 6$.
c. From Equation (3.15), P_{30} is located at the position $(30(14+1))/100 = 4.50$. Thus P_{30} lies half the distance between the fourth and fifth data points. Thus $P_{30} = 7.5$.

Additionally, what is the percentile associated with $X' = 15$? From Equation (3.16),

$$\text{Percentile of } 15 = \frac{10}{14} \times 100 = 71.42.$$

Thus, $X' = 15$ locates the 71st percentile of the X values, that is, approximately 71% of all observations on X lie below 15.

3.10 KURTOSIS

The concept of *kurtosis* refers to how flat or rounded the peak of a distribution happens to be. While there are a variety of ways to measure the degree of kurtosis in a data set, one approach, which is quite straightforward to apply, is the so-called *coefficient of kurtosis*:

$$k = \frac{q.d.}{P_{90} - P_{10}}, \tag{3.17}$$

where the quartile deviation ($q.d.$) is determined from Equation (3.13). The benchmark or point of reference for determining the degree of kurtosis for a data set is the normal distribution (it is continuous, bell-shaped, and symmetrical). In this regard:

a. For a normal curve, $k \approx 0.263$.
b. For a flat curve, $k \approx 0$.
c. For a sharply peaked curve, $k \approx 0.5$.

Typically k is computed when one has a fairly large data set.

An alternative to Equation (3.17) is to calculate *standard kurtosis* as the average of the Z-scores raised to the fourth power or

$$k_4 = \frac{\Sigma_{i=1}^{n} Z_i^4}{n-1} = \frac{1}{s^4}\frac{\Sigma_{i=1}^{n}(X_i - \bar{X})^4}{n-1}. \tag{3.18}$$

For a normal distribution (again used as our benchmark), $k_4 = 3$. Then our estimate of the degree of kurtosis is $k_4 - 3$ (called the *excess*). So if $k_4 - 3 > 0$ (positive kurtosis), the data set has a peak that is sharper than that of a normal distribution;

and if $k_3 - 3 < 0$ (negative kurtosis), the data set has a peak that is flatter than that of a normal distribution. In this latter instance, the distribution might have elongated tails, thus indicating the presence of outliers. For instance, again using the Example 3.8 data set, it can be shown that $\Sigma(X_i - \bar{X})^4 = 98$, $s^4 = 21.76$, and thus

$$k_4 - 3 = \frac{1}{21.76}\left(\frac{98}{3}\right) - 3 = -1.4988.$$

Hence this data set exhibits considerable uniformity—its peak is much flatter than that of a normal distribution.

3.11 DETECTION OF OUTLIERS

We noted above that a measure of central location such as the mean is affected by each observation in a data set and thus by extreme values. Extreme values may or may not be *outliers* (atypical or unusually small or large data values). For instance, given the data set X: 1, 3, 6, 7, 8, 10, it is obvious that 1 and 10 are extreme values for this data set, but they are not outliers. If, instead, we had X: 1, 3, 6, 7, 8, 42, then, clearly, 42 is an extreme value that would also qualify as an outlier.

There are two ways to check a data set for outliers:

1. We can apply the empirical rule—for a data set that is normally distributed (or approximately so, but certainly symmetrical), we can use Z-scores to identify outliers: an observation X_i is deemed an outlier if it is Z-score Z_i lies below –2.24 or above 2.24. Here ± 2.24 are the *cut-off points* or *fences* that leave us with only 2.5% of the data in each tail of the distribution. Both of these cases can be subsumed under the inequality

$$\frac{|X_i - \bar{X}|}{s} > 2.24. \tag{3.19}$$

While Equation (3.19) is commonly utilized to detect outliers, its usage ignores a rather acute problem—the values of both $\bar{X}$ and s are themselves affected by outliers. Why should we use a device for detecting outliers that itself is impacted by outliers? Since the median is not affected by outliers, let us modify Equation (3.19) to obtain the following:

$$\frac{|X_i - \text{med}|}{\text{mad}/0.6745} > 2.24, \tag{3.19.1}$$

where *mad* denotes the *sample median absolute deviation* that is determined as the median of the absolute deviations from the median.

That is, to find *mad*, first determine the set of absolute deviations from the median or

$$|X_1 - \text{med}|, |X_2 - \text{med}|, \ldots, |X_n - \text{med}|$$

and then find their median or middle value.

Example 3.12

Suppose that from a normal population we randomly obtain the set of $n = 10$ sample values:

$$0,\ 1,\ 1,\ 3,\ 4,\ 6,\ 8,\ 10,\ 11,\ 50.$$

Then $med = 5$ and

$$|0-5| = 5,\ |1-5| = 4,\ |1-5| = 4,\ |3-5| = 2, |4-5| = 1, |6-5| = 1,$$
$$|8-5| = 3,\ |10-5| = 5,\ |11-5| = 6,\ |50-5| = 45.$$

Arranging these absolute deviations from the median in an increasing sequence yields

$$1,\ 1,\ 2,\ 3,\ 4,\ 4,\ 5,\ 5,\ 6,\ 45.$$

Then $mad = 4$ and, from Equation (3.19.1)

$$\frac{|50-5|}{4/0.6745} = 7.58 > 2.24.$$

Thus only $X = 50$ can be considered to be an outlier.

2. We can employ quartiles—given the $\text{IQR} = Q_3 - Q_1$, the *fences* are:

$$\begin{aligned}(\text{Lower})\ l &= Q_1 - 1.5\ \text{IQR},\\ (\text{Upper})\ u &= Q_3 + 1.5\ \text{IQR}.\end{aligned} \tag{3.20}$$

So if a value of X is less than the lower fence l or greater than the upper fence u, it is considered to be an outlier.

Example 3.13

Given the $n = 14$ values of X appearing in Example 3.11, use the fences given by Equation (3.20) to determine if this data set exhibits any outliers. We previously found that $Q_1 = 6.75$. Since Q_3 is located at the position $\frac{3(14+1)}{4} = 11.25$ (it is a quarter of the distance between the 11th and 12th data points), we find $Q_3 = 15.25$. Then IQR $=$ $Q_3 - Q_1 = 8.50$ and

$$l = 6.75 - 1.5(8.50) = -6.0,$$
$$u = 15.25 + 1.5(8.50) = 28.0.$$

Since none of the X_i's is less than -6 or greater than 28, this data set does not have any outliers.

A modification of Equation (3.20) that employs departures from the median is: X_i is an outlier if

$$X_i < \text{med} - h\,\text{IQR} \quad \text{or if} \quad X_i > \text{med} + h\,\text{IQR}, \tag{3.20.1}$$

where

$$h = \frac{17.63n - 23.64}{7.74n - 3.71}.$$

What if outliers are detected for some data set? How should they be handled? Since the mean, for example, is affected by the presence of outliers, one could possibly discard any of the offending outlying values. However, if these data points are considered important or essential to a study, then discarding them would obviously not be appropriate. A useful alternative is to compute an *α-trimmed mean*, $0 < \alpha < 0.5$, and use it in place of $\bar{X}$. That is, for the set of n data points arranged in an increasing sequence, the $\alpha-$trimmed mean is determined by deleting the αn smallest observations and the αn largest observations, and then taking the arithmetic mean of the remaining data points. For instance, if $n = 30$ and $\alpha = 0.10$, then we would delete the $\alpha n = 3$ smallest and largest values of X, and then compute the simple average of the remaining 24 data values.

3.12 SO WHAT DO WE DO WITH ALL THIS STUFF?

How are the various descriptive measures developed in this chapter to be used in analyzing a data set? To answer this question, we need only remember that the characteristics of interest of a data set are *shape*, *central location* ("center" for short), and *dispersion* ("spread" for short). In this regard:

1. If a distribution is symmetric (or nearly so) with no outliers, then $\bar{X}$ and s provide us with legitimate descriptions of center and spread, respectively. (This is so because the values of $\bar{X}$ and s are sensitive to both inordinately large or small extreme values of a variable X. And if the distribution lacks symmetry, no single number can adequately describe spread since the two sides of a highly skewed distribution have different spreads.)

2. If a distribution is skewed and/or exhibits outliers, then $\bar{X}$ and s are inadequate for describing center and spread respectively. In this instance, a *boxplot* should be used in order to obtain a summary of both center and spread. To construct a boxplot, we need as inputs the *five number summary*:

$$\min X_i,\ Q_1,\ \text{median}\ (= Q_2),\ Q_3,\ \max X_i$$

along with the fences

$$l = Q_1 - 1.5\,\text{IQR},$$
$$u = Q_3 + 1.5\,\text{IQR},$$

where $\text{IQR} = Q_3 - Q_1$ (the interquartile range). Given this information, let us now undertake the following steps (see Fig. 3.6):

a. Draw a central box spanning Q_1 and Q_3.
b. A vertical line in the box marks the median Q_2.
c. Horizontal lines (called *whiskers*) extend from the box out to $\min X_i$ and $\max X_i$.
d. Find $\text{IQR} = Q_3 - Q_1$ and then plot the lower and upper fences l and u, respectively, on the horizontal lines using brackets "[" and "]" (any data point smaller than l or larger than u is deemed an outlier).

To interpret a boxplot such as the one in Fig. 3.6, first find the median that locates the center of the distribution. Then assess the spread—the span of the box (the IQR) shows the spread in the middle half of the data set while the range ($\max X_i - \min X_i$) yields a measure of the "total spread" of the data. If all data points are located with $[l,u]$, then there are no outliers. Clearly, Fig. 3.6 displays an outlier at the high end of the data values.

Once a box plot has been constructed, it can readily be used to assess the shape of a distribution. For instance:

a. If the median is near the center of the box and the horizontal lines or whiskers have approximately the same length, then the distribution is more or less symmetric (Fig. 3.7a).
b. If the median is located left-of-center of the box or the right horizontal line is substantially longer than its left counterpart, then the distribution is skewed to the right (Fig. 3.7b).

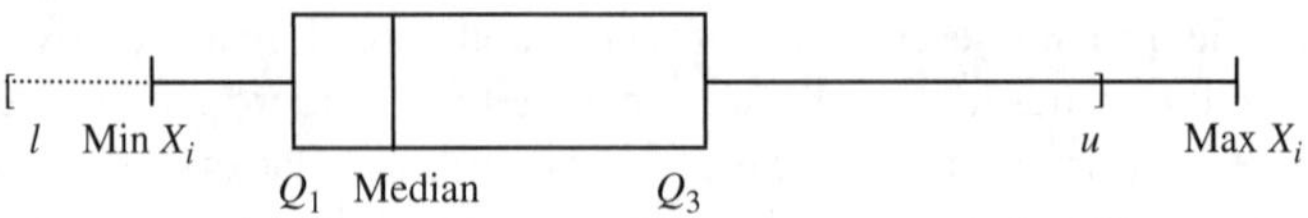

FIGURE 3.6 Boxplot.

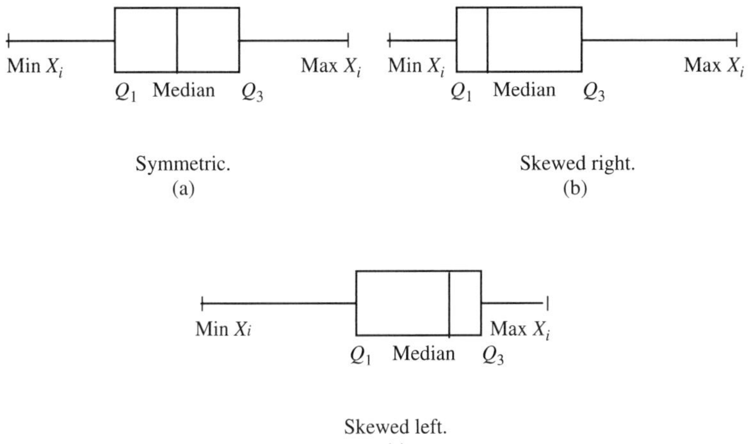

FIGURE 3.7 Boxplots.

c. If the median is located right-of-center of the box or the left horizontal line is substantially longer than the right horizontal line, then the distribution is skewed to the left (Fig. 3.7c).

Example 3.14

Given the following set of $n = 25$ observations on a variable X, discuss the shape, center, and spread of X's distribution.

X: 43, 98, 51, 86, 87, 86, 85, 71, 62, 77, 86, 90, 80, 97, 70, 75, 80, 86, 70, 95, 94, 81, 62, 90, 86

Arranging these X values in an increasing sequence yields

43, 51, 62, 62, 70, 70, 71, 75, 77, 80, 80, 81, 85, 86, 86, 86, 86, 86, 87, 90, 90, 94, 95, 97, 98.

Then:

$$\text{Min}\, X_i = 43, \max X_i = 98,$$

$$Q_1 \text{ has the position } \frac{n+1}{4} = 6.5 \text{ or } Q_1 = 70.5,$$

$$\text{Median} = Q_2 \text{ has the position } \frac{n+1}{2} = 13 \text{ or } Q_2 = 85,$$

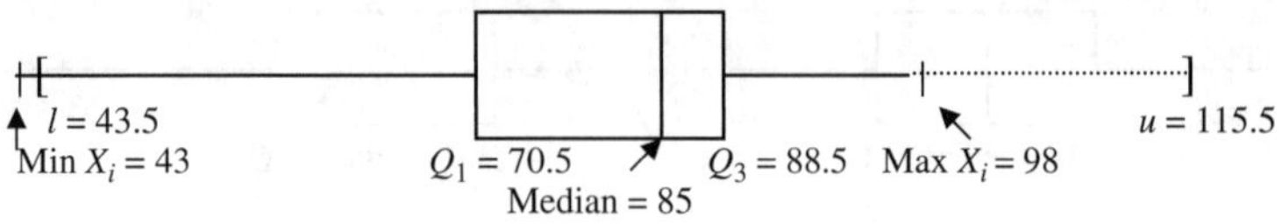

FIGURE 3.8

$$Q_3 \text{ has the position } \frac{3(n+1)}{4} = 19.5 \text{ or } Q_3 = 88.5, \text{ IQR} = 18,$$

$$l = Q_1 - 1.5\,\text{IQR} = 43.5,$$
$$u = 15.25 + 1.5\,\text{IQR} = 115.5.$$

Armed with this information, the resulting boxplot appears in Fig. 3.8. As this diagram indicates, the X distribution is highly skewed to the left and has a low-end outlier corresponding to the value 43.

Example 3.15

A task that one often faces in the area of applied statistics is the comparison of the characteristics of two (or more) data sets. As we have just determined, there are essentially three ways to proceed with any such comparison: (a) absolute frequency distributions and histograms provide us with a general picture of the overall shape of the data sets; (b) comparing their descriptive measures of central location, shape, and dispersion (e.g., mean, median, mode, standard deviation, and so on) adds much more information but may give us a distorted view of their relative properties if the distributions lack symmetry and contain outliers; and (c) boxplots, while showing less detail than, say, histograms, are useful for describing skewed distributions or distributions displaying outliers. Suppose we have $n_1 = 60$ sample observations on a variable X_1 and $n_2 = 60$ sample observations on the variable X_2 (Table 3.6).

TABLE 3.6 Observations on X_1, X_2

X_1						X_2					
17	20	31	38	40	50	10	18	22	35	43	49
15	22	30	38	39	60	16	19	21	37	44	44
17	20	33	41	40	55	12	16	21	37	50	49
22	27	29	39	41	62	17	20	20	37	48	46
23	22	29	35	38	50	22	19	19	39	55	51
27	25	31	34	35	51	19	21	22	38	56	52
20	25	35	35	37	56	20	19	30	39	54	48
19	20	36	41	41	62	13	16	29	40	57	48
15	22	35	38	51	54	17	17	30	45	53	46
16	28	36	40	52	38	17	19	31	43	50	40

TABLE 3.7 Absolute Frequency Distribution for X_1

Classes of X_1	Absolute Frequency f_j
15–22	14
23–30	9
31–38	16
39–46	9
47–54	5
55–62	6
63–70	1
	60

A. Absolute frequency distributions and histograms.
   Tables 3.7 and 3.8 house the absolute frequency distributions for these two variables and Figs. 3.9 and 3.10 display their corresponding histograms. What sort of comparisons emerge?
B. Descriptive measures for X_1 and X_2.

X_1	X_2
Central location	Central location
Mean = 35.1833	Mean = 32.7500
Median = 35	Median = 33
Mode = 35	Mode = 19
Dispersion	Dispersion
Range = 53	Range = 47
IQR = 17	IQR = 27
Standard deviation = 13.2352	Standard deviation = 14.2907
Coefficient of variation = 37.6177	Coefficient of variation = 43.6358
Shape	Shape
Skewness = 0.0139	Skewness = 0.9621
Kurtosis = 0.3471	Kurtosis = −1.5134

TABLE 3.8 Absolute Frequency Distribution for X_2

Classes of X_2	Absolute Frequency f_j
10–17	10
18–25	16
26–33	4
34–41	9
42–49	12
50–57	9
	60

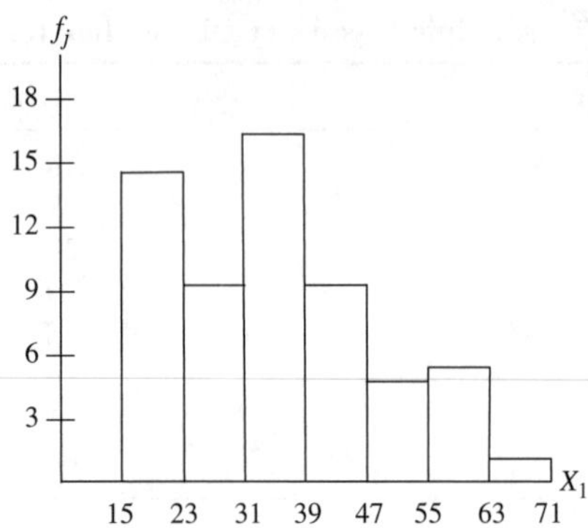

FIGURE 3.9 Absolute frequency histogram for X_1.

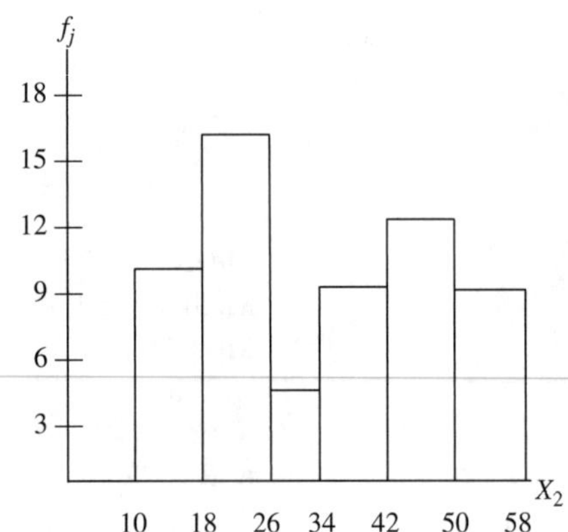

FIGURE 3.10 Absolute frequency histogram for X_2.

A glance at these sets of descriptive measures for X_1 and X_2 reveals the following:

1. The average value for the X_1 data points is slightly larger than the average of the X_2 observations.
2. The distribution for X_1 is much more symmetrical than that for X_2.
3. There is a greater range associated with the X_1 values and the X_2 variable has a higher degree of both absolute variation and relative variation.
4. Both distributions are skewed to the right and X_2 has a much greater degree of skewness relative to that of X_1.
5. Based upon their kurtosis measures, each distribution is a bit flatter than a normal distribution, with X_2 displaying a greater departure from normality than X_1.

(a) Boxplot for X_1.

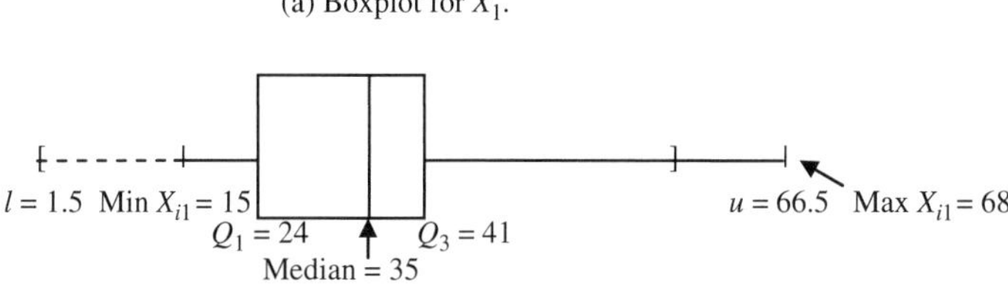

Skewed right, 68 is an outlier.

(b) Boxplot for X_2.

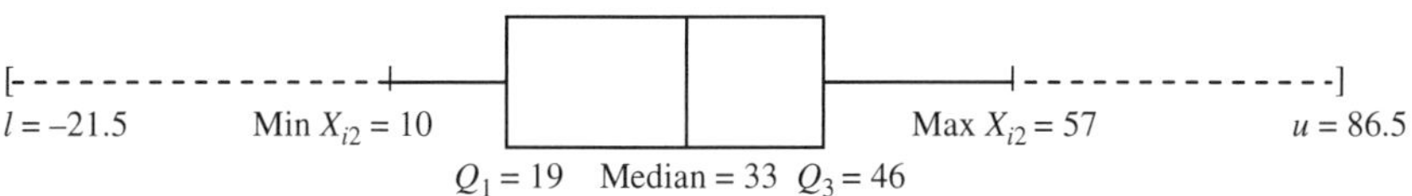

FIGURE 3.11 Boxpots for X_1 and X_2.

6. Looking to their IQR's, X_2 has a greater degree of variation in the middle half of its distribution relative to that corresponding to the X_1 data set.

C. Boxplots for X_1 and X_2.
Boxplots facilitate the "side-by-side comparison" of two (or more) distributions (Fig. 3.11).

EXERCISES

1 Expand the following expressions as sums of the terms involved:

a. $\Sigma_{i=1}^{4} a_i X_i^2$
b. $\Sigma_{i=1}^{4} a^i X_i$
c. $\Sigma_{j=1}^{3} \left(a^j - X_j\right)$
d. $\Sigma_{j=2}^{5} X_j^{j+1}(-1)^{j-1}$

2 Express the following in closed-form summation notation:

a. $X_1 Y_1 + X_2 Y_2 + X_3 Y_3$
b. $a_1 X_1^2 + a_2 X_2^2 + a_3 X_3^2$
c. $X_1 + X_2 + X_3 - Y_1 - Y_2 - Y_3$
d. $-X_1/Y_1 + X_2/Y_2 - X_3/Y_3$

3 Given that $X_1 = 1,\ X_2 = 4,\ X_3 = 6,\ Y_1 = 5,\ Y_2 = 1$ and $Y_3 = -4$, find:

a. $\Sigma_{i=1}^{3} X_i Y_i$
b. $\Sigma_{i=1}^{3} (X_i - Y_i)$
c. $\Sigma_{i=1}^{3} X_i Y_i^2$
d. $\left(\Sigma_{i=1}^{3} X_i\right)\left(\Sigma_{i=1}^{3} Y_i\right)$

4 Are the following valid equalities?

a. $\Sigma_{i=1}^{n} X_i^2 = \left(\Sigma_{i=1}^{n} X_i\right)^2$
b. $\Sigma_{i=1}^{n} X_i Y_i = \left(\Sigma_{i=1}^{n} X_i\right)\left(\Sigma_{i=1}^{n} Y_i\right)$

Verbalize each side of each of these expressions.
What general rules follow?

5 Given the following sample data sets, find the mean, median, mode(s), range, variance, and standard deviation:

a. X: 9, 11, 2, 5, 13
b. X: 8, 4, 14, 10, 12, 8, 4, 10, 2
c. X: 6, 2, 4, 0, −6, 0, −4, −4, 2
d. X: 8, 6, 20, 24, 14, 16, 2, 22

6 For data set (a) in Problem 3.5, verify that $\Sigma_{i=1}^{n} (X_i - \bar{X}) = 0$.

7 John has received his grades for the past semester. He wants to determine his quality-point average (QPA). John made Bs in two three-credit courses, an A in a five-credit course, and a C in a four-credit course. The numerical values assigned to the letter grades are: A = 4, B = 3, C = 2, D = 1 and F = 0. What is John's QPA?

8 Sarah has decided to make her own special trail mix rather than pay \$2.50 per pound at her local grocery store. She purchased 4 lbs of peanuts at \$3.50 per pound, 3 lbs of acorns at \$2.75 per pound, and 2 lbs of sprouts at \$2.25 per pound (yum). How does Sarah's cost per pound compare with her grocer's cost per pound?

9 Suppose a variable X has a bell-shaped distribution with a mean of 100 and a standard deviation of 10.

a. What percentage of X values lies between 90 and 110?
b. What percentage of X values lies between 80 and 120?
c. What percentage of X values lies between 70 and 130?
d. What percentage of X values lies below 80 or above 120?
e. What percentage of X values lies above 120?

10 A variable X is known to have a bell-shaped distribution with a mean of 510 and a standard deviation of 90?

a. What percentage of X values lies between 420 and 600?
b. What percentage of X values is less than 420 or greater than 600?
c. What percentage of X values is less than 330?
d. What percentage of X values is greater than 780?

11 A variable X has a mean of \$1.35 and a standard deviation of \$0.10.

a. What percentage of X values lies within 2.5 standard deviation of the mean? What is the implied range of X values?
b. What percentage of X values lies within 3 standard deviations of the mean? What is the implied range of X values?
c. What is the minimum percentage of X values that lies between \$1.15 and \$1.55?

12 A variable X has a mean of 27 and a standard deviation of 8.

a. What percentage of X values will lie within 1.5 standard deviation of the mean?
b. What are the values of X that lie within 1.5 standard deviations of the mean?
c. What is the minimum percentage of X values that lie between 3 and 51?

13 Transform the data set X: 2, 3, 4, 11, 15 into a set of Z-scores and fully interpret each such score. Verify that these Z-scores have a mean of zero and a standard deviation of 1.

14 Suppose we have the data set X: 1, 5, 7, 15. From X we obtain a new variable $Y = X - 10$ with values –9, –5, –3, and 5. Verify that the variances of the X and Y variables are equal. In general, the variance of a variable X equals the variance of its *linear translation* $Y = X + b$, b arbitrary. Use this fact to simplify the calculation of the variance of X: 1156, 1163, 1113, and 1100. [Hint: Try $b = -1135$.] Note: A more general form of this translation is $Y = aX + b$, $a \neq 0$, b arbitrary.

15 Suppose that over a 6 yr period a series of X values grows by the following year-to-year percentage values: 5%, 7%, 10%, 11%, and 20%. What is the average rate of growth in X over this period? If you decide to calculate $\bar{X} = (5 + 7 + 10 + 11 + 20)/5 = 10.6$ you would be incorrect. Why? Simply because there is such a thing as *compounding* going on. The correct procedure for finding average growth is to use the *geometric mean*

$$\mathrm{GM} = \sqrt[n]{X_1 \cdot X_2 \cdots X_n} = (X_1 \cdot X_2 \cdots X_n)^{\frac{1}{n}}$$

or the nth root of the product of n terms. What is the (geometric) mean growth rate for X? [Note: In general, $GM \leq \bar{X}$.]

16 Which of the following distributions (A or B or C) has more variability associated with it?

a. A: $\bar{X}_A = \$3.00,\ s_A = \$0.50$
B: $\bar{X}_B = \$3.02,\ s_B = \$0.63$

b. A: $\bar{X}_A = 115,\ s_A = 25$
B: $\bar{X}_B = 129,\ s_B = 15$

c. A: $\bar{X}_A = 35,\ s_A = 10$
B: $\bar{X}_B = 41,\ s_B = 15$
C: $\bar{X}_C = 21,\ s_C = 13$

17 For each of the data sets in Problem 3.5, determine and interpret the Pearson coefficient of skewness.

18 Given the following data set of $n = 20$ X values, find:

a. The quartiles
b. The third, sixth, and ninth deciles
c. The quartile deviation
d. The 10th, 65th and 90th percentiles
e. What is the percentile of the values $X = 2.34$ and $X = 5.20$?

0.99 1.16 1.75 2.34 2.50 2.75 3.41 3.48 3.91 3.96
4.00 4.04 4.11 4.75 5.20 5.45 5.79 6.16 6.28 7.75

19 For the following set of $n = 20$ observations on a variable X, find the following:

a. The quartiles
b. The fourth, seventh, and eighth deciles
c. The quartile deviation
d. The 10th, 35th, and 90th percentiles
e. What is the percentile associated with the values $X = 13$ and $X = 41$?

1 7 8 9 13 15 20 21 23 27 30 35 40 41 42
43 52 71 89 92

20 Given the data set in Problem 3.18, calculate and interpret the coefficient of kurtosis (Eq. (3.17)).

21 Given the data set in Problem 3.19, calculate and interpret the coefficient of kurtosis (Eq. (3.17)).

22 For the following data set, calculate and interpret the degree of kurtosis or the excess [use Equation(3.18)].

$$X{:}\ 3,\ 6,\ 7,\ 5,\ 9,\ 8,\ 10,\ 7,\ 6,\ 5$$

23 Check the data set in Problem 3.18 for outliers using Equation (3.20). Recalculate the fences using Equation (3.20.1).

24 Check the data set in Problem 3.19 for outliers using Equation (3.20). Recalculate the fences using Equation (3.20.1).

25 Check the following data set for the presences of outliers. Use Equation (3.19.1).

$$X\colon 1,\ 15,\ 19,\ 14,\ 18,\ 20,\ 17,\ 28,\ 40$$

26 For the data set in Problem 3.25, compare $\bar{X}$ with the α—trimmed mean for $\alpha = 0.20$.

27 Create boxplots for:

a. The data set in Problem 3.18

b. The data set in Problem 3.19

Fully interpret the results.

28 Given the following absolute frequency distribution for a variable X, find: $\bar{X}$, median, mode, s, Q_1, Q_3, D_3, D_6, P_{10} and P_{33}.

Classes of X	f_j
16–21	10
22–27	15
28–33	30
34–39	40
40–45	51
46–51	39
50–57	20
58–63	10
	215

APPENDIX 3.A DESCRIPTIVE CHARACTERISTICS OF GROUPED DATA

Let us assume that a data set has been summarized in the form of an absolute frequency distribution involving classes of a variable X and absolute class frequencies f_j (see Chapter 2). This type of data set appears in Table 3.A.1. Quite often one is faced with analyzing a set of observations presented in this format. (For instance, some data sets contain proprietary information and the owner will only agree to present the data in summary form and not release the individual observation values. Or maybe the data set is just too large to have all of its observations printed in some report or document.) Given that the individual observations lose their identity in the grouping process, can

TABLE 3.A.1 Absolute Frequency Distribution

Classes of X	f_j
20–29	3
30–39	7
40–49	8
20–59	12
60–69	10
70–79	6
80–89	4
	50

we still find the mean, median, mode, standard deviation, and quantiles of X? We can, but not by using the formulas presented earlier in this chapter. In fact, we can only get "approximations" to the mean, median, and so on. However, as we shall soon see, these approximations are quite good. In what follows, we shall assume that we are dealing with a sample of size n.

3.A.1 The Arithmetic Mean

To approximate $\bar{X} = \sum_{i=1}^{n} X_i/n$, let us use the formula

$$\bar{X} \approx \frac{\Sigma_{j=1}^{k} m_j f_j}{n}, \tag{3.A.1}$$

where m_j is the class mark (midpoint) of the jth class and there are k classes. (Note that in Equation (3.A.1) we sum over all k classes instead of over the n data values.) Why will this approximation work? Since m_j serves as a representative observation from the jth class, it follows that $m_j f_j$ gives a good approximation to the sum of the observations in the jth class (i.e., there are f_j data points in the jth class and they are all taken to equal m_j). Hence $\Sigma_{j=1}^{k} m_j f_j \approx \Sigma_{i=1}^{n} X_i$. [Incidentally, Equation (3.A.1) is actually a weighted mean—the class marks are weighted by the class frequencies and $\Sigma_{j=1}^{k} f_j = n$.]

While Equation (3.A.1) is appropriate for approximating $\bar{X}$, it can be modified somewhat to give

$$\bar{X} \approx A + \left(\frac{\Sigma_{j=1}^{k} f_j d_j}{n}\right) c, \tag{3.A.2}$$

where

A = The *assumed mean* (it is the class mark of an arbitrarily selected class near the middle of the distribution)

C_A = The class that houses A

d_j = The *class deviation* for the *j*th class (it is the number of classes away from C_A that any other class finds itself)

c = Common class interval or class length

[Equation (3.A.2) can be derived from Equation (3.A.1) by substituting $m_j = A + c\, d_j$ into the latter and simplifying.] Let us use Equation (3.A.2) to find $\bar{X}$ for the distribution appearing in Table 3.A.1 (see Table 3.A.2). We first arbitrarily select C_A as the class 50–59. Thus $A = 54.5$ (that classes midpoint). Next, class C_A gets the zeroth class deviation, lower classes get negative class deviations (the class 30–39 is two classes below C_A), and higher classes get positive class deviations (the class 80–89 is three classes above C_A). From Equation (3.A.2), with $c = 10$,

$$\bar{X} \approx 54.5 + \left(\frac{3}{50}\right)10 = 55.1.$$

Example 3.A.1

Let us employ Equation (3.A.2) to approximate $\bar{X}$ for the data set presented in Table 3.A.3. Upon arbitrarily selecting the class 30–39 to house the assumed mean A, we find that

$$\bar{X} \approx A + \left(\frac{\Sigma f_j d_j}{n}\right) c = 34.5 + \left(\frac{2}{100}\right) 10 = 34.7.$$

3.A.2 The Median

Suppose we want to determine the median of the grouped data set appearing in Table 3.A.1 (repeated here as Table 3.A.4). Which class contains the median? (Note that the observations are already in an increasing sequence since the classes are ordered from 20–29 to 80–89.) To answer this question, let us first determine the

TABLE 3.A.2 Absolute Frequency Distribution

Classes of X	f_j	A	d_j	$f_j d_j$
20–29	3		−3	−9
30–39	7		−2	−14
40–49	8		−1	−8
C_A: 50–59	12	54.5	0	0
60–69	10		1	10
70–79	6		2	12
80–89	4		3	12
	50			3

TABLE 3.A.3 Absolute Frequency Distribution

Classes of X	f_j	A	d_j	f_jd_j
0–9	2		−3	−6
10–19	13		−2	−26
20–29	31		−1	−31
C_A: 30–39	20	34.5	0	0
40–49	16		1	16
50–59	10		2	20
60–69	4		3	12
70–79	3		4	12
80–89	1		5	5
	100			2

position of the median—it is located at the position $\frac{n+1}{2} = 25.5$, that is, it lies half the distance between the 25th and 26th observations. So which class contains the 25th and 26th observations. Let us start cumulating the absolute frequencies f_j, that is, "up to" the class 50–59 we have 18 observations; and "up through" this class we have 30 data points. Hence the median resides in the class 50–59. Let us denote this class as C_m—"the class containing the median."

We can now apply the formula

$$\text{Median} \approx L + \left(\frac{\frac{n}{2} - CF}{f_{C_m}} \right) c, \qquad (3.A.3)$$

where

$L =$ Lower boundary of C_m
$CF =$ Cumulative frequency of the classes preceding C_m
$f_{C_m} =$ Absolute frequency of C_m
$c =$ Length of C_m

TABLE 3.A.4 Absolute Frequency Distribution

Classes of X	f_j	
20–29	3	} 18
30–39	7	
40–49	8	
C_m: 50–59	12	
60–69	10	
70–79	6	
80–89	4	
	50	

Then from Table 3.A.4 and Equation (3.A.3),

$$\text{Median} \approx 49.5 + \left(\frac{25-18}{12}\right)10 = 55.33.$$

Example 3.A.2

For the distribution presented in Table 3.A.5, use Equation (3.A.3) to approximate the median. Since $\frac{n+1}{2} = 50.5$, the median lies half of the distance between the 50th and 51st observations. Since we have 46 observations "up to" the class 30–39 and 66 observations "up through" it, the median must lie within the class 30–39 (this class is labeled C_m). Then from Equation (3.A.3),

$$\text{Median} \approx 29.5 + \left(\frac{50-46}{20}\right)10 = 31.5.$$

3.A.3 The Mode

To determine the mode for a grouped data set (e.g., Table 3.A.6), let us first identify *the modal class* (denoted C_o) or class with the highest frequency. Clearly, this is the class 50–59. Also, we know that the *premodal class* is 40–49; and the *postmodal class* is 60–69. Armed with these considerations, we now turn to the following approximation formula

$$\text{Mode} \approx L + \left(\frac{\Delta_1}{\Delta_1 + \Delta_2}\right)c, \tag{3.A.4}$$

TABLE 3.A.5 Absolute Frequency Distribution

Classes of X	f_j	
0–9	2	
10–19	13	} 46
20–29	31	
C_m: 30–39	20	
40–49	16	
50–59	10	
60–69	4	
70–79	3	
80–89	1	
	100	

TABLE 3.A.6 Absolute Frequency Distribution

Classes of X	f_j	
20–29	3	
30–39	7	
40–49	8	$\Delta_1 = 12 - 8 = 4$
C_o: 50–59	12	
60–69	10	$\Delta_2 = 12 - 10 = 2$
70–79	6	
80–89	4	
	50	

where

$L =$ Lower boundary of C_o

$\Delta_1 =$ The difference between the frequency of the modal class and the premodal class

$\Delta_2 =$ The difference between the frequency of the modal class and the postmodal class

$c =$ Length of C_o

(Note that if $\Delta_1 = \Delta_2$, then mode $\approx L + \frac{1}{2}c =$ class mark or midpoint of the modal class.) So from Equation (3.A.4) and Table 3.A.6,

$$\text{Mode} \approx 49.5 + \left(\frac{4}{4+2}\right)10 = 56.16.$$

Example 3.A.3

Using the Table 3.A.7 data set and Equation (3.A.4), find the mode. Since C_o is the class 20–29, we can readily obtain

$$\text{Mode} \approx 19.5 + \left(\frac{18}{18+11}\right)10 = 25.71.$$

TABLE 3.A.7 Absolute Frequency Distribution

Classes of X	f_j	
0–9	2	
10–19	13	
C_o: 20–29	31	$\Delta_1 = 31 - 13 = 18$
30–39	20	$\Delta_2 = 31 - 20 = 11$
40–49	16	
50–59	10	

3.A.4 The Standard Deviation

To find the standard deviation for a set of grouped data, we can use the expression

$$s \approx \sqrt{\frac{\Sigma_{j=1}^{k} f_j (m_j - \bar{X})^2}{n-1}}, \tag{3.A.5}$$

where we sum over the k classes. To perform this calculation, squared deviations from the mean $\bar{X}$ (using the class marks or the m_j's) are weighted by the class frequencies f_j. But since $m_j = A + cd_j$ and $\bar{X} = A + (\Sigma f_j d_j/n)\,c$, a substitution of these quantities into Equation (3.A.5) yields the more computationally friendly formula:

$$s \approx c\sqrt{\frac{\Sigma f_j d_j^2}{n-1} - \frac{(\Sigma f_j d_j)^2}{n(n-1)}}. \tag{3.A.6}$$

Here c is the common class interval and d_j is a "class deviation." (We can actually use the same d_j's that were employed in finding $\bar{X}$. Keep in mind, however, that this need not be the case since the location of the zeroth class deviation is arbitrary.) Looking to Table 3.A.1, let us use Equation (3.A.6) to approximate s. Table 3.A.8 houses our calculations.

$$s \approx 10\sqrt{\frac{133}{49} - \frac{(3)^2}{50 \times 49}} = 10\sqrt{2.714 - 0.0037} = 16.46$$

Example 3.A.4

Using the grouped data set presented in Table 3.A.3, use Equation (3.A.6) to approximate s. Table 3.A.10 lays out our basic computations for insertion into Equation (3.A.6). Remember that the location of the zeroth class deviation in the

TABLE 3.A.8 Absolute Frequency Distribution

Classes of X	f_j	d_j	d_j^2	$f_j d_j$	$f_j d_j^2$
20–29	3	−3	9	−9	27
30–39	7	−2	4	−14	28
40–49	8	−1	1	−8	8
50–59	12	0	0	0	0
60–69	10	1	1	10	10
70–79	6	2	4	12	24
80–89	4	3	9	12	36
	50			3	133

TABLE 3.A.9 Absolute Frequency Distribution

Classes of X	f_j	d_j	d_j^2	$f_j d_j$	$f_j d_j^2$
0–9	2	−4	16	−8	32
10–19	13	−3	9	−39	117
20–29	31	−2	4	−62	124
30–39	20	−1	1	−20	20
40–49	16	0	0	0	0
50–59	10	1	1	10	10
60–69	4	2	4	8	16
70–79	3	3	9	9	27
80–89	1	4	16	4	16
	100			−98	362

d_j column is arbitrary.

$$s \approx 10\sqrt{\frac{362}{99} - \frac{(-98)^2}{100 \times 99}} = 10\sqrt{3.65 - 0.97} = 16.371$$

3.A.5 Quantiles (Quartiles, Deciles, and Percentiles)

We previously defined quantiles as positional measures that divide the observations on a variable X into a number of equal portions (given, of course, that the observations are in an increasing sequence.) How are the various quantiles determined when we have a set of grouped data? (Remember that the observations are already in an increasing sequence by virtue of the way the classes are structured.)

1. Quartiles (split the observations on X into four equal parts)
   (a) The first quartile Q_1 (the point below which 25% of the observations on X lie) is located at the position $n/4$; suppose it falls within the class denoted C_{Q_1}. Then

$$Q_1 \approx L + \left(\frac{\frac{n}{4} - \text{CF}}{f_{Q_1}} \right) c \tag{3.A.7}$$

where

$L =$ Lower boundary of C_{Q_1}
$CF =$ Cumulative frequency of the classes preceding C_{Q_1}
$f_{Q_1} =$ Absolute frequency of C_{Q_1}
$c =$ Length of C_{Q_1}

(b) The second quartile Q_2 (the point below which 50% of the observations on X lie) is the median [see Equation (3.A.3)].

(c) The third quartile Q_3 (the point below which 75% of the observations on X lie) is located at the position $3n/4$; suppose it falls within the class labeled C_{Q_3}. Then

$$Q_3 \approx L + \left(\frac{\frac{3n}{4} - CF}{f_{Q_3}} \right) c, \tag{3.A.8}$$

where

L = Lower boundary of C_{Q_3}
CF = Cumulative frequency of the classes preceding C_{Q_3}
f_{Q_3} = Absolute frequency of C_{Q_3}
c = Length of C_{Q_3}

2. Deciles (split the observations on X into 10 equal parts)
The jth decile D_j (the point below which $10j$% of the observations on X lie) is located at the position $jn/10$; suppose it falls within the class C_{D_j}. Then

$$D_j \approx L + \left(\frac{\frac{jn}{10} - CF}{f_{D_j}} \right) c, j = 1, \ldots, 9, \tag{3.A.9}$$

where

L = Lower boundary of C_{D_j}
CF = Cumulative frequency of the classes preceding C_{D_j}
f_{D_j} = Absolute frequency of C_{D_j}
c = Length of C_{D_j}

3. Percentiles (split the observations on X into 100 equal parts)
The jth percentile P_j (the point below which j% of the observations on X lie) is located at the position $jn/100$; let us say it falls within the class C_{P_j}. Then

$$P_j \approx L + \left(\frac{\frac{jn}{100} - CF}{f_{P_j}} \right) c, j = 1, \ldots, 99, \tag{3.A.10}$$

where

L = Lower boundary of C_{P_j}
CF = Cumulative frequency of the classes preceding C_{P_j}
f_{P_j} = Absolute frequency of C_{P_j}
c = Length of C_{P_j}

TABLE 3.A.10 Absolute Frequency Distribution

Classes of X	f_j	$\leq$
5–7	2	2
8–10	5	7
11–13	22	29
14–16	13	42
17–19	7	49
20–22	1	50
	50	

Example 3.A.5

Given the grouped data set presented in Table 3.A.9, find Q_1, D_4, and P_{10}. For convenience, a "$\leq$" cumulative frequency column has been included so that the various CF values can be readily determined.

a. Q_1's position is $n/4 = 50/4 = 12/5$ (it lies half the distance between the 12th and 13th data points). Thus, C_{Q_1} is the class 11–13. Then from Equation (3.A.7),

$$Q_1 \approx 10.5 + \left(\frac{12.5 - 7}{22}\right)3 = 11.25.$$

Thus 25% of all observations on X lie below 11.25.

b. D_4 is found at the position $4n/10 = 200/10 = 20$ (its value corresponds to that of the 20th observation). Thus C_{D_4} is in the class 11–13. Then from Equation (3.A.9),

$$D_4 \approx 10.5 + \left(\frac{20 - 7}{22}\right)3 = 12.27.$$

Thus 40% of all observations on X lie below 12.27.

c. P_{70}*'s position is* $70n/100 = 350/100 = 35$ *(its value corresponds to the value of the 35th data point). Hence* $C_{P_{70}}$ *is the class 14–16. Then from* Equation (3.A.10),

$$P_{70} \approx 13.5 + \left(\frac{35 - 29}{13}\right)3 = 14.88.$$

Thus 70% of all observations on X fall below the value 14.88.

4

ESSENTIALS OF PROBABILITY

4.1 SET NOTATION

This section offers a brief review of set notation and concepts (readers familiar with this material showed proceed to the next section). These (set) notions are important in that they offer the reader a way of "thinking" or a way to "organize information". As we shall see shortly, only "events" have probabilities associated with them, and these events are easily visualized/described and conveniently manipulated via set operations.

Let us define a *set* as a collection or grouping of items without regard to structure or order—we have an amorphous group of items. Sets will be denoted using capital letters ($A, B, C, \ldots$). An *element* is a member of a set. Elements will be denoted using small-case letters ($a, b, c, \ldots$). A set is defined by listing its elements. If forming a list is impractical or impossible, then we can define a set in terms of some key property that the elements, and only the elements, of the set possess. For instance, the set of all odd numbers can be defined as $N = \{n|n + 1 \text{ is even}\}$. Here n is a representative member of the set N and the vertical bar reads "such that."

If an element x is a member of set X, we write $x \in X$(*element inclusion*); if x is not a member of X, then $x \notin X$(we negate the element inclusion symbol). If set A is a subset of B (a *subset* is a set within a set), we write $A \subseteq B$(set inclusion); if A is not a subset of B, then $A \nsubseteq B$ (we negate the set inclusion symbol).[1]

[1] If $A \subseteq B$, then either $A \subset B$ (A is a *proper subset* of B, meaning that B is a larger set) or $A = B$. Actually, *set equality* is formally defined as follows: if $A \subseteq B$, and $B \subseteq A$, then $A = B$.

Statistical Inference: A Short Course, First Edition. Michael J. Panik.

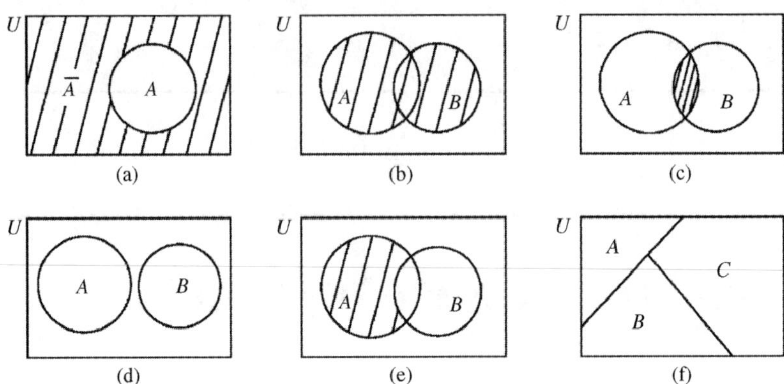

FIGURE 4.1 Venn diagrams for selected sets.

The *null set* (denoted by ∅) is the empty set; it is the set with no elements and thus is a subset of every set. The *universal set* (denoted by *U*) is the set containing all elements under a particular discussion. For instance, how many people on earth are 10 ft tall? Clearly, that constitutes a null set. The universal set, however, does not contain "everything under the sun"—it contains essentially what you want it to contain; you restricted the discussion. In this regard, it could be everyone in a particular room, or everyone in a given city, or everyone in some country, and so on.

The *complement of a set A* (denoted by $\bar{A}$) is defined relative to the universal set; it consists of all elements of *U* lying outside *A* (Fig. 4.1a).[2] It should be apparent that $\bar{\bar{A}} = A$, $\bar{\varnothing} = U$, and $\bar{U} = \varnothing$.

The *union of sets A and B* (denoted as $A \cup B$) is the set of elements in *A* or in *B* or in both *A* and *B* (Fig. 4.1b). (Note that the "or" used here is the "inclusive or.") Also, the *intersection of sets A and B* (written as $A \cap B$) is the set of elements common to both *A* and *B* (Fig. 4.1c). If $A \cap B = \varnothing$, then sets *A* and *B* are said to be *disjoint* or *mutually exclusive* (Fig. 4.1d). The *difference of sets A and B* (denoted by *A* – *B*) is that portion of *A* that lies outside *B* (Fig. 4.1e), that is, $A - B = A \cap \bar{B}$ (it highlights what *A* has in common with everything outside *B*). Finally, a *partition of U* is a collection of mutually exclusive and *collectively exhaustive* sets in *U* (Fig. 4.1f). As Fig. 4.1f reveals,

$$A \cap B = A \cap C = B \cap C = \varnothing \text{ (mutually exclusive)}$$

$$U = A \cup B \cup C \text{ (collectively exhaustive)}.$$

Thus the union of *A*, *B*, and *C* renders all of *U*.

[2] Illustrations such as those in Fig. 4.1 are called *Venn diagrams*; the universal set is depicted as a rectangle and individual sets in *U* are drawn (typically) as circles.

Based upon the preceding definitions, it is readily seen that

$$
\begin{array}{ll}
A \cup A = A & A \cup \bar{A} = U \\
A \cap A = A & A \cap \bar{A} = \varnothing \\
A \cup \varnothing = A & \text{DeMorgan's Laws:} \\
A \cap \varnothing = \varnothing & 1.\ \overline{A \cup B} = \bar{A} \cap \bar{B}; \\
A \cup U = U & 2.\ \overline{A \cap B} = \bar{A} \cup \bar{B} \\
A \cap U = A &
\end{array}
\tag{4.1}
$$

4.2 EVENTS WITHIN THE SAMPLE SPACE

Let us define a *random experiment* as a class of occurrences that can happen repeatedly, for an unlimited number of times, under essentially unchanged conditions. That is, we should be able to perform "similar trials" of this experiment "in the long run." A *random phenomenon* is a phenomenon such that, on any given trial of a random experiment, the outcome is unpredictable. The resulting outcome will be termed an *event*, that is, an event is an outcome of a random experiment.

The *sample space* (denoted by S) is the collection of all possible outcomes of a random experiment; it is a class of events that are mutually exclusive and collectively exhaustive. A *simple event* E_i is a possible outcome of a random experiment that cannot be decomposed into a combination of other events—it is essentially a point in S (Fig. 4.2). A *compound event* A is an event that can be decomposed into simple events (Fig. 4.2). (Here A is made up of five simple events.)

Thus far, we have been dealing exclusively with points (or groups of points) in S. How do we get some numbers out of our random experiment? This is the job of the *random variable* X. Formally, it is a real-valued (point-to-point) function defined on the elements of S. That is, it is a rule or law of correspondence that associates with each $E_i \in S$ some number X_i (Fig. 4.3). Let us now consider an example problem that will pull all of these concepts together.

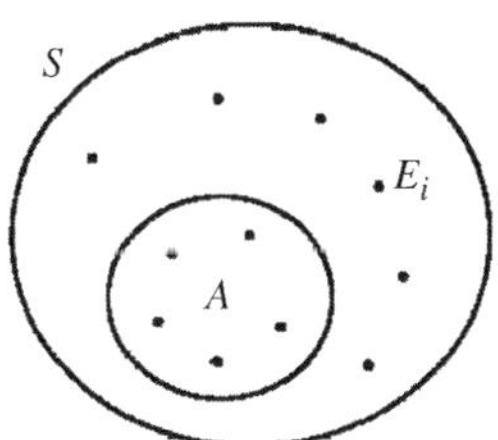

FIGURE 4.2 Events as subsets of S.

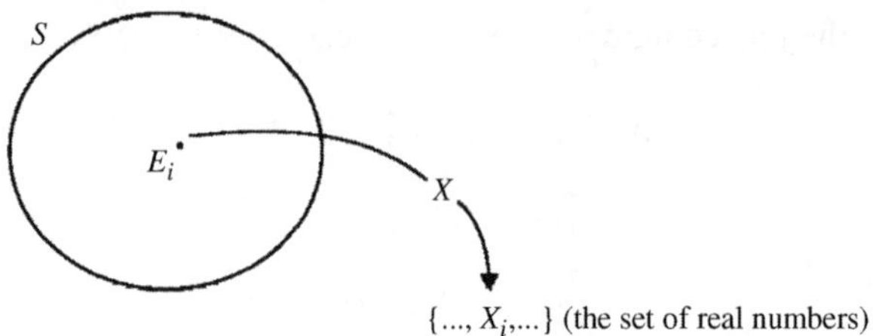

FIGURE 4.3 X connects E_i to the real number X_i.

Example 4.1

Let a random experiment consist of rolling a pair of (six-sided) dice. If the pair of dice is "fair" (i.e., not "loaded"), then, on any one roll, the outcome is a random phenomenon—the faces showing cannot be easily predicted. Figure 4.4 depicts the sample space S and the 36 simple events (points) in S. Let the random variable X be defined as the "sum of the faces showing." Clearly, X: 2, 3, 4, 5, 6, 7, 8, 9, 10, 11, 12 (eleven values or real numbers).[3]

Next, let us specify the (compound) event A as "getting a sum of the faces equal to seven." Is there more than one way to get a sum equal to 7? Clearly, the answer is yes since X will map each of the simple events E_6 or E_{11} or E_{16} or E_{21} or E_{26} or E_{31} into 7. Hence $A = \{E_6, E_{11}, E_{16}, E_{21}, E_{26}, E_{31}\}$ (see Fig. 4.4). In fact, a glance back at all of the "ors" connecting these simple events enables us to see immediately that any compound event (such as A) is the union of the simple events comprising it. When will event A occur? Event A occurs when, on any given trial of our random experiment, any one of the simple events comprising A occurs, for example, A occurs if E_6 occurs or if E_{11} occurs, and so on.

4.3 BASIC PROBABILITY CALCULATIONS

How do we associate probabilities with events? Since events are subsets of S, let us employ a *set function*[4] P from the sample space to the *unit interval* **[0,1]**. Here P is a rule that can be ascribed to any event $A \subseteq S$ some number $P(A)$, $0 \leq P(A) \leq 1$, called its *probability* (Fig. 4.5). Note that if $P(A) = 0$, then A is the *impossible event*; and if

[3] Actually, the random variable can be anything that one wants it to be—you are only limited by the physical structure of the random experiment. For instance, we could take X to be the "product of the faces showing." Now the lowest value of X is 1 and the highest value of X is 36. Also, what if one die was red and the other was blue. Then X could be defined as, say, the outcome on the red die divided by the outcome on the blue die. In this instance, the lowest value of X is $1/6$ and the highest value is 6. Other possibilities for defining X abound. What is important to remember is that it is the random variable that defines the possible outcomes of the random experiment—it tells us how to get some real numbers from the random experiment.

[4] Set functions should be familiar to everyone—although you probably never referred to them as such. For instance, "length" is a set function: the length of the line segment from a to b (it is a set) is determined by the rule "length $= b - a$." The area of a rectangle with height h and base b (it is also a set) is calculated according to the rule "area $= b \times h$." In each case, a rule associated with a set yields some nonnegative number called its length or area. Is volume a set function?

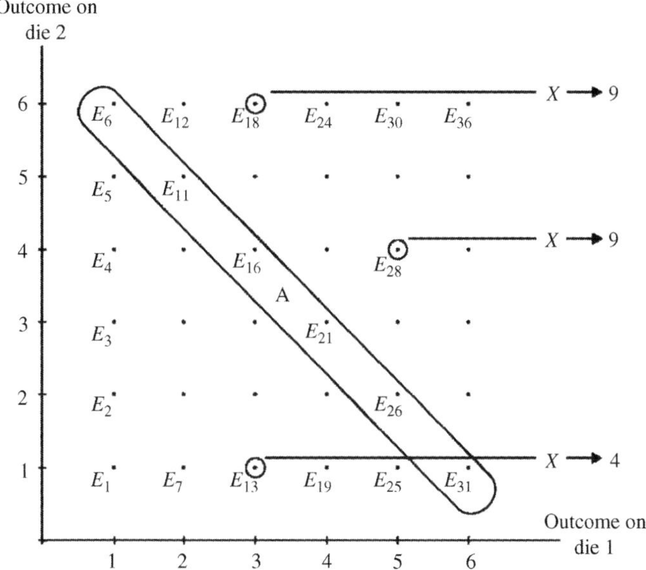

FIGURE 4.4 Sample space S and its simple events E_i, $i = 1, \ldots, 36$.

$P(A) = 1$, then A is the *certain event*. Thus probability is a number between 0 and 1 inclusive, where the larger is $P(A)$, the more likely it is that event A will occur.

What are the characteristics of P? To answer this question, we start with a set of *probability axioms* (an *axiom* is an accepted or established principle—a self-evident truth):

Axiom 1. $P(A) \geq 0$ for any event $A \subseteq S$.

Axiom 2. $P(S) = 1$.

Axiom 3. If A, B are mutually exclusive events in S, then $P(A \cup B) = P(A) + P(B)$, where "$\cup$" is interpreted as "or".

Thus the union of disjoint events in S is the sum of their individual probabilities.

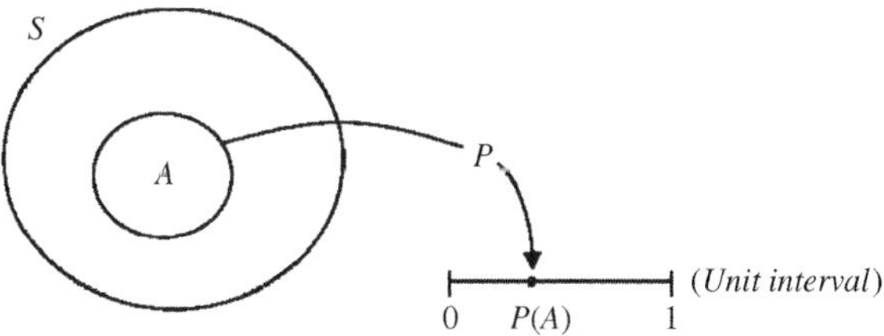

FIGURE 4.5 *P* is a set function on *S*.

These axioms lead to a set of *corollaries* (a *corollary* is a consequence—something that naturally follows):

Corollary 1. $P(\bar{A}) = 1 - P(A), A \subseteq S$. Thus the probability that "event A does not occur" equals one minus the probability that it does occur.

Corollary 2. $0 \leq P(A) \leq 1, A \subseteq S$.

Corollary 3. $P(\varnothing) = 0$.

In what follows, these axioms and corollaries will expedite or facilitate the calculation of probabilities of events.

Let us return to the event $A = \{E_6 \cup E_{11} \cup E_{16} \cup E_{21} \cup E_{26} \cup E_{31}\}$ (the sum of the faces showing on the roll of a pair of dice is seven). What is $P(A)$? To answer this, we first note that the simple events E_i, $i = 1, \ldots, 36$, in S are *equiprobable* or $P(E_i) = \frac{1}{36}$ for all i. (Thus Axioms 1 and 2 above hold.) Then,

$$\begin{aligned}
P(A) &= P(\cup E_i \text{ comprising } A) \\
&= \textstyle\sum_{E_i \in A} P(E_i) \text{ (by Axiom 3)} \\
&= P(E_6) + P(E_{11}) + P(E_{16}) + P(E_{21}) + P(E_{26}) + P(E_{31}) \\
&= \frac{1}{36} + \frac{1}{36} + \frac{1}{36} + \frac{1}{36} + \frac{1}{36} + \frac{1}{36} = \frac{6}{36} = \frac{1}{6}.
\end{aligned}$$

Thus the probability of an event A equals the sum of the probabilities of the individual simple events comprising A.

Can we take a shortcut to this final answer? We can by applying the

Classical definition of probability: if a random experiment has n equiprobable outcomes $[P(E_i) = \frac{1}{n}$ for all $i]$ and if n_A of these outcomes constitute event A, then

$$P(A) = \frac{n_A}{n}. \tag{4.2}$$

Hence the classical approach to probability has us count equiprobable points or simple events within S: we take the number of points in A and divide that value by the total number of simple events in S. Then according to Equation (4.2), for A specified as "the sum of the faces is seven," we simply take $P(A) = \frac{6}{36}$ (Fig. 4.4).

Axiom 3 dealt with the instance in which events A and B cannot occur together, that is, $A \cap B = \varnothing$. If events A and B can occur together, then $A \cap B \neq \varnothing$ so that $P(A \text{ and } B) = P(A \cap B) \neq 0$. In this latter instance, how do we determine $P(A \text{ or } B) = P(A \cup B)$? Here we look to the

General addition rule for probabilities: for events $A, B \subseteq S$ with $A \cap B \neq \varnothing$,

$$P(A \cup B) = P(A) + P(B) - P(A \cap B). \tag{4.3}$$

Why net out $P(A \cap B)$ from $P(A) + P(B)$? Since $A \cap B$ is in A, we count its probability of occurrence once in $P(A)$; and since $A \cap B$ is also in B, its probability of occurrence is counted a second time in $P(B)$. To avoid double counting $P(A \cap B)$, we net it out once on the right-hand side of Equation (4.3). (To convince yourself that this is so, see Fig. 4.1b.)

Example 4.2

Given $P(A) = 0.6$, $P(B) = 0.3$, and $P(A \cap B) = 0.15$ [events A and B can occur together since obviously $A \cap B \neq \varnothing$ given that $P(A \cap B)$is a positive number], determine the following set of probabilities:

a. What is the probability that event B does not occur?

$$P(\bar{B}) = 1 - P(B) = 1 - 0.3 = 0.7$$

b. What is the probability that event Aor event B occurs?

$$P(A \cup B) = P(A) + P(B) - P(A \cap B) = 0.6 + 0.3 - 0.15 = 0.75$$

c. What is the probability that event A occurs and B does not occur?

$$P(A \cap \bar{B}) = P(A - B) = P(A) - P(A \cap B) = 0.6 - 0.15 = 0.45$$

d. What is the probability that event A does not occur or B does not occur?

$$P(\bar{A} \cup \bar{B}) = P(\overline{A \cap B}) = 1 - P(A \cap B) = 1 - 0.15 = 0.85 \quad \text{[See Equation (4.1).]}$$

e. What is the probability that event A does not occur and B does not occur?

$$P(\bar{A} \cap \bar{B}) = P(\overline{A \cup B}) = 1 - P(A \cup B) = 1 - 0.75 = 0.25$$

Example 4.3

Suppose we have a group of 30 students of which 15 are chemistry majors, five are athletes, and two chemistry majors are athletes. How can this information be organized so that various probabilities concerning these students can be determined? Let:

$S = \{\text{All students (30)}\}$
$A = \{\text{All chemistry majors (15)}\}$
$B = \{\text{All athletes (5)}\}$
$A \cap B = \{\text{All chemistry majors who are also athletes (2)}\}$

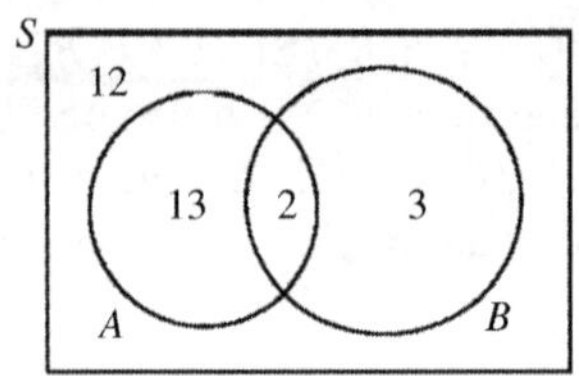

FIGURE 4.6 Students allocated to S.

Figure 4.6 depicts the apportioning of points or simple events to S. Then we can readily find:

a. The probability of picking a chemistry major who is an athlete:

$$P(A \cap B) = {}^{2}/_{30}.$$

b. The probability of selecting a student who is either a chemistry major or an athlete:

$$P(A \cup B) = P(A) + P(B) - P(A \cap B) = \frac{15}{30} + \frac{5}{30} - \frac{2}{30} = {}^{18}/_{30}.$$

c. The probability of selecting a student who is neither a chemistry major nor an athlete:

$$P(\bar{A} \cap \bar{B}) = 1 - P(A \cup B) = 1 - \frac{18}{30} = {}^{12}/_{30}.$$

d. The probability of picking a chemistry major who is not an athlete:

$$P(A \cap \bar{B}) = P(A - B) = P(A) - P(A \cap B) = \frac{15}{30} - \frac{2}{30} = {}^{13}/_{30}.$$

(Note that these probabilities could have been computed in an alternative fashion by a direct application of the classical definition of probability (Eq. (4.2)). Here one just counts points in the events under consideration relative to the points in S.)

4.4 JOINT, MARGINAL, AND CONDITIONAL PROBABILITY

Suppose $S = \{E_i, i = 1, \ldots, n\}$ with $P(E_i) = \frac{1}{n}$ for all i. Let us cross-classify the points or simple events in S by forming a row partition ($A_1, A_2 \subseteq S$ with $A_1 \cap A_2 = \varnothing$ and $S = A_1 \cup A_2$) and a column partition ($B_1, B_2 \subseteq S$ with $B_1 \cap B_2 = \varnothing$ and $S = B_1 \cup B_2$) as indicated in Fig. 4.7. Here n_{ij} is the number of simple events in

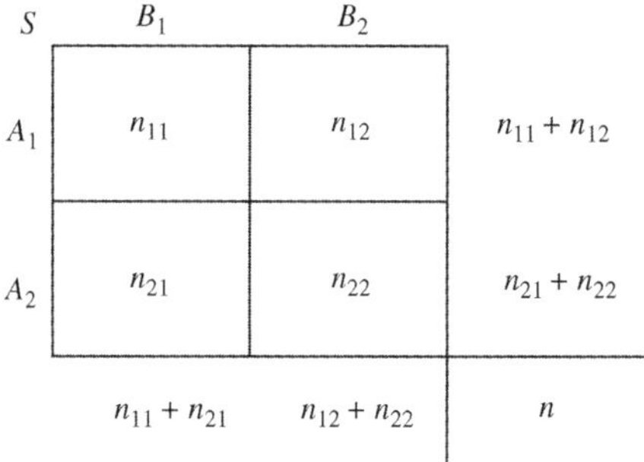

FIGURE 4.7 The cross-partitioning of S.

row i and column j, with $n_{11} + n_{12} + n_{21} + n_{22} = n$. We can now define the following probabilities:

a. *Joint Probability*: Considers the probability of the simultaneous occurrence of two events, that is,

$$P(A_1 \cap B_1) = \frac{n_{11}}{n}, P(A_1 \cap B_2) = \frac{n_{12}}{n} \text{ and so on.}$$

b. *Marginal Probability*: Used whenever one or more criteria of classification are ignored (we can ignore either the row or column partition), that is,

$$P(A_1) = \frac{n_{11} + n_{12}}{n} \quad \text{(we ignore the column partition),}$$

$$P(B_2) = \frac{n_{12} + n_{22}}{n} \quad \text{(we ignore the row partition), and so on.}$$

c. *Conditional Probability*: We are interested in the probability of occurrence of one event, given that another event has definitely occurred. (This type of probability can be calculated as the ratio between a joint and a marginal probability.) For instance, given that event B_1 has occurred, what is the probability that event A_2 also occurs? Let us write this probability as

$$P(A_2|B_1) = \frac{P(A_2 \cap B_1)}{P(B_1)},$$

where the vertical bar is read "given" and the probability of the given event "always appears in the denominator." Then

$$P(A_2|B_1) = \frac{P(A_2 \cap B_1)}{P(B_1)} = \frac{n_{21}/n}{(n_{11} + n_{21})/n} = \frac{n_{21}}{n_{11} + n_{21}}.$$

Similarly,

$$P(B_2|A_1) = \frac{P(A_1 \cap B_2)}{P(A_1)} = \frac{n_{12}/n}{(n_{11} + n_{12})/n} = \frac{n_{12}}{n_{11} + n_{12}}, \text{ and so on.}$$

Let us take a shortcut to these two results. That is, we can apply the classical definition of probability as follows. For $P(A_2|B_1)$, since event B_1 has occurred, we can ignore the B_2 column above. Hence the B_1 column becomes our *effective sample space*. Then, all we have to do is divide the number of points in $A_2 \cap B_1$ by the number of points in the given event B_1 or in the effective sample space. Thus, we can go directly to the ratio $n_{21}/(n_{11} + n_{21})$. Similarly, to find $P(B_2|A_1)$, simply divide the number of elements in $A_1 \cap B_2$ by the number of elements in the given event A_1 (thus the first column of the above partition serves as the effective sample space), to which we have $n_{12}/(n_{11} + n_{12})$.

Since the notion of a conditional probability is of considerable importance in its own right, let us generalize its definition. Specifically, let $A, B \subseteq S$ with $P(B) \neq 0$. Then

$$P(A|B) = \frac{P(A \cap B)}{P(B)}. \tag{4.4}$$

[It is important to note that Equation (4.4) is an "evidentiary" probability; it is the probability of A given that we "observe" B. Hence this conditional probability has nothing to do with cause and effect.] And if $P(A) \neq 0$,

$$P(B|A) = \frac{P(A \cap B)}{P(A)}. \tag{4.5}$$

Given that $P(A \cap B)$ appears in the numerator of both Equations (4.4) and (4.5), we can solve for this term and form the

General multiplication rule for probabilities:

$$\begin{aligned} P(A \cap B) &= P(B) \cdot P(A|B) \\ &= P(A) \cdot P(B|A). \end{aligned} \tag{4.6}$$

Armed with the concept of a conditional probability, we can now introduce the notion of independent events. Specifically, two events are *independent* if the occurrence of one of them in no way affects the probability of occurrence of the other. So if events $A, B \subseteq S$ are independent, then $P(A|B) = P(A)$; and $P(B|A) = P(B)$. Clearly, these two equalities indicate that A and B must be "mutually independent." Under independence, Equation (4.6) yields the following:

Test criterion for independence: if events $A, B \subseteq S$ are independent, then

$$P(A \cap B) = P(A) \cdot P(B). \tag{4.7}$$

How can we generate independent events? Suppose some random experiment has multiple trials. For example, we flip a coin five times in succession and we let event A_1 be "the coin shows heads on the first flip," event A_2 is "the coin shows heads on the second flip," and so on. Since the outcome (heads or tails) on any one flip or trial does not affect the outcome on any other trial, the trials are considered independent. But if the trials are independent, then the events $A_1, \ldots, A_5$ must be independent. Hence $P(A_1 \cap A_2) = P(A_1) \cdot P(A_2) = \frac{1}{2} \cdot \frac{1}{2} = {}^1/_4$, $P(A_1 \cap A_2 \cap A_3) = P(A_1) \cdot P(A_2) \cdot P(A_3) = \left(\frac{1}{2}\right)^3 = {}^1/_8$ and so on. In this regard, independent trials will generate independent events.

Similarly, we can generate independent events by *sampling with replacement*. For instance, suppose a jar contains three white and seven black marbles. Two marbles are drawn at random with replacement (meaning that the first marble drawn is returned to the jar before the second marble is drawn). What is the probability that both are black? On draw 1, $P(B_1) = {}^7/_{10}$; on draw 2, $P(B_2) = {}^7/_{10}$. Hence

$$P(B_1 \cap B_2) = P(B_1) \cdot P(B_2) = \left(\frac{7}{10}\right)^2 = 0.49.$$

But what if we sampled without replacement? (Now, the first marble drawn is not returned to the jar before the second marble is selected.) Obviously, the probability of getting a black (or white) marble on the second draw depends on the color obtained on the first draw. Clearly, there is a conditional probability calculation that must enter the picture. Now, from Equation (4.6) we have the following:

$$\begin{aligned} P(B_1 \cap B_2) &= P(B_1) \cdot P(B_2|B_1) \\ &= \frac{7}{10} \cdot \frac{6}{9} = 0.4667. \end{aligned}$$

Here the events B_1 and B_2 are not independent; the probability that B_1 and B_2 occurs is conditional on B_1 occurring. Figure 4.8 describes a *decision tree* diagram under independent trials (events) or sampling with replacement. The appropriate probabilities are listed beneath each *node*. Note that as we move along each *branch*, we simply multiply the node probabilities encountered. What is the sum of the four joint probabilities?

Figure 4.9 offers a similar diagram in which the events are not independent because of sampling without replacement.

(Here too the sum of the four joint probabilities is 1.)

What if a third trial of this experiment is undertaken? Under sampling with replacement, we can readily find, say, the following:

$$P(W_1 \cap B_2 \cap W_3) = P(W_1) \cdot P(B_2) \cdot P(W_3) = \frac{3}{10} \cdot \frac{7}{10} \cdot \frac{3}{10} = 0.042.$$

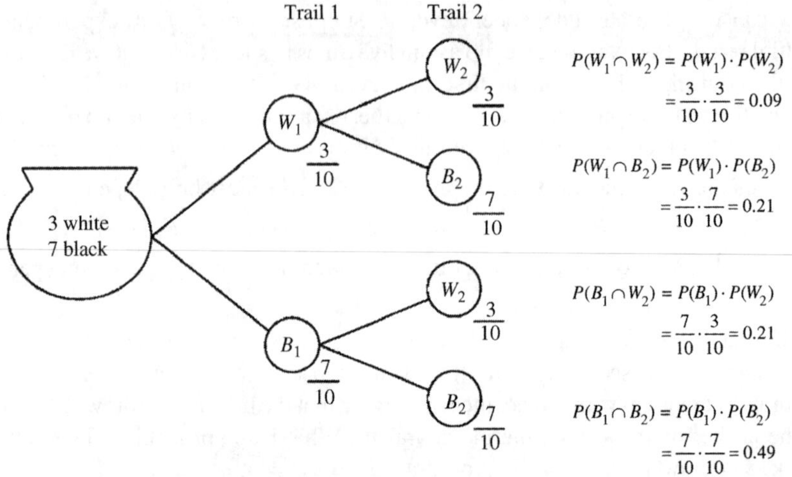

FIGURE 4.8 Sampling with replacement.

But if we sample without replacement,

$$P(W_1 \cap B_2 \cap W_3) = P(W_1) \cdot P(B_2|W_1) \cdot P(W_3|W_1 \cap B_2) = \frac{3}{10} \cdot \frac{7}{9} \cdot \frac{2}{8} = 0.0583.$$

What is the distinction between events that are mutually exclusive and events that are independent? Mutually exclusive events cannot occur together, that is, $A \cap B$ cannot occur. Hence $A \cap B = \varnothing$ and thus $P(A \cap B) = P(\varnothing) = 0$. In fact, mutually exclusive events are "dependent events"—the occurrence of one of them depends on the other

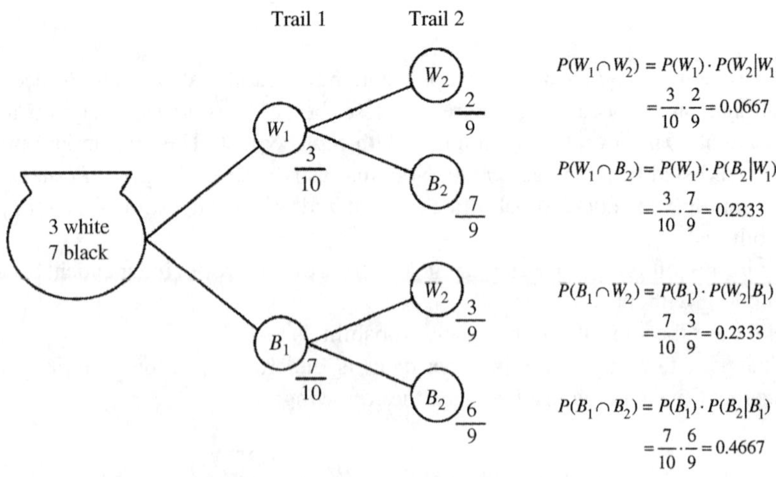

FIGURE 4.9 Sampling without replacement.

not occurring. However, independent events can occur together ($A \cap B$ can occur)—it is just that the occurrence of one of them does not affect the probability of occurrence of the other. For example, $P(A|B) = P(A)$ so that $P(A \cap B) = P(A) \cdot P(B)$.

4.5 SOURCES OF PROBABILITIES

Let us now examine the various ways in which probabilities are determined. Two broad categories of probabilities are typically recognized: *objective probability* versus *subjective probability*. Moreover, under the objective heading, we have the *classical approach* versus the *empirical determination* of probability.

The classical approach utilities *a priori* (before the fact) knowledge, that is, we can determine a *theoretical probability* by counting points or simple events within S. Clearly, this approach utilizes the physical structure of an implied experiment and is used when we have n equiprobable simple events in S.

An empirical probability utilizes *a posteriori* (after the fact) knowledge; it makes use of accumulated or historical experience or is determined by actually performing a random experiment. That is, we can compute the relative frequency of an event via a random experiment. Specifically, if A is an event for a random experiment and if that experiment is repeated n times and A is observed in n_A of the n repetitions, then the *relative frequency of event* $\boldsymbol{A}$ is n_A/n. What is the connection between relative frequency and probability? The link is provided by the *frequency limit principle*: if in a large number n of independent repetitions of a random experiment the relative frequencies n_A/n of some event A approach a number $P(A)$, then this number is ascribed to A as its probability, that is,

$$\lim_{n \to \infty} \frac{n_A}{n} = P(A) \text{ or } \frac{n_A}{n} \to P(A) \text{ as } n \to \infty.$$

Here $P(A)$ is a long-run stable value that the relative frequencies approach—it is the expected limit of n_A/n (Fig. 4.10). (The reader may want to flip a coin, say, 30 times and, for the event A: the coin shows heads, calculate n_A/n. While $n = 30$ is certainly not the " long run," one can get the "flavor" of the frequency limit principle by performing this rather trivial random experiment.)

Sometimes, the frequency limit principle is termed as the *law of large numbers*: if $P(A) = p$ and n independent repetitions of a random experiment are made, then the probability that n_A/n differs from p by any amount, however small, tends to one as n increases without bound, That is,

$$P\left(\left|\frac{n_A}{n} - p\right| < \varepsilon\right) \to 1 \text{ as } n \to \infty,$$

where ε is an arbitrary small positive number. It is important to note that the law of large numbers does not state that n_A/n is necessarily getting closer and

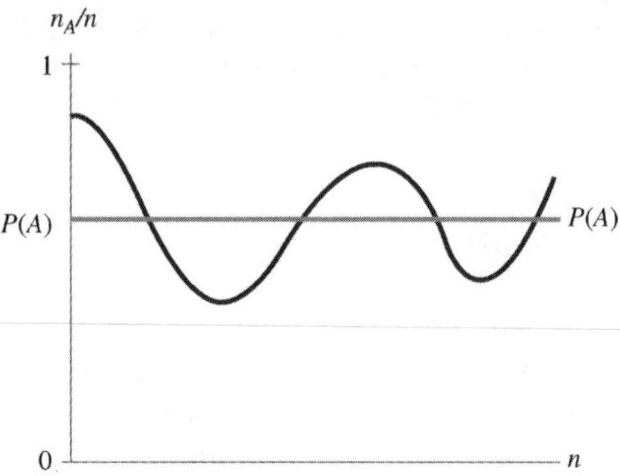

FIGURE 4.10 n_A/n converges to $P(A)$for sufficiently large n.

closer to p as n gets larger and larger—it does state, however, that as n increases without bound, the probability tends to 1 that n_A/n will not differ from p by more than ε.

At times, the frequency limit principle and the law of large numbers are erroneously referred to as the so-called "law of averages," which posits that, for n large enough, the relative frequency n_A/n somehow "balances or evens out" to approximate $P(A)$. At the foundation of the law of averages is the notion (the *gambler's fallacy*) that if we get a run of, say, 100 heads in 100 flips of a coin, then certainly "tails is due to occur." (This is not a legitimate conclusion since the probability of tails on any one trial is always ½; and, in reality, the coin may actually be biased toward heads.) That is, a run of tails is due in order to get even so that the total number of tails will eventually equal the total number of heads. Hence, according to the law of averages, in the long run "chances" tend to even out—given enough flips of a coin, the total number of heads thrown will be equal to the total number of tails obtained so that the probability of heads will be 0.5. Actually, just the opposite occurs: as the number of tosses of a coin increases, so does the probability that the percentage of heads (or tails) gets near 0.5, but the difference between the realized number of heads (or tails) and the number needed to get to a percentage of 0.5 increases. The law of large numbers eventually compensates for the 100 heads in a row by a flood of tails—subsequent tosses resulting in tails swamp or overwhelm the 100 heads even if we have 50% heads.

To summarize, the law of large numbers stipulates that: the values of $\frac{n_A}{n} - p$ concentrate about zero as n increases or, for sufficiently large n, $\frac{n_A}{n}$ has a high probability of being near p; however, it does not require that the individual values of $\frac{n_A}{n} - p$ will necessarily "all" be near zero or that $\frac{n_A}{n}$ will remain near p if n is increased further.

Finally, a *subjective probability* is a "best-guess" probability or a quantified degree of belief—it relates to experiments that "cannot be repeated." For instance, at 9 a.m. one Wednesday one may "feel" that the "market" will take a bit of business news very favorably and that this will be reflected in the DOW gaining at least 200 points by the end of the trading day. Clearly, this is a "one-shot" phenomenon.

EXERCISES

1 Which of the following numbers could not be the probability of an event? 0.0, 1.2, 0.8, 0.0001, −0.7, −1.4, 1.0

2 Describe the following situations:

a. Events A and B are disjoint.
b. Events C and D are complements. (Are C and D also disjoint?)

3 Suppose for events A and B, we have $P(A)=0.25$ and $P(B)=0.45$. Find the following:

a. $P(A \cup B)$ if $A \cap B = \emptyset$
b. $P(A \cup B)$ if $P(A \cap B) = 0.15$
c. $P(\bar{B})$
d. $P(A \cap B)$ if $P(A \cup B) = 0.65$

4 Suppose that for events A and B, we have $P(B)=0.3$, $P(A \cup B) = 0.65$, and $P(A \cap B)=0.12$. Find $P(A)$.

5 For a random experiment, let the sample space appear as $S=\{1,2,3,4,5,6,7,8\}$ and suppose the simple events in S are equiprobable. Find the following:

a. The probability of event $A=\{1, 2, 3\}$
b. The probability of event $B=\{2, 3, 6, 7\}$
c. The probability of event $C=\{$an even integer$\}$
d. The probability of event $D=\{$an odd integer$\}$
e. The probability that A and B occurs
f. The probability that A or B occurs
g. $P(S)$
h. The probability that C and D occurs
i. The probability that A does not occur
j. The probability that C or D occurs
k. The probability that C does not occur
l. The probability that A does not occur and B does not occur
m. The probability that A does not occur or B does not occur

6 Suppose a marble is randomly selected from a jar containing 10 red, 3 black, and 7 blue marbles. Find the probability of the following:

a. The marble is red or black

b. The marble is black or blue

c. The marble is not blue

d. The marble is red or not blue

7 Find the probability that a card randomly selected from an ordinary deck of 52 cards will be a face card (event A) or a spade (event B) or both.

8 Find the probability that a card randomly selected from an ordinary deck of 52 cards will be an ace (event A) or a king (event B).

9 If we toss a fair coin five times in succession, what is the probability of getting all heads (event H)? [Hint: The tosses are independent.]

10 Suppose five coins are tossed simultaneously. Find the probability of getting at least one head (event A) and at least one tail (event B).

11 Suppose A and B are two events with $P(A \cap B) = 0.4$ and $P(A) = 0.6$. Find $P(B|A)$.

12 Given that A and B are two events with $P(A) = 0.7$ and $P(B|A) = 0.4$, find $P(A \cap B)$.

13 For events A and B, if $P(A) = 0.5$ and $P(B|A) = 0.45$, are A and B independent?

14 For events A and B, if $P(A) = 0.2$, $P(B) = 0.3$, and $P(A \cap B) = 0$, are A and B independent? How would you characterize events A and B?

15 A company's sales force consists of four men and three women. The CEO will randomly select two people to attend the upcoming trade show in Las Vegas. Find the probability that both are men.

16 A box contains 10 red tokens numbered 1 through 10 and 10 white tokens numbered 1 through 10. A single token is selected at random. What is the probability that it is as follows:

a. Red

b. An odd number

c. Red and odd

d. Red and even

17 Suppose A and B are independent events with $P(A) = 0.3$, $P(B) = 0.2$. What is the probability that the following will occur:

a. Both A and B will occur

b. Neither one will occur

c. One or the other or both will occur

18 Two marbles are randomly drawn from a jar containing three white and seven red marbles. Find the probability of obtaining two white marbles if we do the following:

a. We draw with replacement

b. We draw without replacement

19 Let A and B be events with $P(A)=0.3$, $P(B)=0.7$, and $P(A \cap B)=0.1$. Find the following:

a. $P(\bar{B})$

b. $P(\bar{A})$

c. $P(\bar{A} \cap \bar{B})$

d. $P(A \cup B)$

e. $P(A|B)$

f. $P(\bar{B}|A)$

g. $P(A \cap \bar{B})$

h. $P(\bar{A} \cap B)$

20 Suppose events A, B, and C are mutually exclusive and collectively exhaustive with $P(A) = \frac{1}{2}$, $P(B \cup C)$ and $P(B) = \frac{1}{2}P(C)$. Find $P(A)$, $P(B)$, and $P(C)$.

21 Suppose we conduct a random survey of 10,000 individuals in order to investigate the connection between smoking and death from lung cancer. The results appear in Table E.4.21. Find the following:

TABLE E.4.21 Smoking and Lung Cancer

	Died from lung cancer (event D)	Did not die from lung cancer (event $\bar{D}$)
Never smoked (event A)	45	6500
Former smoker (event B)	110	1400
Current smoker (event C)	250	1695

a. The probability that a person who died from lung cancer was a former smoker.

b. The probability that a person who was a former smoker died from lung cancer.

c. The probability that a person who died from lung cancer was never a smoker.

d. The probability that a person who did not die from lung cancer was a former smoker.

22 A bag contains 20 red marbles, 8 green marbles, and 7 white marbles. Suppose two marbles are drawn at random without replacement. Find the following:

a. The probability that both are red.

b. The probability that the first marble selected is red and the second is green.

c. The probability that the first marble selected is green and the second is red.

d. The probability that one marble selected is red and the other is green. [Hint: Use a tree diagram to solve this problem.]

TABLE E.4.23 Gender and College Graduation

	Gender	
College graduate	Male (M)	Female (F)
Yes (Y)	40	30
No (N)	20	10

23 Job applicants at the local mall were asked if they had graduated college. The responses are given in Table E.4.23. What is the probability that a randomly selected individual will be as follows:

a. Male and a college graduate
b. Either female or a college graduate
c. A college graduate
d. Male, given that the person chosen is a college graduate
e. Not a college graduate, given that the person chosen is female
f. Are the categories of college graduate and gender independent?

24 A jar contains six red and four blue marbles. Three are drawn at random without replacement. What is the probability that all are blue? What is the probability of a blue marble on the first draw and a red one on the second and third draws? Construct a tree diagram for this experiment and calculate all terminal probabilities. Also, express each terminal probability symbolically.

25 Suppose we draw a card at random from an ordinary deck of 52 playing cards. Consider the events:

$A = \{\text{A black card is drawn}\}$
$B = \{\text{An ace is drawn}\}$
$C = \{\text{A red card is drawn}\}$

a. Are events A and C mutually exclusive?
b. Are events A and B mutually exclusive?
c. Are events B and C mutually exclusive?
d. Are events A and C collectively exhaustive?
e. Does $P(A \cap B) = P(A) \cdot P(B)$?
f. Does $P(A \cap C) = P(A) \cdot P(C)$?
g. Does $P(B \cap C) = P(B) \cdot P(C)$?
h. Based upon your answers to parts (e) and (f), how can we characterize the pairs of events A and B and A and C?

TABLE E.4.26 Employment Status and Opinion

	Opinion		
Employment status	Yes (Y)	No (N)	Indifferent (I)
Production (P)	57	49	5
Development (D)	22	10	1
Sales (S)	37	55	2
Clerical (C)	14	38	5

26 The employees at XYZ Corp. fall into one of four mutually exclusive categories: production, development, sales, and clerical. The results of a recent company poll pertaining to the issue of favoring wage/salary increases rather than getting dental benefits are presented in Table E.4.26. Given that an employee is to be chosen at random, find the following probabilities:

a. $P(Y \cap I)$
b. $P(D \cup N)$
c. $P(D \cap N)$
d. $P(C \cap \bar{Y})$
e. $P(S \cup \bar{N})$
f. $P(D|Y)$
g. $P(S|\bar{Y})$
h. Are employment status and opinion statistically independent?

27 What is the connection between probability and odds? *Odds* relate to the chances in favor of an event A to the chances against it, for example, if the odds are two to one that event A will occur, then this means that there are two chances in three that A will occur. In general, the odds of x to y is typically written x:y—x successes to y failures.

How are odds converted to probabilities? Specifically, if the odds are x:y in favor of an event A, then

$$P(A) = \frac{x}{x+y}.$$

So if the odds are 2:1 in favor of event A, then $P(A) = \frac{2}{2+1} = \frac{2}{3}$. In addition, if the odds are x:y in favor of event A, then the odds are y:x against A. Hence

$$P(\bar{A}) = \frac{y}{x+y}.$$

Given that the odds are 1:2 against A, we thus have $P(\bar{A}) = \frac{1}{1+2} = \frac{1}{3}$.

What about converting probabilities to odds? If the probability that event A will occur is $P(A)$, then the odds in favor of A are $P(A) : P(\bar{A})$. So if the probability of event A is $\frac{1}{2}$, then the odds in favor of the occurrence of A are 1:1. Also, if the odds in favor of the occurrence of event A are $P(A) : P(\bar{A})$, then the odds against A are $P(\bar{A}) : P(A)$. This said,

a. If the odds are 6:4 that Mary's boyfriend will propose marriage tonight, what is the probability that a proposal occurs?

b. What is the probability that a proposal will not be made? [Hint: The odds are 4:6 against the proposal.]

c. If Mary awoke with the feeling that the probability of her boyfriend proposing marriage that evening is $\frac{4}{5}$, what are the odds that a proposal will occur?

d. What are the odds that a proposal will not occur?

5

DISCRETE PROBABILITY DISTRIBUTIONS AND THEIR PROPERTIES

5.1 THE DISCRETE PROBABILITY DISTRIBUTION

A probability distribution is termed *discrete* if the underlying random variable X is discrete. (Remember that a random variable X is a real-valued function defined on the elements of a sample space S—its role is to define the possible outcomes of a random experiment.) In this regard, the number of values that X can assume forms a *countable set*: either there exists a finite number of elements or the elements are *countably infinite* (can be put in one-to-one correspondence with the positive integers or counting numbers).

For instance, let X represent the outcome (the face showing) on the roll of a single six-sided die. Then $X = \{1, 2, 3, 4, 5, 6\}$—the X values form a countable set. However, what if we roll the die repeatedly until a five appears for the first time? Then if X represents the number of rolls it takes to get a five for the first time, clearly, $X = \{1, 2, 3, \ldots\}$, that is, the X values form a countably infinite set.

This said, a *discrete probability distribution* is a distribution exhibiting the values of a discrete random variable X and their associated probabilities. In order to construct a discrete probability distribution, we need three pieces of information:

Statistical Inference: A Short Course, First Edition. Michael J. Panik.

1. The *random experiment*: this enables us to obtain the sample space $S = \{E_i, i = 1, \ldots, n\}$, a countable collection of simple events.
2. A *discrete random variable* X defined on S, which also results in a countable set of values (real numbers) $\{X_1, X_2, \ldots, X_p\}$.
3. A *probability mass function* f—a rule for assigning probabilities, that is, $P(X = X_j) = f(X_j)$ is termed the *probability mass* at X_j. (If S has n equiprobable simple events, then $P(E_i) = \frac{1}{n}$ for all $i = 1, \ldots, n$. In this instance, $f(X_j)$ is simply the relative frequency of X_j.) For f to constitute a legitimate probability mass function, we must have
   (a) $f(X_j) \geq 0$ for all j,
   (b) $\sum f(X_j) = 1$.

What is the role of a discrete probability distribution? It indicates how probabilities are distributed over all possible values of X. In this regard, a discrete probability distribution is completely determined once the sequence of probabilities $f(X_j)$, $j = 1, \ldots, p$, is given.

How is a discrete probability distribution depicted? Typically, one should "think table" (see Table 5.1).

Example 5.1

Let a random experiment consist of rolling a single (fair) six-sided die. The random variable X is the value of the face showing. Determine the implied discrete probability distribution. Since $S = \{E_1, \ldots, E_6\}$ and $P(E_i) = \frac{1}{6}$ for all i (we have a set of six equiprobable outcomes), it follows that X:1, 2, 3, 4, 5, 6 with $f(X_j) = \frac{1}{6}$ for all j ($\frac{1}{6}$ is the relative frequency for each X_j). Table 5.2 illustrates the resulting discrete probability distribution. The probability mass function is given in Fig. 5.1. Since the probabilities are equal, we shall term this type of discrete distribution a *uniform probability distribution*.

Let our random experiment involve the rolling of a pair of (fair) six-sided dice. The random variable X is taken to be the sum of the faces showing. What is the implied discrete probability distribution? Now, $S = \{E_i, i = 1, \ldots, 36\}$ and $P(E_i) = \frac{1}{36}$ for all i,

TABLE 5.1 A Discrete Probability Distribution

X	$f(X)$
X_1	$f(X_1)$
X_2	$f(X_2)$
.	.
.	.
.	.
X_p	$f(X_p)$
	1

TABLE 5.2 Discrete Probability Distribution

X	$f(X)$
1	1/6
2	1/6
3	1/6
4	1/6
5	1/6
6	1/6
	1

and X assumes the eleven values {2, 3, 4, 5, 6, 7, 8, 9, 10, 11, 12} (see Example 4.1). Since the E_i are equiprobable, the probabilities $f(X_j), j=1,\ldots,11$, are determined on a relative frequency basis. Table 5.3 illustrates the discrete probability distribution, and Fig. 5.2 displays the associated probability mass function.

Given Table 5.3, we can determine the following for instance:

$$P(X<4)=P(X=2)+P(X=3)=3/36$$
$$P(X\geq 9)=P(X=9)+P(X=10)+P(X=11)+P(X=12)=10/36$$
$$P(6\leq X\leq 8)=P(X=6)+P(X=7)+P(X=8)=16/36$$
$$P(6<X\leq 8)=P(X=7)+P(X=8)=11/36$$
$$P(6<X<8)=P(X=7)=6/36.$$

At this point in our discussion of discrete probability distributions, one additional concept merits our attention. Specifically, let us consider what is called the *cumulative distribution function*:

$$F(X_j)=P(X\leq X_j)=\sum_{k\leq j} f(X_k) \tag{5.1}$$
(we sum over all subscripts $k\leq j$).

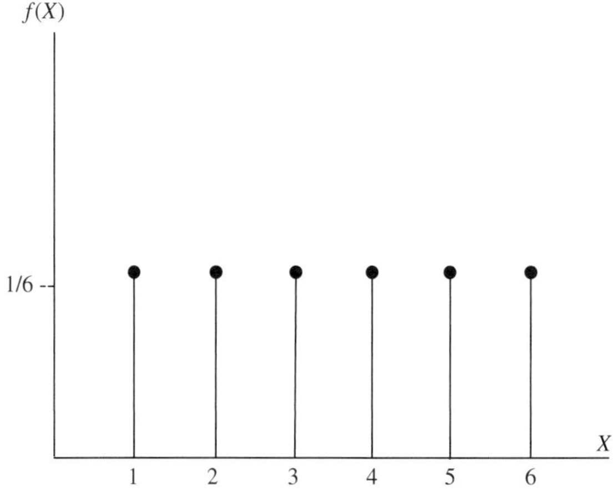

FIGURE 5.1 Probability mass function.

TABLE 5.3 Discrete Probability Distribution

X	$f(X)$
2	1/36
3	2/36
4	3/36
5	4/36
6	5/36
7	6/36
8	5/36
9	4/36
10	3/36
11	2/36
12	1/36
	1

Here $F(X_j)$ renders the probability that the random variable X assumes a value "$\leq$" X_j, where j is given in advance, for example,

$$F(X_1) = P(X \leq X_1) = \sum_{k \leq 1} f(X_k) = f(X_1) = \frac{1}{36}$$

$$F(X_2) = P(X \leq X_2) = \sum_{k \leq 2} f(X_k) = f(X_1) + f(X_2) = \frac{3}{36}$$

$$F(X_3) = P(X \leq X_3) = \sum_{k \leq 3} f(X_k) = f(X_1) + f(X_2) + f(X_3) = \frac{6}{36}$$

$$\vdots$$

$$F(X_{11}) = P(X \leq X_{11}) = \sum_{k \leq 11} f(X_k) = f(X_1) + f(X_2) + \ldots + f(X_{11}) = 1.$$

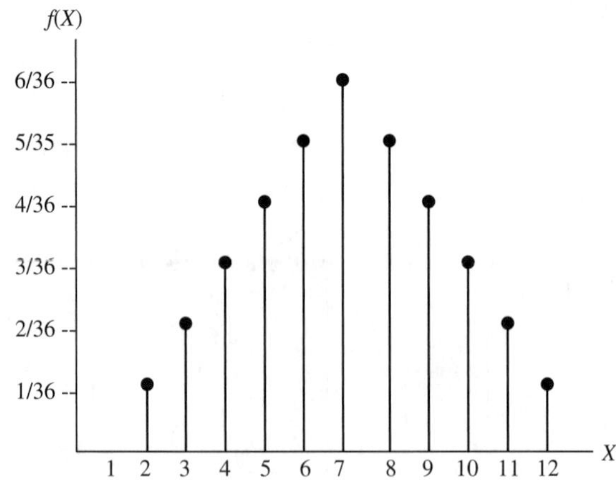

FIGURE 5.2 Probability mass function.

TABLE 5.4 Cumulative Distribution Function

X	$f(X)$	$F(X)$
2	1/36	1/36
3	2/36	3/36
4	3/36	6/36
5	4/36	10/36
6	5/36	15/36
7	6/36	21/36
8	5/36	26/36
9	4/36	30/36
10	3/36	33/36
11	2/36	35/36
12	1/36	36/36 = 1
	1	

Note that the entries in the $F(X)$ column (column 3) of Table 5.4 represent a "running total" of the $f(X)$ values in column two of the same.

The distinction between $f(X)$ and $F(X)$ is critical. For $X = 8$, $f(8) = 5/36$ gives the probability that $X = 8$ exactly while $F(8) = 26/36$ gives the probability that X assumes a value of ≤ 8 (see Table 5.4).

5.2 THE MEAN, VARIANCE, AND STANDARD DEVIATION OF A DISCRETE RANDOM VARIABLE

The starting point for determining the mean and variance of a discrete random variable X (or of X's discrete probability distribution) is Table 5.1. Looking first to the mean of X, convention dictates that the mean be termed the *expectation of X* (or the *expected value of X*) and denoted as follows:

$$E(X) = \sum_j X_j f(X_j) = X_1 f(X_1) + X_2 f(X_2) + \cdots + X_p f(X_p). \tag{5.2}$$

Here $E(X)$ serves as a measure of central location (the center of gravity) of a discrete probability distribution. For instance, given the discrete probability distribution appearing in Table 5.5, let us find and interpret $E(X)$.

As we shall now see, $E(X)$ is the "long-run" average value of X—it represents the mean outcome we expect to obtain if a random experiment is repeated a large number of times in the long run. In this regard, from Table 5.5, we get the following:

$$E(X) = 10\,(0.30) + 40\,(0.70) = 31,$$

that is, if we repeat our random experiment over and over a large number of times, then, in the long run, since 10 occurs 30% of the time and 40 occurs 70% of the time,

TABLE 5.5 A Discrete Probability Distribution

X	$f(X)$
10	0.30
40	0.70
	1.00

the average value of X will be 31. So if the long run consists of, say, 100 trials, then the average of 30 tens and 70 fortys would be 31.

Why does $E(X)$ have the form provided by Equation (5.2)? We noted in Chapter 3 that if some observations were deemed more important than others in calculating the mean, then, to take account of any differing relative importance among the data points, we should employ the weighted mean formula (3.3). In Table 5.5, it should be intuitively clear that the value of X with the larger probability should carry relatively more weight (here any differing relative importance among the X values is determined by the magnitudes of their probability values) in the calculation of the average of X. Hence $E(X)$ is actually a weighted mean, where the weights are probabilities. But since the sum of the probabilities must always be unity ($\sum f(X_j) = 1$), we are left with only the numerator of the weighted mean formula or (5.3)—the denominator is understood to always be 1.

Next, let us find the variance of a discrete random variable X. Let σ^2 (the variance of a nonrandom population variable) serve as our model or guide. How is σ^2 defined? It is the average of the squared deviations from the mean of X. If we mirror this definition, then the *variance of the random variable* ***X*** will be calculated as the "expectation of the squared deviations from the expectation" or

$$V(X) = E[(X - E(X))^2] = \sum_j (X_j - E(X))^2 f(X_j), \tag{5.3}$$

with $E(X)$ determined from Equation (5.2). Once $V(X)$ is found, the *standard deviation of the random variable X* is

$$S(X) = +\sqrt{V(X)}. \tag{5.4}$$

Here $S(X)$ serves as an index of the dispersion of X about its mean value—it indicates average long-run variability about the mean or expectation.

Example 5.2

Given the following discrete probability distribution (incidentally, how do we know that it is a legitimate discrete probability distribution?), find $E(X)$ and $V(X)$. Then determine $S(X)$ (Table 5.6). First,

$$E(X) = 1(0.1) + 2(0.1) + 3(0.3) + 4(0.2) + 6(0.2) + 10(0.1) = 4.2,$$

the long-run average value of X. To find $V(X)$, let us set up a work table (Table 5.7).

TABLE 5.6 Discrete Probability Distribution

X	$f(X)$
1	0.1
2	0.1
3	0.3
4	0.2
6	0.2
10	0.1
	1.0

Then, via Equation (5.3), it is readily seen that $V(X) = 5.96$ and thus $S(X) = 2.44$—in the long run and on the average, the X values will be about 2.44 units away from the mean.

As an alternative to Equation (5.3), let us calculate the variance of X is

$$V(X) = E(X^2) - E(X)^2, \tag{5.5}$$

that is, the variance of X equals "the expectation of X^2" less "the expectation of X itself squared," where

$$E(X^2) = \sum_j X_j^2 f(X_j). \tag{5.6}$$

(We may view Equation (5.3) as a "definitional" equation and Equation (5.5) as a "computational" equation.)

Example 5.3

Given the Example 5.2 discrete probability distribution (Table 5.6), recalculate $V(X)$ using Equation (5.5). Table 5.8 serves as our work table for determining $E(X^2)$ from Equation (5.6).

Given that $E(X^2) = 23.6$, $V(X) = 23.6 - (4.2)^2 = 5.96$ as anticipated.

TABLE 5.7 Calculations for $V(X)$

$X - E(X)$	$(X - E(X))^2$	$(X - E(X))^2 f(X)$
$1 - 4.2 = -3.2$	10.24	1.024
$2 - 4.2 = -2.2$	4.84	0.484
$3 - 4.2 - -1.2$	1.44	0.432
$4 - 4.2 = -0.2$	0.04	0.008
$6 - 4.2 = 1.8$	3.24	0.648
$10 - 4.2 = 5.8$	33.64	3.364
		5.96

TABLE 5.8 Calculations for $E(X^2)$

X^2	$X^2 f(X)$
1	0.1
4	0.4
9	2.7
16	3.2
36	7.2
100	10.0
	23.6

Example 5.4

The local casino offers a high-stakes weekly lottery game in which only 1000 tickets are sold for \$100.00 each. The prizes appear in Table 5.9. What is a player's expected winnings?

To use Equation (5.2), we need some probabilities. To determine these, we first need to define the random variable X. Since the question asked called for the "expected winnings" (a dollar amount), we see that the values of the random variable X must be the dollar amounts in Table 5.9. Thus the associated probabilities are determined as follows:

$P(X = 25{,}000) = 1/1000$ (only one out of 1000 tickets pays \$25,000),
$P(X = 5{,}000) = 4/1000$, and so on. Then from Table 5.10,

$$E(X) = 25{,}000\,(0.001) + 5{,}000\,(0.004) + 500\,(0.005) + 0\,(0.945) = \$70.00.$$

So if you played this lottery game over and over, week after week, for an extended period of time, then your long-run average winnings would be \$70. Clearly, your expected loss for a single lottery ticket would be \$30.

TABLE 5.9 Weekly Lottery Prizes

Number of Prizes	Dollar Amount
1	25,000
4	5,000
50	500
945	0

TABLE 5.10 Prize Probabilities

X	$f(X)$
25,000	0.001
5,000	0.004
500	0.050
0	0.945
	1.000

5.3 THE BINOMIAL PROBABILITY DISTRIBUTION

While there are many discrete probability distributions, by far the most important one is the binomial distribution. To set the stage for a discussion of this distribution, we will need to cover two preliminary notions: (1) counting issues; and (2) the Bernoulli probability distribution.

5.3.1 Counting Issues

We may generally view the concept of a *permutation* as an ordered set of objects. If any two of the objects are interchanged, we have a new permutation. For instance, the letters "a, b, c" form a particular permutation of the first three letters of the alphabet. If we interchange b and c, then "a, c, b" gives us another distinct permutation of these letters. More specifically, a permutation is any particular arrangement of r objects selected from a set of n distinct objects, $r \leq n$.

What is the total number of permutations of r objects selected from a set of n distinct objects? To answer this, let us find

$$_nP_r = \frac{n!}{(n-r)!}, \tag{5.7}$$

the number of permutations of n different objects taken r at a time, where

$$n! = n(n-1)(n-2)\cdots 2\cdot 1.$$

For instance,

$$3! = 3\cdot 2\cdot 1 = 6$$

$$7! = 7\cdot 6\cdot 5\cdot 4\cdot 3\cdot 2\cdot 1 = 5040.$$

Note that $0! \equiv 1$ (the notation "$\equiv$" reads "is defined as"). As a special case of Equation (5.7), if $r = n$, then

$$_nP_r = n!, \tag{5.8}$$

the number of permutations of n different objects taken all together.

Example 5.5

How many ABC Corp. telephone extensions of four different digits each can be made from the digits 0, 1, ..., 9? Here we want the following:

$$_{10}P_4 = \frac{10!}{6!} = \frac{10\cdot 9\cdot 8\cdot 7\cdot 6!}{6!} = 10\cdot 9\cdot 8\cdot 7 = 5{,}040,$$

(the 6! terms cancel).

Next, four tires on a car (not including the "doughnut" spare) are to be rotated so that all will be worn equally. How many ways are there of putting all the four tires on the car? We now want

$$ {}_4P_4 = 4! = 24. $$

If the spare is a fully functional tire, then we would find

$$ {}_5P_5 = 5! = 120. $$

If the order of the distinct items in an arrangement is not important, then we have a *combination*—an arrangement of items without regard to order (clearly, "a, b, c" and "a, c, b" represent the same combination of letters). We are now interested in "what" objects are selected regardless of their order.

What is the total number of combinations of n distinct items taken r at a time? The answer is provided by the following:

$$ {}_nC_r = \frac{n!}{r!(n-r)!}, \tag{5.9} $$

the number of combinations of n distinct objects taken r at a time. As a special case of Equation (5.9), for $r=n$,

$$ {}_nC_n = 1, \tag{5.10} $$

the number of combinations of n different items taken all together. Which is larger, ${}_nP_r$ or ${}_nC_r$? Since

$$ {}_nC_r = \frac{{}_nP_r}{r!}, $$

it follows that for given n and r values, there are fewer combinations than permutations.

Example 5.6

How many committees of six people can be chosen from a slate of 10 people? Since order is irrelevant (one is either a member of a committee or not), we need to find

$$ {}_{10}C_6 = \frac{10!}{6!4!} = \frac{10\cdot 9\cdot 8\cdot 7\cdot 6!}{6!4!} = \frac{10\cdot 9\cdot 8\cdot 7}{4!} = \frac{10\cdot 9\cdot 8\cdot 7}{4\cdot 3\cdot 2\cdot 1} = 5\cdot 3\cdot 2\cdot 7 = 210. $$

So should permutations or combinations be computed in any counting situation? To answer this, we need only ask—"is order important?" If the answer is "yes," then we compute permutations; if "no" is forthcoming, then we determine combinations.

5.3.2 The Bernoulli Probability Distribution

Let us define a random experiment as a *simple alternative or Bernoulli experiment*, that is, there are two possible outcomes—success or failure. For instance:

> tossing a coin (getting heads is termed a success); quality control inspection (passing inspection is deemed a success); and, in general, some attribute (color, style, and so on) is either present (call it a success) or absent.

In this regard, the sample space has only two points or simple events within it: $S = \{E_1, E_2\}$, where E_1 represents success and E_2 represents failure. Additionally, the *Bernoulli random variable* ***X*** has only two possible values that we shall code as $X_1 = 1$ (if a success occurs) and $X_2 = 0$ (for a failure). Suppose $P(X = 1) = p$ depicts *Bernoulli probability*. Then, since $P(S) = 1$, it follows that $P(X = 0) = 1 - p$. Hence the Bernoulli probability distribution can be depicted ("think table") as Table 5.11.

What is the source of the Bernoulli probability distribution? It arises from a single trial of a simple alternative experiment. Moreover, this distribution is completely determined once p, the probability of a success, is given. Note also that:

$$\begin{aligned} E(X) &= \sum_{j=1}^{2} X_j f(X_j) = 1 \cdot p + 0(1-p) = p; \text{ and} \\ V(X) &= E(X^2) - E(X)^2 = \sum_{j=1}^{2} X_j^2 f(X_j) - p^2 \\ &= (1)^2 p + (0)^2 (1-p) - p^2 = p(1-p). \end{aligned}$$

Thus the mean of a Bernoulli probability distribution is the probability of a success and its variance is the probability of a success times the probability of a failure.

5.3.3 The Binomial Probability Distribution

Having discussed the so-called counting issues and the Bernoulli probability distribution, let us generalize the latter by performing some simple alternative experiment n (a fixed number) times in succession. Hence, the source of the binomial distribution is the following: it arises as a series of n repeated independent trials of a simple alternative experiment, with p constant from trial to trial. (Note that the trials will be independent if the outcome on any one trial does not affect, or is not affected by, the outcome on any other trial; and p will be constant from trial to trial if sampling is done with replacement.)

TABLE 5.11 Bernoulli Probability Distribution

X	$f(X)$
1	p
0	$1-p$
	1

As the n independent trials of our simple alternative experiment commence, let us count the number of successes obtained. We shall call this number k. Then the question that naturally arises is: what is the probability of obtaining k successes in the n independent trials? The answer is provided by the *binomial probability formula*:

$$P(k,n,p) = \frac{n!}{k!(n-k)!} p^k (1-p)^{n-k}, \tag{5.11}$$

where the left-hand side of this equation is read "the probability of k successes in n trials." [Don't let this expression intimidate you—we shall soon "strip it naked" by employing some familiar probability concepts. In particular, please review Equation (4.7).]

How are we to rationalize Equation (5.12)? Let the random variables $X_1, \ldots, X_n$ denote the outcomes of the n independent trials of the simple alternative experiment. Each X_i, $i=1, \ldots, n$, is a Bernoulli random variable with the same probability p for $X_i=1$ and $1=p$ for $X_i=0$. (If the trials are independent, then we expect these n Bernoulli random variables to be independent.) Assume the following:

a. On the first k trials, we get k successes. Since the probability of one success is p, the probability of k successes must be p^k, that is, under independence,

$$P(k\text{ successes}) = P(\underbrace{1 \cap 1 \cap \cdots \cap 1}_{k\text{ times}}) = \underbrace{P(1)\cdot P(1)\cdots P(1)}_{k\text{ times}} = \underbrace{p\cdot p\cdots p}_{k\text{ times}} = p^k.$$

b. On the last $n-k$ trials we get $n-k$ failures. Since the probability of one failure is $1-p$, the probability of $n-k$ failures must be $(1-p)^{n-k}$, that is, again under independence,

$$\begin{aligned} P(n-k\text{ failures}) &= P(\underbrace{0 \cap 0 \cap \cdots \cap 0}_{n-k\text{ times}}) = \underbrace{P(0)\cdot P(0)\cdots P(0)}_{n-k\text{ times}} \\ &= \underbrace{(1-p)(1-p)\cdots(1-p)}_{n-k\text{ times}} = (1-p)^{n-k}. \end{aligned}$$

We can now combine parts a and b to obtain:

c. The probability of k successes on the first k trials "and" $n-k$ failures on the last $n-k$ trials is, under independence of these events,

$$\begin{aligned} P(k\text{ successes } \cap n-k\text{ failures}) &= P(k\text{ successes})\cdot P(n-k\text{ failures}) \\ &= p^k(1-p)^{n-k}. \end{aligned}$$

d. But k successes on the first k trials is only one way to get k successes in the n independent trials. Since the k successes can actually occur in any fashion whatever (their order of occurrence is not important), the total number of ways to get k successes in the n trials is ${}_nC_k = n!/k!(n-k)!$. Thus

$$P(k,n,p) = \frac{n!}{k!(n-k)!}p^k(1-p)^{n-k}.$$

We are not quite finished yet. Having just discussed binomial probability, we now turn to the definition of the binomial probability distribution. Remember that, in order to define a discrete probability distribution, we need three pieces of information: the random experiment, the discrete (or "count") random variable X, and the rule for assigning probabilities. To this end: let a random variable X be specified as the "number of successes" in n independent trials of a "simple alternative experiment." Then X *is distributed binomially* with "probability mass function" $P(X; n, p)$, where

$$P(X;n,p) = \frac{n!}{X!(n-X)!}p^X(1-p)^{n-X}, \quad X = 0, 1, \ldots, n. \tag{5.12}$$

Hence the binomial probability distribution is a probability model for the "count" of successful outcomes of a simple alternative experiment.

How should the binomial probability distribution actually be represented? Think table (see Table 5.12).

A few additional comments concerning the binomial probability distribution are in order. First, a binomial probability distribution is completely determined once p and n are given. Second, the binomial distribution is symmetrical if $p = 0.5$; otherwise it is skewed (in particular, as $p \rightarrow 0$, it is skewed to the right; and as $p \rightarrow 1$, it is skewed to the left). Finally, it can be shown that, for a binomial random variable X:

$$E(X) = np \tag{5.13a}$$

$$V(X) = np(1-p). \tag{5.13b}$$

Note that the Bernoulli mean and variance are special cases of these expressions when $n = 1$.

TABLE 5.12 Binomial Probability Distribution

X	$P(X; n, p)$ [from (5.13)]
0	$P(0; n, p)$
1	$P(1; n, p)$
2	$P(2; n, p)$
.	.
.	.
.	.
n	$P(n; n, p)$
	1

To summarize, the characteristics of a binomial experiment are as follows:

1. We have a fixed number n of independent trials of a simple alternative experiment
2. The random variable X, defined as the number of successes, is discrete
3. p, the probability of a success, is constant from trial to trial (which occurs under sampling with replacement).

Example 5.7

Suppose we flip a "fair" coin five times and let a success be defined as "heads occurs" on any toss. Clearly, we have $n = 5$ independent trials of this simple alternative experiment, the probability of a success $p = 1/2$ is constant from trial to trial, and X, the number of heads obtained in the five flips of a coin, is a discrete random variable. The implied binomial probability distribution is given in Table 5.13. For instance:

$$P\left(0; 5, \frac{1}{2}\right) = \frac{5!}{0!5!}\left(\frac{1}{2}\right)^0\left(\frac{1}{2}\right)^5 = \frac{1}{32}$$

$$P\left(1; 5, \frac{1}{2}\right) = \frac{5!}{1!4!}\left(\frac{1}{2}\right)^1\left(\frac{1}{2}\right)^4 = \frac{5}{32}$$

$$P\left(2; 5, \frac{1}{2}\right) = \frac{5!}{2!3!}\left(\frac{1}{2}\right)^2\left(\frac{1}{2}\right)^3 = \frac{10}{32}$$

Since $p = \frac{1}{2}$, the distribution is symmetrical so that the remaining probabilities must be as indicated. To determine the mean and variance of X we need only find the following:

$$E(X) = np = 5/2$$

$$V(X) = np(1 - p) = 5/4.$$

Also, it is readily determined that:

TABLE 5.13 Binomial Probability Distribution When $n = 5$, $p = \frac{1}{2}$

X	$P(X; n, p)$ [from Equation (5.12)]
0	$p(0; 5, ½) = 1/32 = 0.0313$
1	$p(1; 5, ½) = 5/32 = 0.1563$
2	$p(2; 5, ½) = 10/32 = 0.3125$
3	$p(3; 5, ½) = 10/32 = 0.3125$
4	$p(4; 5, ½) = 5/32 = 0.1563$
5	$p(5; 5, ½) = 1/32 = 0.0313$
	1.0000

$P(X \geq 2) = P(2; 5, ½) + P(3; 5, ½) + P(4; 5, ½) + P(5; 5, ½) = \frac{26}{32}$ or
$P(X \geq 2) = 1 - P(X \leq 1) = 1 - P(0; 5, ½) - P(1, 5, ½) = \frac{26}{32}$
$P(2 < X \leq 4) = P(3; 5, ½) + P(4; 5, ½) = \frac{15}{32}$
$P(2 < X < 4) = P(3; 5, ½) = \frac{10}{32}$
P (at least one head) $= 1 - P$ (no heads) $= 1 - \frac{1}{32} = \frac{31}{32}$
$P(X > 2) = P(3; 5, ½) + P(4; 5, ½) + P(5; 5, ½) = \frac{16}{32}$ or
$P(X > 2) = 1 - P(X \leq 2) = 1 - P(0; 5, ½) - P(1; 5, ½) - P(2; 5, ½) = \frac{16}{32}$
$P(2 \leq X \leq 4) = P(2; 5, ½) + P(3; 5, ½) + P(4; 5, ½) = \frac{25}{32}$ or
$P(2 \leq X \leq 4) = P(X \leq 4) - P(X \leq 1) = \frac{31}{32} - \frac{6}{32} = \frac{25}{32}$

One final point concerning the binomial distribution is in order. The reader may have noticed that the decimal values of the probabilities in Table 5.13 have been provided. This is because, for specific values of n and p, a table of binomial probabilities has been provided in the Appendix at the end of this book (Table A.5). Specifically, this table gives the probability of obtaining X successes in n trials of a binomial or simple alternative experiment when the probability of a success is p. So for $n = 5$, and $p = 0.5$, the probabilities appearing in this table are the same as those provided in Table 5.13. A companion table, Table A.6, gives the cumulative probability of X successes in n trials of a binomial experiment when the probability of a success is p. For instance, for $n = 5$ and $p = 0.5$, the probability that $P(X \leq 3) = 0.8125$.

Example 5.8

A company claims that 25% of the people who chew gum buy its brand of chewing gum, "My Cud." If we obtain a random sample of four individuals who chew gum, and, if the company's claim is true, what is the probability that at least one of them will be found to chew "My Cud." Let $n = 4$ and $p = 0.25$. Also, define a success as finding a person who chews "My Cud." Then the random variable X is defined as the number of persons who chew this product found in a sample of size 4. The associated probability distribution appears in Table 5.14, with the individual probabilities obtained from Table A.5. Then $P(X \geq 1) = 0.6836$. Note that this probability could also have been calculated as $P(X \geq 1) = 1 - P(X \leq 0)$.

TABLE 5.14 Binomial Probability Distribution When $n = 4$, $p = 0.25$

X	$P(X; 4, 0.25)$
0	0.3164
1	0.4219
2	0.2109
3	0.0469
4	0.0039
	1.0000

EXERCISES

1 Which of the following random variables (X) is discrete? Why? [Hint: List the appropriate values for each X.]

a. X is the number of light bulbs on a chandelier containing eight light bulbs that need to be replaced this week.

b. X is the number of lobsters eaten at a picnic on the beach.

c. X is the number of customers frequenting a convenience store during business hours.

d. X is the number of points scored by the hometown football team.

e. X is the number of free-throw attempts before the first basket is made.

f. X is the number of senior citizens attending the next town council meeting.

2 Table E.5.2 contains data pertaining to the results of a survey in which $n = 300$ people were randomly selected and asked to indicate their preference regarding the candidates for the upcoming mayoral election. Transform this absolute frequency distribution to a discrete probability distribution. What key process allows us to do this?

TABLE E.5.2 Survey Results

Candidate	Absolute Frequency
Mary Smith	100
William Jones	50
Charles Pierce	40
Harold Simms	110

3 Which of the following tables represents a legitimate discrete probability distribution? Why?

a.		b.		c.		d.	
X	$f(X)$	X	$f(X)$	X	$f(X)$	X	$f(X)$
1	0.5	−1	0.1	1	0.2	1	0.1
2	0.1	2	0.1	2	0.3	2	0.4
3	0.1	3	0.3	3	0.3	3	0.1
4	0.2	4	0.2	4	−0.2	4	0.3
5	0.1	5	0.3	5	0.4	5	0.3

4 Given the following discrete probability distribution, find the following:

a. $P(X \leq 20)$ **c.** $P(X < 35)$

b. $P(X \geqslant 25)$ **d.** $P(X > 15)$

X	10	15	20	25	30	35
$f(X)$	0.1	0.3	0.1	0.3	0.1	0.1

5 The accompanying table provides an absolute frequency distribution obtained under random sampling from a large population of district managers for the ABC Corp. The random variable X depicts job satisfaction score ("1" is the lowest score and "5" is the highest score). Convert this absolute frequency distribution to a discrete probability distribution. Graph the probability mass function. Then find

a. $P(X > 3)$ **b.** $P(X \leq 2)$ **c.** $E(X)$

X	1	2	3	4	5
Absolute frequency	7	13	5	51	24

6 Given the following discrete probability distribution, find: $E(X)$; $V(X)$ [use Equation (5.3) and then use Equation (5.5) as a check]; and $S(X)$. Graph the associated probability mass function and locate $E(X)$.

X	0	1	2	3	4
$f(X)$	0.15	0.15	0.35	0.25	0.10

7 Given the following discrete probability distribution, find: $E(X)$; $V(X)$ (use Equation (5.5)); and $S(X)$. Graph the associated probability mass function.

X	−1	0	1	2
$f(X)$	0.1	0.2	0.3	0.4

8 Suppose that in calculating $S(X)$ for some discrete random variable X you find that $S(X) < 0$. What should you conclude? Suppose you are told that a discrete random variable X has $S(X) = 0$. What can you conclude?

9 A stock broker asserts that if the economy grows at 4% per annum, then your investment will yield a profit of $50,000. However, if the economy grows at only 2% per annum, your investment will render $15,000 in profit. And if the economy slips into a recession, you will lose $50,000 in profit. The U.S. Chamber of Commerce predicts that there is about a 20% chance that the economy will grow at 4%, a 70% chance of 2% growth, and a 10% chance of recession. What is the expected profit from your investment?

10 Suppose we toss a fair coin three times in succession. Let the random variable X be defined as the number of heads obtained in the three flips of the coin. Determine X's discrete probability distribution and graph its probability mass function. [Hint: Each outcome point or simple event is a point in three dimensions, that is, each point appears as (outcome on flip 1, outcome on flip 2, outcome on flip 3). There are eight such points in the three-dimensional sample space.] Also, find $E(X)$ and $S(X)$.

11 Let the random variable X represent the number of children per married couple after 10 years of marriage. Give X's discrete probability distribution (appearing below), find and interpret $E(X)$. Does a fractional part of a child make sense? How should this result be reported?

X	0	1	2	3	4	5	6
$f(X)$	0.07	0.15	0.30	0.25	0.15	0.05	0.03

12 Last week, 100,000 tickets were sold in a special state lottery program. Each ticket costs $3. The number of prizes and their dollar amounts appear in Table E.5.12. What is your expected gain or winnings? What is the expected gain/loss for a single lottery ticket? Interpret your results.

TABLE E.5.12 Lottery Prizes

Number of Prizes	Prize Amount ($)
1	10,000
3	5,000
1,000	100
10,000	3
88,996	0

13 Find the following:

a. ${}_7P_3$

b. ${}_6P_6$

c. ${}_8P_2$

d. ${}_{14}P_{13}$

e. ${}_7C_4$

f. ${}_{52}C_4$

g. ${}_4C_2$

h. ${}_6C_2$

14 Find the number of ways a chairman, vice chairman, and secretary could be elected from a group of seven candidates.

15 How many committees of five people can be formed from a group of 10 people?

16 Determine which of the following random experiments depicts a binomial experiment. If the random experiment is not a binomial experiment, indicate why. [Hint: Review the characteristics of a binomial experiment.]

a. A new drug for seasickness is taken by 50 randomly chosen individuals, with the number of individuals responding favorably recorded.

b. A random sample of 100 registered democrats is chosen and the respondents are asked if they favor gun control.

c. Three cards are randomly selected from a deck of 52 playing cards with replacement. The number of kings chosen is recorded.

d. A random sample of 30 high school seniors is taken and the students are asked to state their heights.

e. A random sample of 20 vehicles is taken from a state motor pool parking lot and the mileage is noted.

f. Two cards are randomly selected from a deck of 52 playing cards without replacement. The number of clubs chosen is recorded.

g. A field-goal kicker has a 20% chance of making a successful kick from 50 yards away. He is asked to keep kicking from that distance until he makes his first field goal.

17 A binomial probability distribution is completely determined when n and p are given. For each of the following combinations of n and p, compute the probability of X successes:

a. $n=8, p=0.35, X=3$

b. $n=10, p=0.40, X=3$

c. $n=15, p=0.85, X=12$

d. $n=20, p=0.60, X=17$

18 For each of the following combinations of n and p, determine the associated binomial probability distribution. [Hint: Use Table A.5.]

a. $n=8, p=0.35$

b. $n=10, p=0.40$

c. $n=15, p=0.85$

d. $n=20, p=0.60$

19 Using the binomial probability distribution obtained in Exercise 18.c, find the following:

a. $P(X \leq 3)$

b. $P(X<9)$

c. $P(X>10)$

d. $P(X\geqslant 8)$

e. P (at least one success)

20 For each of the binomial probability distributions obtained in Exercise 18, find $E(X)$ and $V(X)$. [Hint: Use Equations (5.14a and b).]

21 According to a Washington, DC think tank, 55% of all Americans support some form of so-called gun control. Suppose 12 citizens are randomly selected and the number in favor of gun control is recorded. Find the following:

a. The probability that exactly seven of them are gun-control advocates.
b. The probability that at most eight of them favor gun control.
c. The probability that at least five of them favor gun control.
d. The probability that at least seven, but no more than 10, favor gun control.

22 In clinical trials of an experimental drug for the treatment of tree-nut allergies, 5% of the patients experienced a headache as a side effect. Suppose a random sample of 20 subjects is taken and the number of them reporting a headache is recorded. Find the following:

a. The probability that exactly three subjects experienced a headache as a side effect.
b. The probability that no more than three subjects experienced a headache as a side effect.
c. The probability that at least one, but no more than four, subjects experienced a headache as a side effect.
d. The probability that at least half of the subjects experienced a headache as a side effect.

6

THE NORMAL DISTRIBUTION

6.1 THE CONTINUOUS PROBABILITY DISTRIBUTION

A probability distribution is termed *continuous* if its underlying random variable X is continuous, that is, if X can assume an infinite or uncountable number of real values over some interval. Such variables are "measured" in terms of distance, temperature, weight, time, and so on. For instance, $X=$ time between arrivals of vehicles at a toll booth is a continuous random variable. Also, if two points A and B are chosen at random on a real line segment of length L (Fig. 6.1), then the distance $X = |A - B|$ (the absolute value[1] of A minus B), $0 \leq X \leq L$, can assume an infinite number of values and thus is a continuous random variable. [How about this definition of a continuous random variable? A random variable X is *continuous* if $P(X = a) = 0$ for all real a. Does it work?]

The expression representing a continuous random variable X is called a *probability density function*. With X being continuous, its probability density function can be drawn without lifting one's pen from a sheet of paper. Hence the graph of X's probability density function is without breaks. More formally, a real-valued function f is a *probability density function* for a continuous random variable X if (a) $f(x) \geq 0$ for all real $x \in (-\infty, +\infty)$ and (b) the total area beneath $f(x)$ is one.

[1] The absolute value function is defined as follows:

$$|x| = \begin{cases} x, & x \geq 0 \\ -x, & x < 0 \end{cases}$$

Statistical Inference: A Short Course, First Edition. Michael J. Panik.

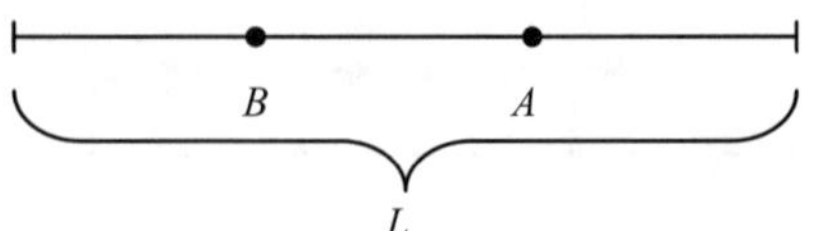

FIGURE 6.1 Points *A* and *B* on line segment *L*.

In fact, a continuous probability distribution is completely determined once its probability density function $f(x)$ is given. For example, let $f(x)$ be a *uniform probability distribution* defined by the following:

$$f(x) = \begin{cases} \dfrac{1}{\beta - \alpha}, & \alpha < x < \beta \\ 0 & \text{elsewhere} \end{cases}$$

(Fig. 6.2a). For $\alpha = 0$ and $\beta = 1$,

$$f(x) = \begin{cases} 0, & x \le 0 \\ 1, & 0 < x < 1 \\ 0, & x \ge 1 \end{cases}$$

(Fig. 6.2b). Here,

$$P(a \le X \le b) = b - a \quad \text{for } 0 \le a \le b \le 1.$$

Note that now an event is an interval on the *X*-axis (more on this point later). Clearly, $P(0 \le X \le 1) = 1 - 0 = 1$ as required. (The reader should verify that $P(0.25 \le X \le 0.75) = 0.50$.) In general, if *X* is a continuous random variable with probability density function $f(x)$, then $P(a \le X \le b)$ is the area under $f(x)$ from *a* to *b*.

6.2 THE NORMAL DISTRIBUTION

A continuous random variable *X* has a *normal distribution* if its probability density function is bell-shaped, symmetrical about its mean μ, and asymptotic to the *X* or horizontal axis (Fig. 6.3) In what follows, we shall refer to the normal probability density function as simply the "normal curve." *Asymptotic* to the horizontal axis

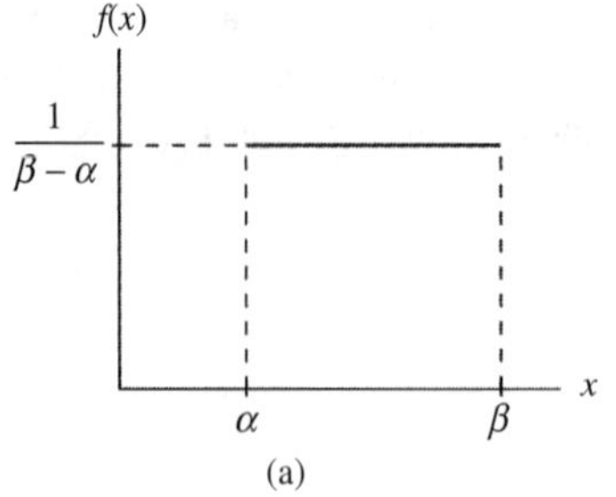

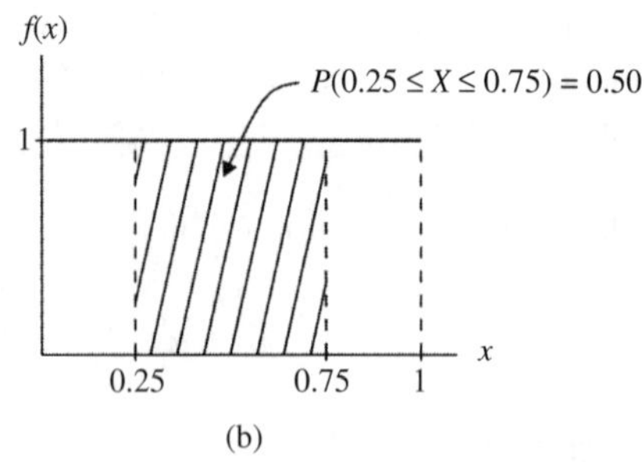

FIGURE 6.2 A continuous uniform probability distribution.

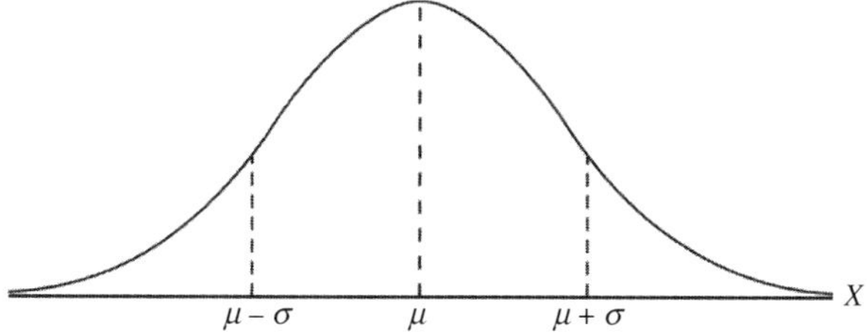

FIGURE 6.3 The normal curve.

means that the normal curve extends from $-\infty$ to $+\infty$ with its tails gradually approaching, but never touching, the horizontal axis. The normal probability distribution is completely determined once its mean μ and standard deviation σ are given. In this regard, if X follows a normal distribution with a known mean and standard deviation, then we will write: X is $N(\mu, \sigma)$. So if X is normally distributed with a mean of 50 and a standard deviation of 10, then this information is summarized as: X is $N(50, 10)$.

Some additional properties of the normal density curve are:

1. The normal curve attains its maximum at $X = \mu$ (Fig. 6.3). Hence μ = mode of X = median of X.
2. The normal curve has points of inflection[2] at $\mu - \sigma$ and $\mu + \sigma$ (Fig. 6.3).
3. The normal curve is constructed so that the total area beneath it is 1. In this regard, the *empirical rule* holds (Fig. 6.4):

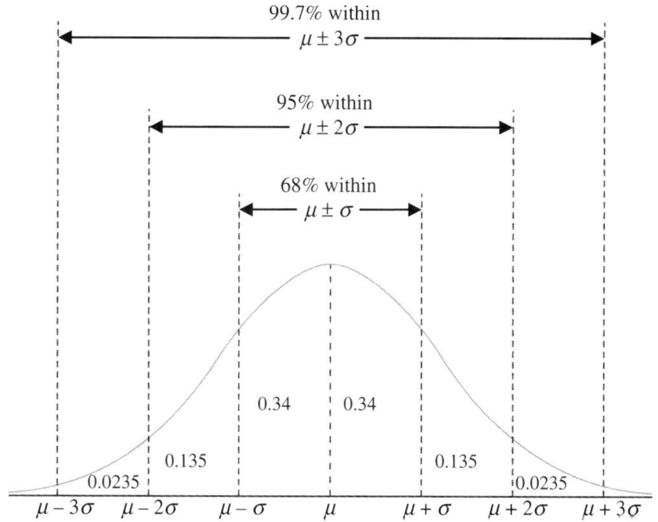

FIGURE 6.4 The empirical rule.

[2] A *point of inflection* occurs where a curve crosses its tangent line and changes the direction of its concavity from upward to downward or vice versa.

a. About 68% of the total area under the normal curve lies within 1 standard deviation of the mean or within the interval from $\mu - \sigma$ to $\mu + \sigma$.

b. 95% of the total area under the normal curve lies within 2 standard deviations of the mean or within the interval from $\mu - 2\sigma$ to $\mu + 2\sigma$.

c. 99.7% of the total area under the normal curve lies within 3 standard deviations of the mean or within the interval from $\mu - 3\sigma$ to $\mu + 3\sigma$.

Property 3 above now sets the stage for a segue to the next section.

6.3 PROBABILITY AS AN AREA UNDER THE NORMAL CURVE

Suppose X is $N(\mu, \sigma)$. If we are interested in finding $P(a \leq X \leq b)$, then we can do so by determining the area under the normal density curve from a to b (Fig. 6.5).

(Note that when dealing with a continuous univariate probability density function, the notion of an "event" is simply an interval $a \leq x \leq b$ on the X-axis.) While integral calculus techniques can be applied to find an area under the normal curve, we shall judiciously not attempt to do so. Instead, we shall read the area under the normal curve directly from a table. However, before we do so, let us note that there actually exists an infinite number of potential normal curves—one for each possible (allowable) value for μ and σ. For instance, as μ increases in value from μ_0 with σ fixed, the normal curve is translated a distance to the right of μ_0 (Fig. 6.6a).

But if we increase the value of σ from σ_0 with μ fixed, the normal curve changes its shape—it becomes wider and flatter (as it must since the total area beneath it is to be preserved at 1) (Fig. 6.6b). In fact, as Fig. 6.6b reveals, the area under each of these normal curves from a to b is different (the area decreases as σ increases). So if there exists an infinite number of possible normal curves and the area, say, from a to b, beneath them can vary, how come we do not have an infinite number of tables in the back of the book instead of just a single "one size fits all" table? (An obviously rhetorical question.) Apparently, some sort of transformation is being made.

To see exactly how the transformation works, let us assume that X is $N(\mu, \sigma)$ with $\sigma \neq 0$. Then:

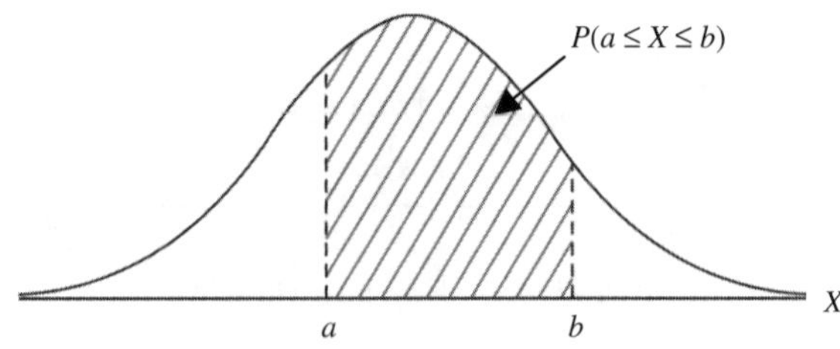

FIGURE 6.5 Probability as area under $N(\mu, \sigma)$.

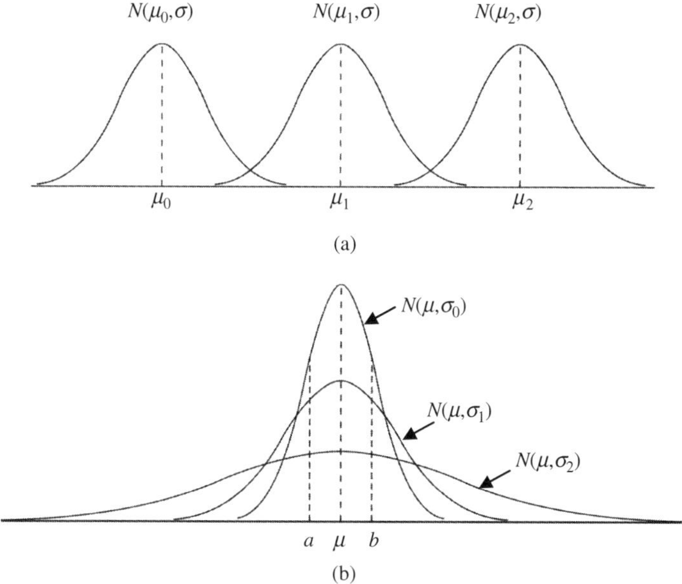

FIGURE 6.6 (a) Increasing the mean μ $(\mu_0 < \mu_1 < \mu_2)$. (b) Increasing the standard deviation σ $(\sigma_0 < \sigma_1 < \sigma_2)$.

1. $Z = \frac{X-\mu}{\sigma}$ is $N(0, 1)$, that is, Z follows a *standard normal distribution* (Fig. 6.7). The properties of the standard normal distribution are:
   (a) It has a mean of 0 and a standard deviation of 1. Since it attains its maximum at $Z = 0$, it follows that mean of Z = median of Z = mode of Z.
   (b) It is symmetrical about 0 with points of inflection at $Z = -1$ and $Z = 1$.
   (c) Z is expressed in standard deviation units, that is, $Z_0 = 2.5$ is located 2.5 standard deviations above the mean of zero. $Z_1 = -1.6$ is located 1.6 standard deviations below the mean of zero.
   (d) The area under $N(0, 1)$ is 1.
   (e) $N(0, 1)$ is asymptotic to the horizontal axis because $N(\mu, \sigma)$ is.

2. $$P(a \le X \le b) = P\left(\frac{a-\mu}{\sigma} \le \frac{X-\mu}{\sigma} \le \frac{b-\mu}{\sigma}\right) = P(Z_1 \le Z \le Z_2) \quad (6.1)$$
   (Fig. 6.8).

3. Let us consider $P(a \le X \le b) = P(X = a) + P(a<X<b) + P(X = b)$. What is the value of $P(X = a)$ and $P(X = b)$? Since there is no area under a curve at a single point (a point has no width), it must be the case that $P(X = a) = P(X = b) = 0$ so that $P(a \le X \le b) = P(a<X<b)$. Hence it does not matter

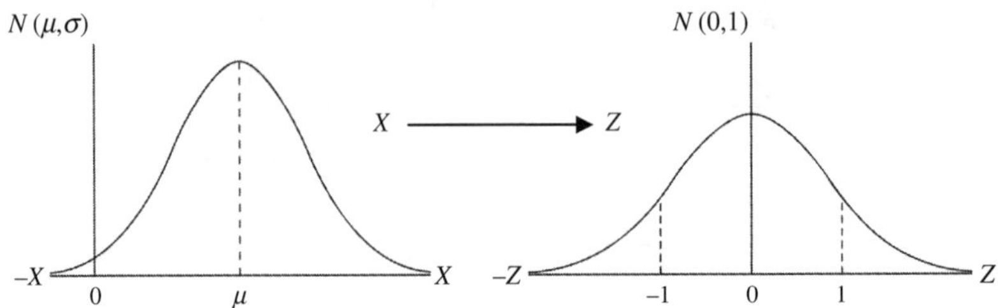

FIGURE 6.7 Transformation of normal X to standard normal $Z = \frac{X-\mu}{\sigma}$.

whether or not the interval end points a and b are included in the above probability statement; we get the same area. (Note that it does matter if X is a discrete random variable.)

4. Once we transform X to Z, the *empirical rule* still holds (Fig. 6.9)
   (a) About 68% of the total area under $N\ (0,\ 1)$ lies within 1 standard deviation of the mean or within the interval from -1 to 1.
   (b) 95% of the total area under $N\ (0,\ 1)$ lies within 2 standard deviations of the mean or within the interval from –2 to 2.
   (c) 99.7% of the total area under $N\ (0,\ 1)$ lies within 3 standard deviations of the mean or within the interval from -3 to 3.

The normal curve area table (Table A.1 of the Appendix A) is termed the *Standard Normal Area Table* since it relates to Z. How is this table read? Specifically, Z values are given in the far left (Z) column to one decimal point, for example, 0.1, 0.9, 2.1, 2.6, and so on. If we are interested in $Z = 1.85$, then we move down the Z column to 1.8 and then go across to the 0.05 column heading. Probabilities or areas or percentages appear in the body of the table. While areas cannot be negative, negative Z-values are allowed.

So what does the standard normal area table give us? It provides us with $A(Z_0)$—the proportion of the total area beneath the $N\ (0,\ 1)$ curve between 0 and any other point Z_0 on the "positive" Z-axis (Fig. 6.10). (Note that, by symmetry, the total area between $-Z_0$ and 0 is also $A(Z_0)$.) For instance, the area under $N\ (0,\ 1)$ from 0 to $Z_0 = 1.35$ is $A(Z_0) = 0.4115$ (Fig. 6.11). As an aid in finding $A(Z_0)$, one should always draw the graph of the desired area.

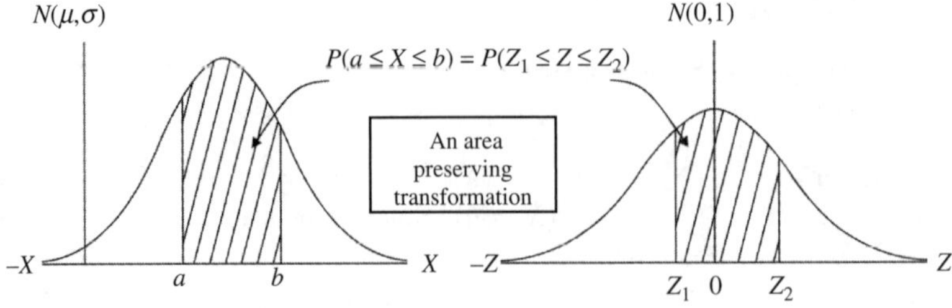

FIGURE 6.8 The area under $N(\mu, \sigma)$ from a to b equals the area under $N(0, 1)$ from Z_1 to Z_2.

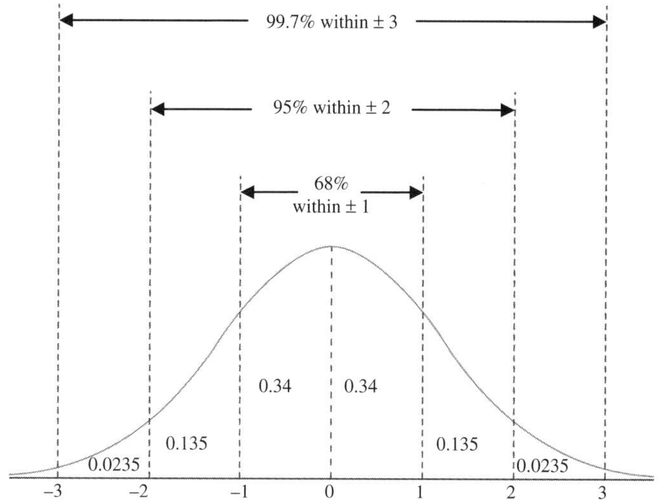

FIGURE 6.9 The empirical rule in terms of Z.

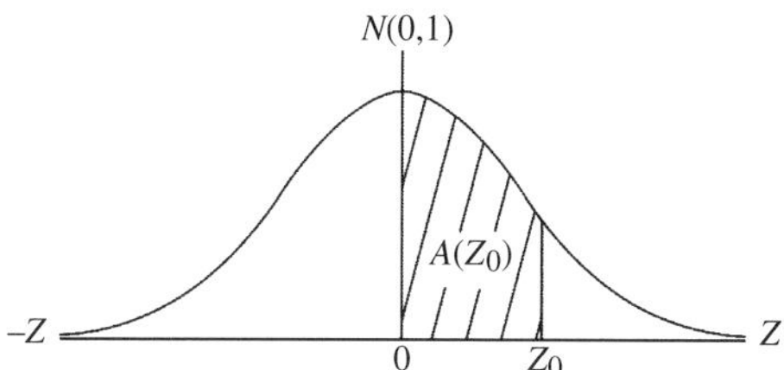

FIGURE 6.10 Proportion of the total area between 0 and Z_0.

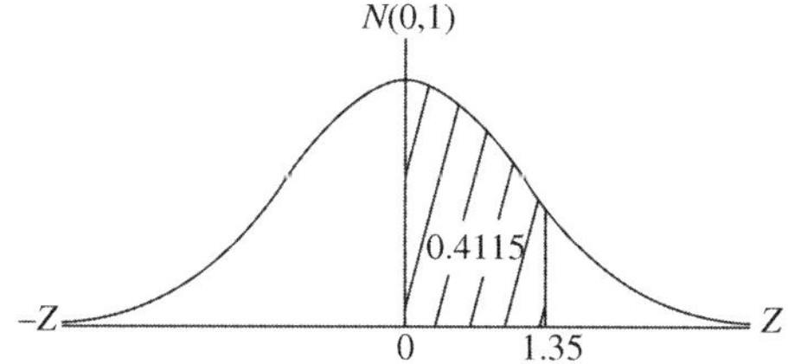

FIGURE 6.11 Area under $N(0, 1)$ from 0 to 1.35.

Example 6.1

Let us consider finding the following areas under $N(0,\ 1)$. (Hint: The total area from $-\infty$ to $+\infty$ is 1; the area from $-\infty$ to 0 is 0.5, as is the area from 0 to $+\infty$.)

a. Area from 0 to 1.61? (See Fig. 6.12.)
b. Area from –0.09 to 0? (See Fig. 6.13.)
c. Area from $-\infty$ to 1.70? (See Fig. 6.14.)
d. Area from –0.56 to $+\infty$? (See Fig. 6.15.)
e. Area from 0.79 to $+\infty$? (See Fig. 6.16.)
f. Area from $-\infty$ to –1.53? (See Fig. 6.17.)
g. Area from 0.87 to 1.45? (See Fig. 6.18.)

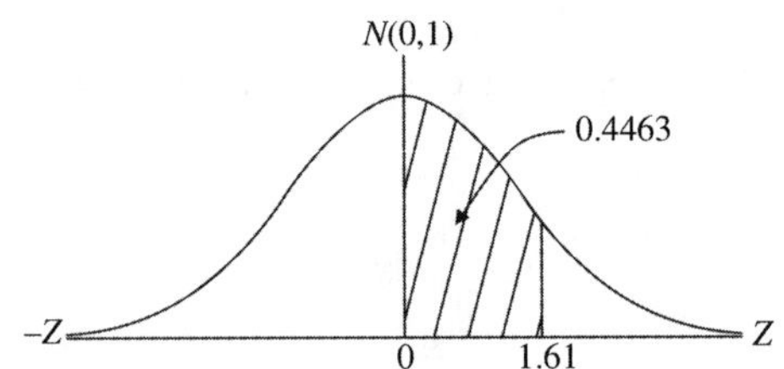

FIGURE 6.12 Area under $N(0,\ 1)$ from 0 to 1.61.

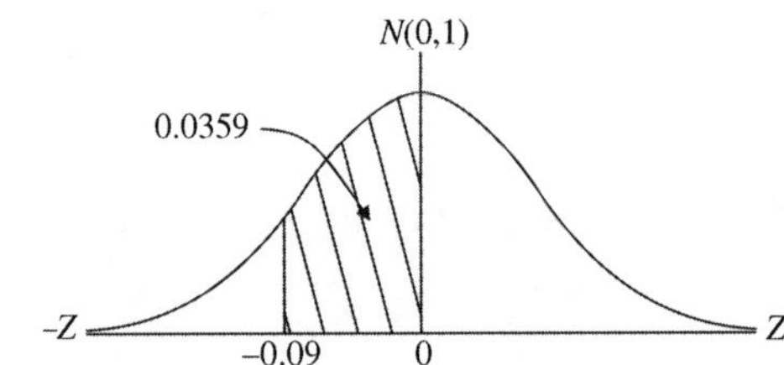

FIGURE 6.13 Area under $N(0,\ 1)$ from –0.09 to 0.

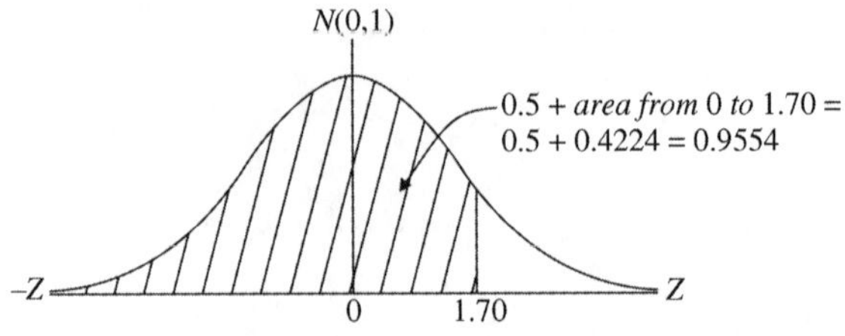

FIGURE 6.14 Area under $N(0,\ 1)$ from $-\infty$ to 1.70.

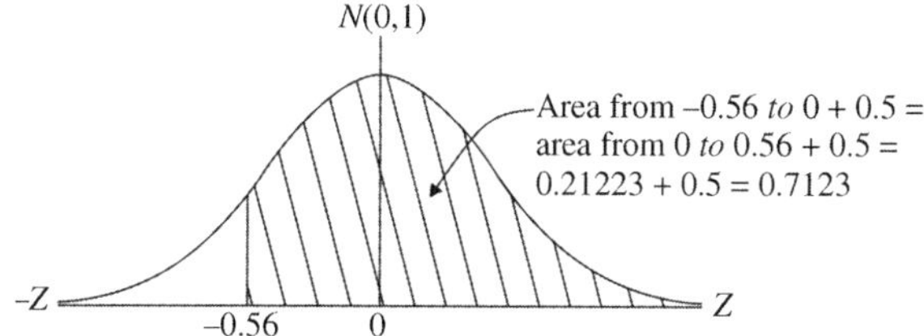

FIGURE 6.15 Area under $N(0, 1)$ from -0.56 to $+\infty$.

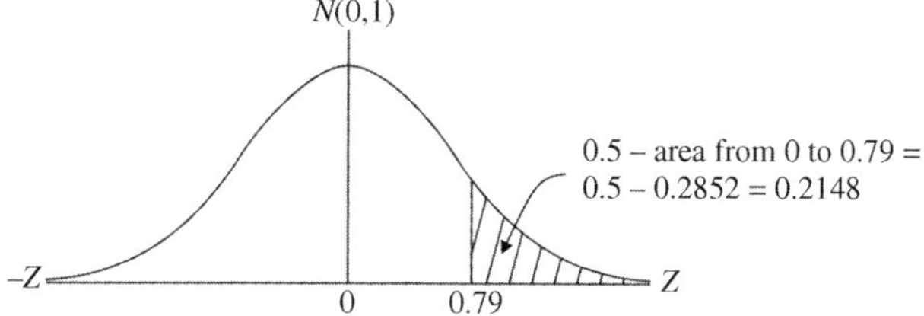

FIGURE 6.16 Area under $N(0, 1)$ from 0.79 to $+\infty$.

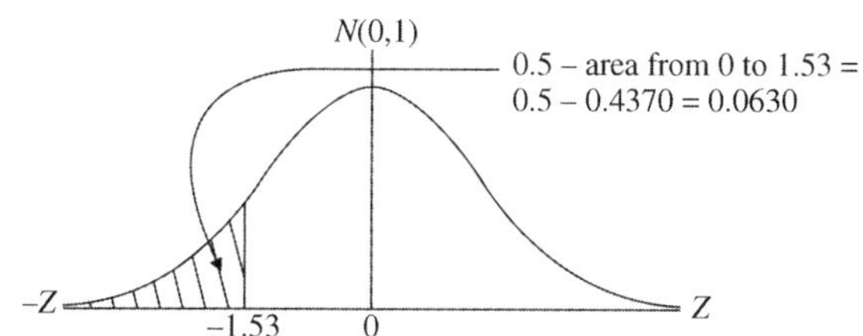

FIGURE 6.17 Area under $N(0, 1)$ from $-\infty$ to -1.53.

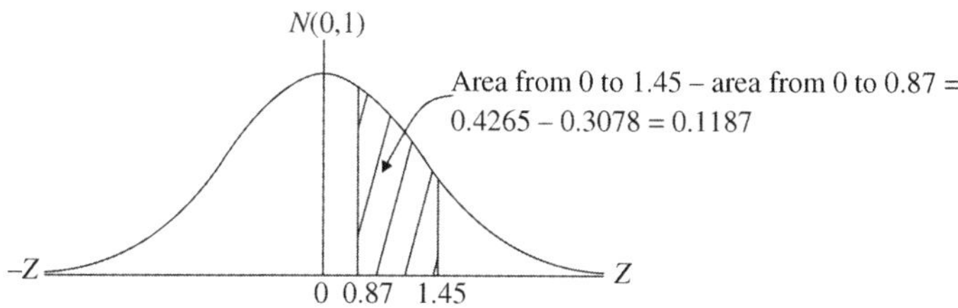

FIGURE 6.18 Area under $N(0, 1)$ from 0.87 to 1.45.

Example 6.2

Suppose the random variable X is $N(\mu, \sigma) = N(50, 10)$. Let us find the following:

a. $P(X \leq 66) = ?$ Our procedure will be to transform the given probability statement in terms of X to an equivalent probability statement in terms of Z

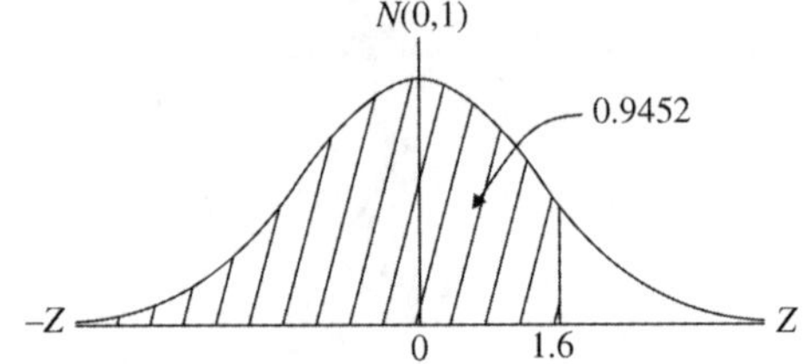

FIGURE 6.19 Area under $N(0, 1)$ from $-\infty$ to 1.6.

[see Equation (6.1)]. How is this accomplished? From "each" term in parentheses subtract the mean of X and then divide each resulting difference by the standard deviation of X, where $X \to Z = \frac{X-\mu}{\sigma}$. To this end, we have $P(\frac{X-\mu}{\sigma} \leq \frac{66-50}{10}) = P(Z \leq 1.6) = 0.5 + 0.4452 = 0.9452$ (Fig. 6.19).

b. $P(X \geq 42) = ?$

$$P\left(\frac{X-\mu}{\sigma} \geq \frac{42-50}{10}\right) = P(Z \geq -0.8) = 0.2881 + 0.5 = 0.7881$$

(Fig. 6.20).

c. $P(40 \leq X \leq 60) = ?$

$$P\left(\frac{40-50}{10} \leq \frac{X-\mu}{\sigma} \leq \frac{60-50}{10}\right) = P(-1 \leq Z \leq 1) = 2(0.3413) = 0.6826$$

(Fig. 6.21).

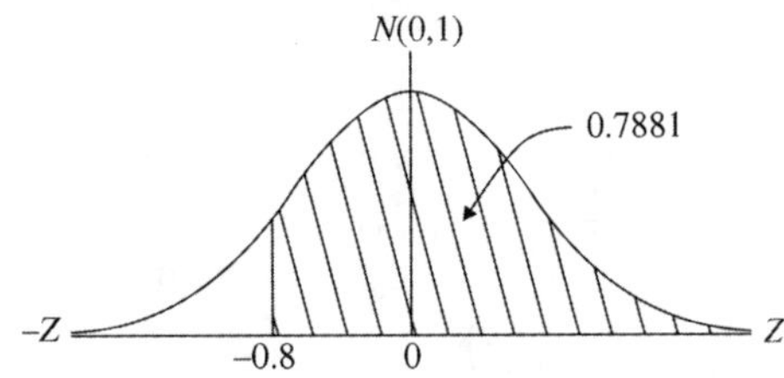

FIGURE 6.20 Area under $N(0, 1)$ from -0.8 to $+\infty$.

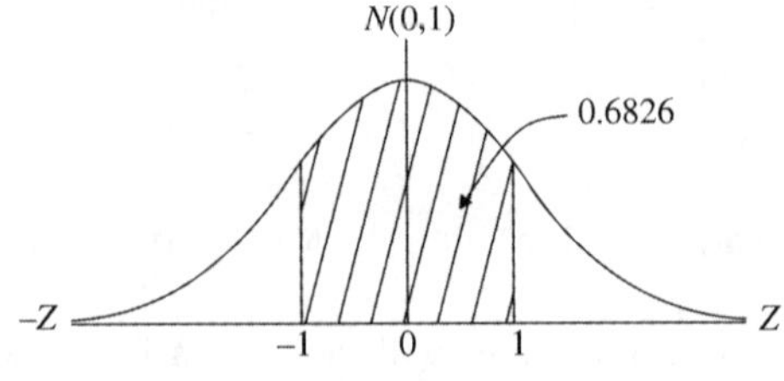

FIGURE 6.21 Area under $N(0, 1)$ from -1 to 1.

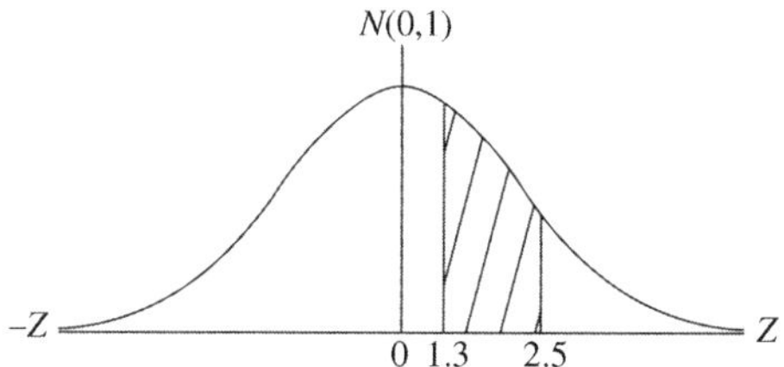

FIGURE 6.22 Area under $N(0, 1)$ from 1.3 to 2.5.

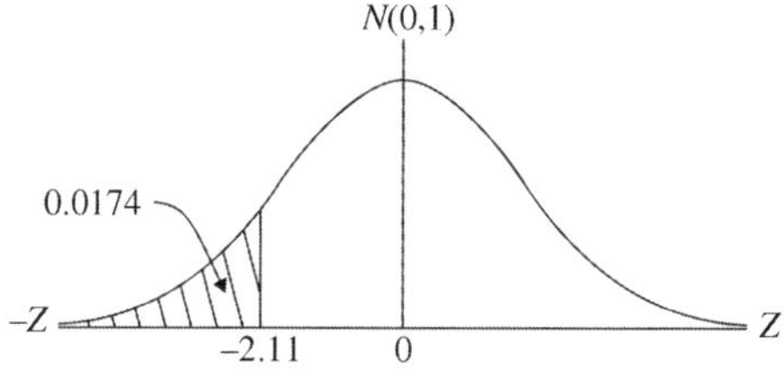

FIGURE 6.23 Area under $N(0, 1)$ from $-\infty$ to -2.11.

d. $P(63 \leq X \leq 75) = ?$

$$P\left(\frac{63-50}{10} \leq \frac{X-\mu}{\sigma} \leq \frac{75-50}{10}\right) = P(1.3 \leq Z \leq 2.5)$$
$$= 0.4938 - 0.4032 = 0.0906$$

(Fig. 6.22).

e. $P(X = 48) = ?$ The probability that X equals a single specific value (there is no interval of positive length here) is "always" zero. Hence $P(X = 48) = 0$.

Example 6.3

Let X be $N\,(2.9,\ 0.9)$, where the random variable X represents the length of life (in years) of an electric can opener. If this can opener has a 1 yr warranty, what fraction of original purchases will require replacement? We need to determine the following:

$$P(X \leq 1) \leq P\left(Z \leq \frac{1-2.9}{0.9}\right) = P(Z \leq -2.11) = 0.5 - 0.4826 = 0.0174$$

(Fig. 6.23).
Hence about 1.74% of original sales will require replacement.

Example 6.4

Suppose X is $N(5000,\ 1500)$, where the random variable X depicts the annual amount spent (in dollars) by customers in the service department at a local car dealership. The service manager wants to sponsor a special promotional program that will offer

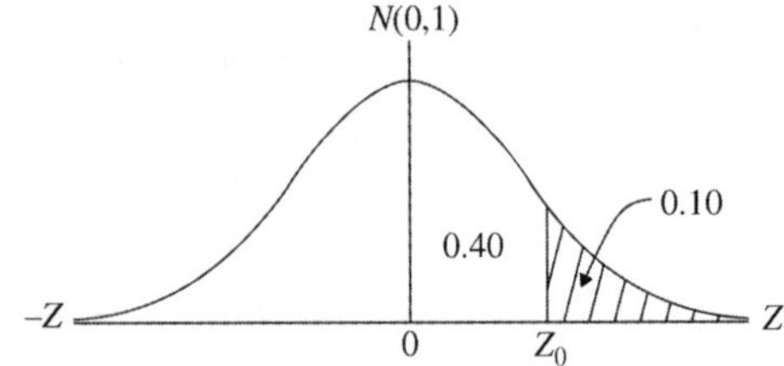

FIGURE 6.24 Area under $N(0, 1)$ from Z_0 to $+\infty$.

discount coupons to those customers whose annual expenditure level puts them in the "top 10%" of the service customer expenditure distribution. How much does a customer have to spend annually to qualify for a book of discount coupons?

Note that we are given the probability level or area in the right-hand tail of the $N(0, 1)$ distribution—it is 0.10 (Fig. 6.24). Hence we ask: "What is the value of Z (call it Z_0) that cuts off 10% of the area in the right-hand tail of the standard normal distribution?" To answer this question, let us start with the probability statement:

$$P(Z \geq Z_0) = P(Z \geq \frac{X_0 - 5000}{1500}) = 0.10.$$

Hence we must "untransform" Z_0 (which is in standard deviation units) to find X_0 (which is in dollars). But what is the value of Z_0?

Since the area from 0 to Z_0 is 0.40, we need to find 0.40 in the body of the $N(0, 1)$ table and then read the table in reverse to obtain Z_0. This renders $Z_0 = 1.28$ (approximately). Hence

$$Z_0 = \frac{X_0 - 5{,}000}{1{,}500} = 1.28.$$

Then solving for X_0 yields

$$X_0 = 5000 + 1500\,(1.28) = \$6920.00.$$

Hence anyone who has spent at least this amount is sent a book of discount coupons.

In general, since the "transformed" X_0 is

$$\frac{X_0 - \mu}{\sigma} = Z_0,$$

we can "untransform" Z_0 and solve for the underlying X_0 value as

$$X_0 = \mu + \sigma Z_0. \tag{6.2}$$

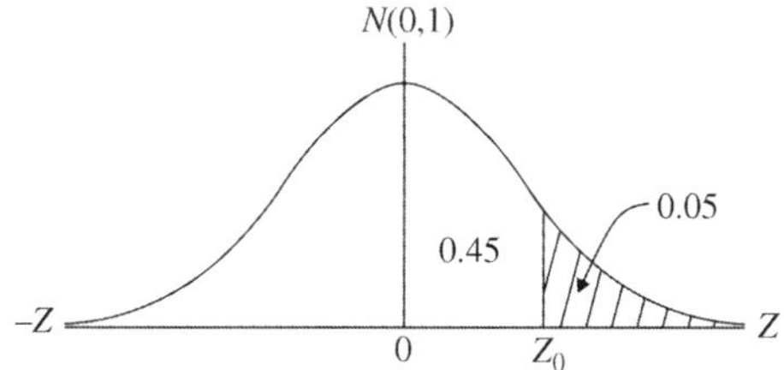

FIGURE 6.25 Area under N (0, 1) from Z_0 to $+\infty$.

Example 6.5

Suppose that weekly production (in number of units assembled) of the *XYZ* Corp. is $N(2000,\ 35)$. To increase output, management decides to pay a bonus on the upper 5% of the distribution of weekly production. How many units must a worker assemble per week in order to qualify for a bonus? (Incidentally, from a practical standpoint, are there any pitfalls associated with such a plan?)

Since we are given an area or probability of 0.05 in the upper tail of the $N(0,\ 1)$ distribution at the outset, let us form the probability statement

$$P(Z \geq Z_0) = P\left(Z \geq \frac{X_0 - 2000}{35}\right) = 0.05$$

(see Fig. 6.25). Since the remaining area under $N\ (0,\ 1)$ to the right of 0 is 0.45, we find 0.45 in the body of the $N\ (0,\ 1)$ table and read it in reverse to obtain $Z_0 = 1.645$ (approximately).

Then since

$$Z_0 = \frac{X_0 - 2000}{35} = 1.645,$$

we obtain, via Equation (6.2), $X_0 = 2000 + 35\,(1.645) = 2057.58 \approx 2058$ units. So any worker assembling at least 2,058 units per week qualifies for a bonus.

Example 6.6

We noted in Chapter 3 that a $Z-$ score $(Z = (X - \mu)/\sigma)$ expresses distance from the mean in standard deviation units (if $Z = -0.75$, then the associated X value lies $3/4$ of a standard deviation below μ); and it can be used to determine the relative location of an X value in a set of data. This being the case, suppose you take a "spatial relations" exam that is given nationally to thousands of (college) students. Research has shown that such exam scores (along with I.Q. scores, and so on) tend to be normally distributed. In this regard, let us assume that the distribution of these scores or the random variable X is $N(1050,\ 157)$. Weeks pass and finally your score of 1305 arrives in the mail. How did you fare relative to all others that took the exam? For your score,

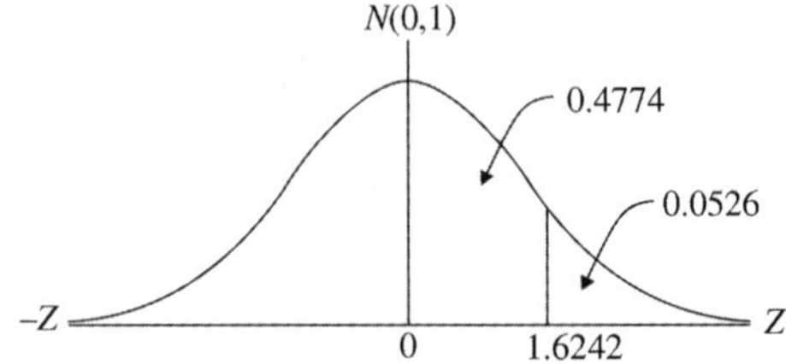

FIGURE 6.26 Relative standing for a test score of 1305.

$$Z = \frac{X - \mu}{\sigma} = \frac{1305 - 1050}{157} = 1.6242.$$

Hence your score of 1305 lies about 1.62 standard deviations above the mean. But this is not the whole story, as Fig. 6.26 reveals. Since the area from 0 to 1.62 is 0.4474, the area in the right-hand tail of the $N(0,\ 1)$ distribution is 0.0526. Hence only about 5.26% of the students had a score higher than your score. Congratulations!

6.4 PERCENTILES OF THE STANDARD NORMAL DISTRIBUTION AND PERCENTILES OF THE RANDOM VARIABLE *X*

Suppose a random variable X is $N(\mu,\ \sigma)$. To determine the pth percentile of X, we first need to find the pth percentile of Z. To this end, the *value of Z at the pth percentile*, denoted Z_p, has 100 p% of the area under $N(0,\ 1)$ below it, that is, Z_p satisfies the equation

$$P(Z \le Z_p) = p. \tag{6.3}$$

For instance, to find the 80th percentile of the $N\ (0,\ 1)$ distribution, let us work with the equation

$$P(Z \le Z_{0.80}) = 0.80$$

(Fig. 6.27a). Then, since $0.80 = 0.50 + 0.30$, let us find 0.30 in the body of the $N\ (0,\ 1)$ table and read it in reverse to find $Z_{0.80} = 0.84$. Hence 80% of the values of Z lie below 0.84. Similarly, to find the 30th percentile of values of Z, we work with the equation

$$P(Z \le Z_{0.30}) = 0.30.$$

Since $0.30 = 0.50 - 0.20$, let us get as close as possible to 0.20 in the body of the $N(0,\ 1)$ table and read it in reverse to obtain $Z_{0.30} = -0.52$ (Fig. 6.27b).

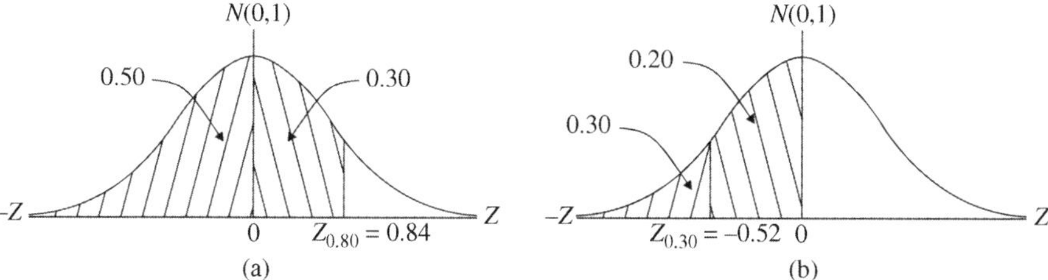

FIGURE 6.27 Percentiles of Z.

Having determined Z_p, we can untransform $Z_p = (X_p - \mu)/\sigma$ [see Equation (6.2)] so as to obtain

$$X_p = \mu + \sigma Z_p. \tag{6.4}$$

Example 6.7

Given that X is N (50, 10), find the 90th percentile of X. Let us start with the 90th percentile of Z or $P(Z \leq Z_{0.90}) = 0.90$ (Fig. 6.28). Given that $0.90 = 0.50 + 0.40$, let us find 0.40 in the body of the N (0, 1) table and read it in reverse to find $Z_{0.90} = 1.28$. Hence 90% of the values of Z lie below 1.28. Turning to Equation (6.4), we can now determine $X_{0.90}$ as

$$X_{0.90} = 50 + 10(1.28) = 62.8,$$

that is, 90% of the X values lie below 62.8.

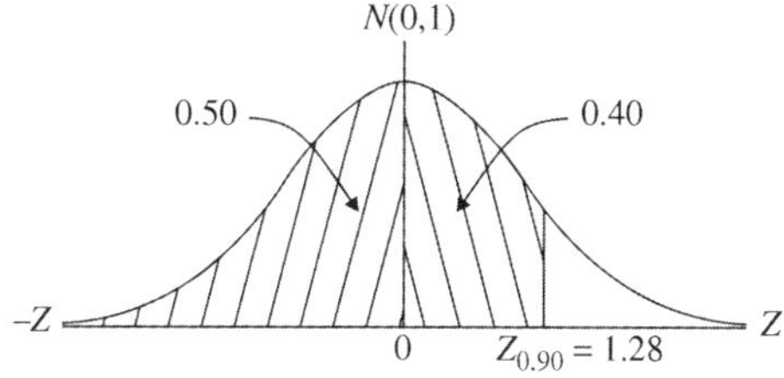

FIGURE 6.28 90th percentile of Z.

To close this chapter, let us summarize some of our thoughts concerning how the $N(0, 1)$ distribution is employed in problem solving. To this end:

1. If we want to determine a probability (area) or percent value, we start with an X − **value**, transform it to a Z − **value**, and then find the answer in the body of the N (0, 1) table (see Example 6.3).
2. But if we need to find an X − **value**, we start with an area (a probability) or percent value, find the associated Z − **value**, and then untransform Z to determine the underlying X (Examples 6.4 and 6.5).

EXERCISES

1 Which of the following random variables is continuous? Explain why?

a. X is the speed of a race car as it passes the finish line.
b. X is the time it takes to fly from London to New York City.
c. X is the capacity of a car's trunk in cubic feet.
d. X is the weight of your roommate.
e. X is the distance traveled by a race horse in 10 seconds.
f. X is the time it takes a race horse to travel 100 ft.

2 Suppose the random variable X has the probability density function

$$f(x) = \begin{cases} \frac{1}{6}, & 0 < x < 6, \\ 0 & \text{elsewhere} \end{cases}$$

find:

a. $P(X = 5)$ **b.** $P(X \leq 2)$ **c.** $P(X > 4)$

3 Given the probability density function for X,

$$f(x) = \begin{cases} \frac{2-x}{2}, & 0 \leq x \leq 2, \\ 0 & \text{elsewhere} \end{cases}$$

find:

a. $P(X > 1)$ **b.** $P(X \leq 1)$

c. $P(0 < X < 2)$ **d.** $P(0.5 < X < 1.5)$

4 Given that X's probability density function appears in Fig. E.6.4, find the following:

a. c. **b.** $P(-1 \leq X \leq 2)$ **c.** $P(X \geq 1.5)$

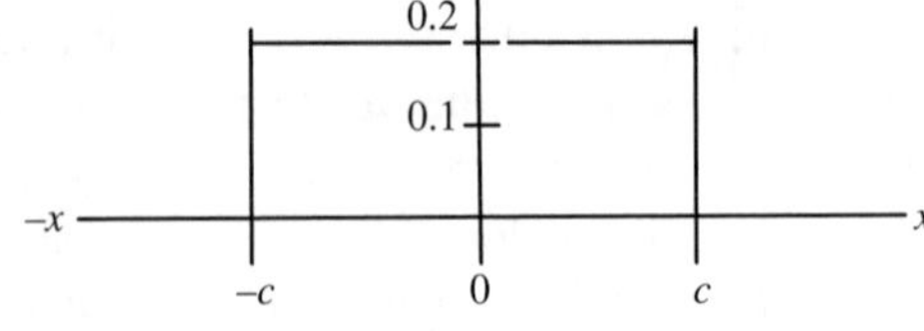

FIGURE E.6.4 Uniform probability density over $[-c, c]$.

5 Given that a random variable X is $N(50,10)$, transform X to a standard normal variable Z. What is the shape or form of Z?

6 Suppose a random variable X is $N(10, 3)$. Find the following:

a. Z_1 for $X_1 = 8$
b. Z_2 for $X_2 = 12$
c. Given that the area under $N(10, 3)$ from $X_1 = 8$ to $X_2 = 12$ is 0.495, what is the area under $N(0,1)$ from $Z_1 = -0.67$ to $Z_2 = 0.67$?

7 Find the area under the standard normal curve that lies to the left of the following:

a. $Z = -2.43$
b. $Z = -0.41$
c. $Z = 1.33$
d. $Z = 0$
e. $Z = 3.31$

8 Find the area under the standard normal curve that lies to the right of the following:

a. $Z = -3.45$
b. $Z = -0.55$
c. $Z = 2.21$
d. $Z = 0$
e. $Z = 3.35$

9 Find the area under the standard and normal curve that lies between:

a. -2.11 and 2.62
b. -0.51 and 0
c. -1.13 and -0.41
d. 1.14 and 2.16
e. 0 and 6.4

10 Find the Z-score such that:

a. The area under the standard normal curve to the left is 0.2.
b. The area under the standard normal curve to the right is 0.25.
c. The area under the standard normal curve to the left is 0.5.
d. The area under the standard normal curve to the right is 0.65.
e. The area under the standard normal curve to the left is 0.85.

11 Find the Z-scores that separate the middle (a) 70%, (b) 99%, and (c) 80% of the standard normal distribution from the area in its tails.

12 A Z-score expresses distance from the mean in terms of standard deviations. How much of the area under the standard normal curve lies within one standard deviation of the mean, that is, between $Z = \pm 1$? Between $Z = \pm 2$? Between $Z = \pm 3$? [Hint: Review the empirical rule.]

13 For each of the following probability statements, find c. [Hint: Read the standard normal area table in reverse, that is, start with an area from the body of the table and move to a point on the horizontal or Z-axis.]

a. $P(Z \leq c) = 0.9938$
b. $P\left(Z > c\right) = 0.4404$
c. $P\left(Z < c\right) = 0.1492$
d. $P(Z \geq c) = 0.5989$
e. $P(-c \leq Z \leq c) = 0.8740$

14 Given that a random variable X is $N(50, 7)$, find the following:

a. $P(X \geq 65)$
b. $P(X < 45)$
c. $P(X = 55)$
d. $P(40 \leq X \leq 65)$
e. $P(55 \leq X \leq 70)$
f. $P(38 < X < 55)$

15 A certain variety of a wood-composite beam has a breaking strength that is $N(1500\text{ lbs}, 100\text{ lbs})$. What percentage of all such beams has a breaking strength between 1450 lbs and 1600 lbs?

16 Given that the random variable X is $N(\mu, \sigma)$, find the value of θ such that:

a. $P(\mu - \theta\sigma \leq X \leq \mu + \theta\sigma) = 0.95$
b. $P(\mu - \theta\sigma \leq X \leq \mu + \theta\sigma) = 0.90$
c. $P(\mu - \theta\sigma \leq X \leq \mu + \theta\sigma) = 0.99$

17 Given that the random variable X is $N(\mu, 5)$ and $P(X \geq 78.5) = 0.0668$, find μ.

18 Given that the random variable X is $N(19, \sigma)$ and $P(X \leq 22.5) = 0.6368$, find σ.

19 Suppose that the overall gasoline mileage for a particular type of vehicle is $N(56, 3.2)$. Find:

a. The proportion of these vehicles getting in excess of 60 mpg.
b. The proportion of these vehicles getting at most 50 mpg.
c. The proportion of these vehicles getting between 58 mpg and 62 mpg.
d. The proportion of these vehicles getting less than 45 mpg.

20 A random variable X is $N(30, 8)$. Find the indicated percentile for X:

a. The 10th percentile
b. The 90th percentile
c. The 80th percentile
d. The 50th percentile

21 The research arm of the ABC Corp. has determined that the use time (in hours) of its microbattery is $N(324, 24)$. Find:

a. The proportion of time that the microbattery will last at least 300 hours.
b. The proportion of time it will last between 310 hr and 350 hr.
c. The use time that is in the top 20% of the distribution.
d. The range of use time that constitutes the middle 90% of the distribution.

22 The average stock price for companies constituting the Standard & Poor's (S&P) 500 Index is $30 and the standard deviation of the stock prices is $8.20. If the implied distribution of stock prices is assumed to be normal, find the following:

a. The probability that an S&P 500 company will have a stock price not higher than $20.
b. The probability that an S&P 500 company will have a stock price of at least $40.
c. How high does a stock price have to be in order to put an S&P 500 company in the top 10% of the distribution of S&P 500 stock prices?
d. The range of stock prices that make up the middle 80% of the distribution of S&P 500 stock prices.

23 A college basketball player must have a points-per-game average in the upper 2% of all college basketball players in order to qualify for *All American* status. If points-per-game values are normally distributed with a mean of 25 and a standard deviation of 15, what points-per-game average must a player have in order to qualify for *All American*?

24 For a large group of trade school applicants the distribution of scores on a mechanical aptitude test is $N(1000, 50)$. You are given an opportunity to take the test and you score 1310. (Wow!) How did you fare relative to all the others that took the test?

25 A binomial random variable X has $n = 100$ and $p = 0.20$. Can the normal approximation to binomial probabilities be legitimately applied? Explain. Find the following:

a. The probability of at most 15 successes.
b. The probability of 18 to 22 successes.

c. The probability of at least 25 successes.
d. The probability of exactly 20 successes.

26 If the unemployment rate in a certain city is 6% and $n = 100$ people are randomly selected at a bus station over a 10 hr period, find the following:

a. The expected number of people who are unemployed.
b. The standard deviation of those who are unemployed.
c. The probability that exactly eight are unemployed.
d. The probability that at least four are unemployed.
e. The probability that between six and ten are unemployed.

APPENDIX 6.A THE NORMAL APPROXIMATION TO BINOMIAL PROBABILITIES

As we shall now see, normal probabilities can serve as useful approximations to binomial probabilities. This is because, as n increases without bound, the normal distribution is the limiting from of the binomial distribution provided p, the probability of a success, is not too close to either 0 or 1. So if X is a binomial random variable with mean $E(X) = np$ and $\sqrt{V(X)} = \sqrt{np(1-p)}$, then, for a and b integers,

$$P(a \leq X \leq b) \approx P\left(\frac{a - np}{\sqrt{np(1-p)}} \leq Z \leq \frac{b - np}{\sqrt{np(1-p)}}\right). \tag{6.A.1}$$

This normal approximation works well provided $np(1-p) > 10$.

To improve the approximation provided by Equation (6.A.1), let us introduce the so-called *continuity correction* (which is needed because we are using a continuous distribution to approximate a discrete one).[3] Then under the continuity correction, Equation (6.A.1) becomes

$$P(a \leq X \leq b) \approx P\left(\frac{a - 0.5 - np}{\sqrt{np(1-p)}} \leq Z \leq \frac{b + 0.5 - np}{\sqrt{np(1-p)}}\right). \tag{6.A.2}$$

[3] Consider integers a, b, and c, where $a < b < c$. Since binomial X is discrete, it has no probability mass between a and b, and between b and c. However, with X continuous, it has non- zero probability "only over intervals." For instance, given $X = a$, the binomial probability is $P(X = a; n, p)$. But if we employ the continuity correction of adding and subtracting 0.5 to account for the gaps between integer values (where binomial probabilities are undefined) and calculate the $N(0,\ 1)$ probability over the interval $a \pm 0.5$, we get $P(X = a; n, p) \approx P(a - 0.5 \leq Z \leq a \pm 0.5)$. Here we are assuming that the binomial probability $P(X;\ n,\ p)$ is the probability associated with an interval by length one and centered on X.

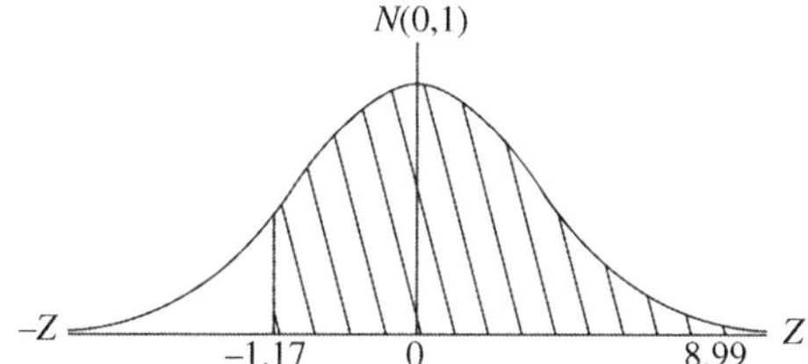

FIGURE 6.A.1 Area under $N(0, 1)$ from -1.17 to 8.99.

Example 6.A.1

Suppose a random experiment consists of tossing a fair pair of dice $n = 1000$ times in succession. Let a success be defined as the sum of the faces showing equals five. In addition, let the random variable X represent the number of successes obtained in the n trials. Clearly, X is distributed binomially with the following:

$$n = 1000,\ p = 4/36 = 0.1111,\ 1 - p = 0.8889$$
$$E(X) = np = 111.1, \text{ and } V(X) = np(1-p) = 98.757.$$

What is the probability that $100 \leq X \leq 200$? Given the above information, our normal approximation Equation (6.A.2) to this binomial probability appears as follows:

$$\begin{aligned} P(100 \leq X \leq 200) &= P\left(\frac{100 - 0.5 - 111.1}{\sqrt{98.757}} \leq Z \leq \frac{200 + 0.5 - 111.1}{\sqrt{98.757}}\right) \\ &= P(-1.17 \leq Z \leq 8.99) = 0.3790 + 0.5 \\ &= 0.8790 \end{aligned}$$

(Fig. 6.A.1). (Note that this approximation is legitimate since $np(1-p) = 98.757 > 10$.)

7

SIMPLE RANDOM SAMPLING AND THE SAMPLING DISTRIBUTION OF THE MEAN

7.1 SIMPLE RANDOM SAMPLING

The fundamental issue addressed by the notion of sampling is to determine when and under what conditions a sample permits a reasonable generalization about a population. While there are many different ways of extracting a sample from a population, we shall primarily engage in simple random sampling. This is because, under random sampling, we can apply the rules of probability theory to calculate the errors associated with using a sample statistic as an estimator for a population parameter. In what follows, we shall always sample "without replacement," that is, once an element of the population has been selected for inclusion in a given sample, it is no longer eligible to appear again in that sample as the sampling process commences.

How many samples of a given size may be drawn from a given population? Suppose the population is of size N and we are interested in taking samples of size n. Then, since the order of the items within a sample is irrelevant (i.e., an item in the population is either chosen or it is not—we do not care if it was picked first or last), we can theoretically find ${}_NC_n = N!/n!(N-n)!$ possible samples. Hence there are ${}_NC_n$ possible ways to obtain a random sample of size n from a population of size N.

How is a random sample defined? A *sample is random* if each element of the population has the same chance of being included in the sample, that is, each element

Statistical Inference: A Short Course, First Edition. Michael J. Panik.

of the population has a known and equal probability of being included in the sample. Hence the probability of picking any one element of the population on the first draw is $1/N$; the probability of picking any element remaining in the population on the second draw is $1/(N-1)$, and so on.

What is the probability that any one element of the population will be included in the sample? To answer this, we need only find n/N.

What is the probability that any particular sample will be selected? This is just $1/{}_NC_n$.

Now, what do we mean by *simple random sampling*? This is a process of sampling that gives each possible sample combination an equal chance of being selected.

Example 7.1

Six marbles (numbered 1 through 6) are in a container. How many samples of size three can be selected? Here $N=6$ and $n=3$. Thus

$$ {}_6C_3 = \frac{6!}{3!3!} = 20 \text{ possible samples.} $$

The probability of picking any one marble on draw one is 1/6; the probability of selecting any of the remaining marbles on draw two is 1/5; and the probability of picking any of the remaining marbles on draw three is $^1\!/_4$. The probability that any one marble will be included in the sample is $n/N = 3/6 = \frac{1}{2}$. And the probability that any particular sample will be chosen is $1/{}_6C_3 = \frac{1}{20}$.

As a practical matter, how do we actually extract a random sample from a (finite) population? Those readers interested in the details of this process can consult Appendix 7.A (using a table of random numbers). In addition, if we are provided with a data set, how do we know that it has been drawn randomly? A test to assess the randomness of a particular sample appears in Appendix 10.A.

7.2 THE SAMPLING DISTRIBUTION OF THE MEAN

Suppose μ is some unknown population mean. We want to determine its "true value." How should we proceed to find μ? A reasonable approach is to take a simple random sample and make an inference from the sample to the population. How is this inference made? We use an *estimator* for μ. Think of an estimator as a function of the sample values used to estimate μ. A logical estimator for μ is the sample mean $\bar{X} = \sum_i X_i/n$. (Specifying an estimator is a form of *data reduction*—we summarize the information about μ contained in a sample by determining some essential characteristic of the sample values. So for purposes of inference about μ, we employ the realization of $\bar{X}$ rather than the entire set of observed data points.) Under random sampling, $\bar{X}$ is a random variable and thus has a probability distribution that we shall term the *sampling distribution of the mean*—a distribution showing the probabilities

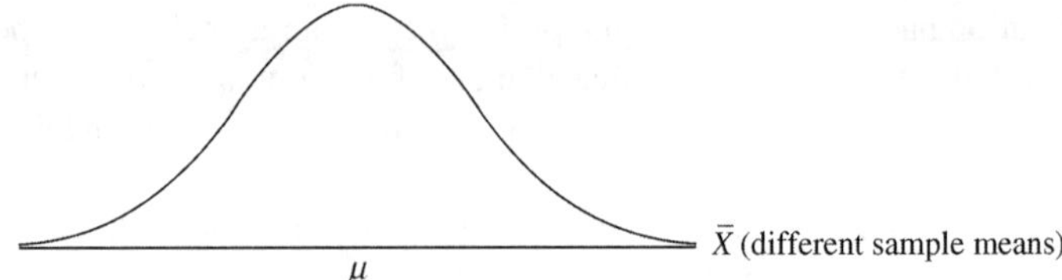

FIGURE 7.1 Sampling distribution of the mean.

of obtaining different sample means from random samples of size n taken from a population of size N (Fig. 7.1).

We shall call $\bar{X}$ a *point estimator* for μ since it reports a single numerical value as the estimate of μ. The value of $\bar{X}$ typically varies from sample to sample (since different samples of size n have different sample values). Hence, we cannot expect $\bar{X}$ to be exactly on target—an error arises that we shall call the *sampling error of* $\bar{X}$. It amounts to $\bar{X} - \mu$. For some samples, the sampling error is positive while for others it will be negative; rarely will it be zero. Hence the values of $\bar{X}$ will tend to cluster around μ, that is, the sampling distribution of the mean is concentrated about μ (Fig. 7.1).

Is $\bar{X}$ a "good" estimator for μ? It will be if $\bar{X}$ has certain desirable properties (which we will get to as our discussion progresses). Interestingly enough, these so-called desirable properties are expressed in terms of the mean and variance of the sampling distribution of $\bar{X}$.

Let us now turn to the process of constructing the sampling distribution of the mean:

1. From a population of size N let us take "all possible samples" of size n. Hence we must repeat some (conceptual) random experiment ${}_NC_n$ times since there are ${}_NC_n$ possible samples that must be extracted.
2. Calculate $\bar{X}$ for each possible sample: $\bar{X}_1, \bar{X}_2, \ldots, \bar{X}_{{}_NC_n}$.
3. Since each sample mean is a function of the sample values, different samples of size n will typically display different sample means. Let us assume that these differences are due to chance (under simple random sampling). Hence the various means determined in step 2 can be taken to be observations on some random variable $\bar{X}$, that is, $\bar{X}$ varies under random sampling depending upon which random sample is chosen. Since $\bar{X}$ is a random variable, it has a probability distribution called the *sampling distribution of the mean* $\bar{X}$: a distribution showing the probabilities (relative frequencies) of getting different sample means from random samples of size n taken from a population of size N.

Why study the sampling distribution of the mean? What is its role? Basically, it shows how means vary, due to chance, under repeated random sampling from the same population. Remember that inferences about a population are usually made on the basis of a single sample—not all possible samples of a given size. As noted above, the various sample means are scattered, due to chance, about the true population mean to be estimated. By studying the sampling distribution of the mean, we can learn something about the error arising when the mean of a single sample is used to estimate the population mean.

Example 7.2

For the population X:1,2,3,4,5 (here $N = 5$), construct the sampling distribution of the mean for samples of size $n = 3$. (Note that it is easily determined that $\mu = 3$ and $\sigma^2 = \left(\sum X_i^2/N\right) - \mu^2 = (55/5) - 9 = 2$.) Since ${}_5C_3 = 10$, there are 10 possible samples of size $n = 3$ (Table 7.1), with the probability of obtaining any particular sample equal to 1/10. A glance at Table 7.1 reveals that some of the sample means occur more than once. Hence we must collect the "different" sample means and thus treat them as observations on the random variable $\bar{X}$ (Table 7.2). The probabilities (or relative frequencies) attached to these $\bar{X}$ values also appear in Table 7.2, for example, the probability of getting a mean of $2\frac{1}{3}$ equals the probability of obtaining a sample that has a mean of $2\frac{1}{3}$. Obviously, it is $\frac{1}{10}$. And the probability of getting $\bar{X} = 3$ is $\frac{2}{10}$ since there are two samples displaying a mean of 3.

What about the mean and variance of the sampling distribution of the mean (or the mean and variance of the random variable $\bar{X}$)? Since $\bar{X}$ is a random variable, its mean is

$$\begin{aligned} E(\bar{X}) &= \sum \bar{X}_j P(\bar{X} = \bar{X}_j) \\ &= 2\left(\frac{1}{10}\right) + 2\frac{1}{3}\left(\frac{1}{10}\right) + 2\frac{2}{3}\left(\frac{2}{10}\right) + \cdots + 4\left(\frac{1}{10}\right) = 3. \end{aligned}$$

TABLE 7.1 All Possible Samples and Their Means

Sample Values	Sample Mean
1, 2, 3	2
1, 2, 4	$2\frac{1}{3}$
1, 2, 5	$2\frac{2}{3}$
1, 3, 4	$2\frac{2}{3}$
1, 3, 5	3
1, 4, 5	$3\frac{1}{3}$
2, 3, 4	3
2, 3, 5	$3\frac{1}{3}$
2, 4, 5	$3\frac{2}{3}$
3, 4, 5	4

TABLE 7.2 Sampling Distribution of $\bar{X}$

$\bar{X}$	$P(\bar{X} = \bar{X}_j)$
2	1/10
$2\frac{1}{3}$	1/10
$2\frac{2}{3}$	2/10
3	2/10
$3\frac{1}{3}$	2/10
$3\frac{2}{3}$	1/10
4	1/10
	1

But $\mu = 3$! Hence we have the interesting result that

$$E(\bar{X}) = \mu,$$

the mean of all the sample means always equals the population mean (regardless of the form of the underlying population). Hence the sampling distribution of the mean (of $\bar{X}$) is centered right on the population mean (Fig. 7.2).

In general, if the expectation of an estimator equals the population value to be estimated, then that estimator is said to be *unbiased*. Thus $\bar{X}$ is an unbiased estimator for μ since $E(\bar{X}) = \mu$, that is, on the average, $\bar{X}$ equals μ or, on the average, the sampling error is zero; there is no systematic tendency for $\bar{X}$ to underestimate or overestimate μ in repeated sampling from the same population.

Stated alternatively, since $E(\bar{X})$ is the mean of $\bar{X}$, we see that as the "number" of samples increases, $\bar{X}$ is on target as an estimator for μ. [Suppose T is an alternative estimator for μ. If T is biased, then the sampling distribution of T is concentrated at $\bar{T} \neq \mu$ (Fig. 7.3).]

Looking to the variance of the sampling distribution of the mean or the variance of the random variable $\bar{X}$, two cases present themselves:

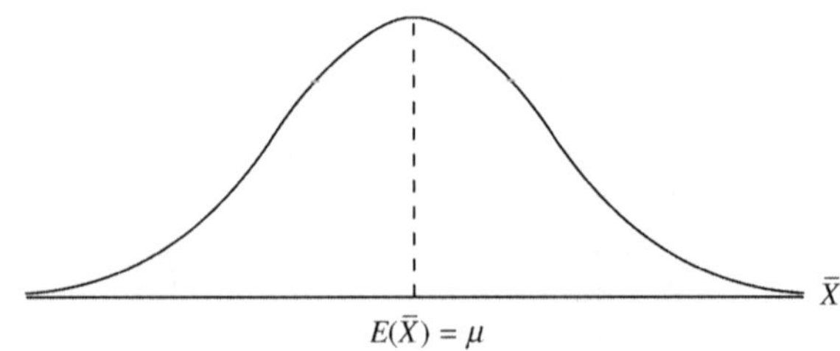

FIGURE 7.2 $\bar{X}$ is a unbiased estimator for μ.

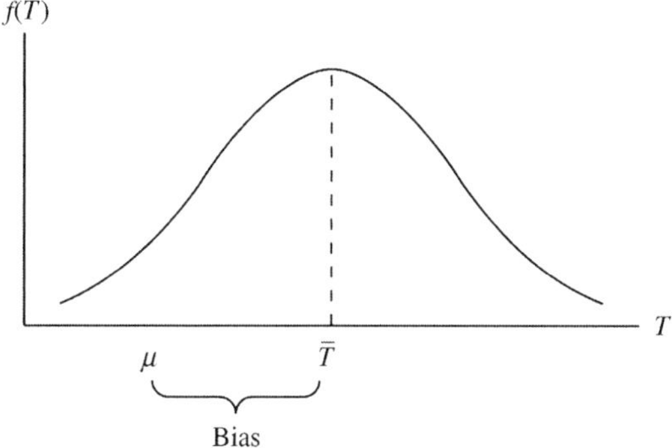

FIGURE 7.3 T is a biased estimator for μ.

1. We sample from an infinite population. Then

$$V(\bar{X}) = \frac{V(X)}{n} = \frac{\sigma^2}{n}, \tag{7.1}$$

   (remember that X is the underlying population variable).

2. We sample from a finite population. Then

$$V(\bar{X}) = \frac{V(X)}{n}\frac{N-n}{N-1} = \frac{\sigma^2}{n}\frac{N-n}{N-1}, \tag{7.2}$$

   where the term $\frac{N-n}{N-1}$ is called the *finite population correction factor*.

For instance, given the results of Example 7.2, it is easily determined from Equation (7.2) that

$$V(\bar{X}) = \frac{2}{3}\frac{5-3}{5-1} = 0.33.$$

Clearly, $V(\bar{X}) = 0.33 < V(X) = \sigma^2 = 2$, that is, there is less variability associated with $\bar{X}$ relative to X. If the population is finite, under what circumstances can we ignore the finite population correction term? We can ignore the said correction if n is small relative to N. In this regard, the cut-off point is usually 5%, that is, if n is less than or equal to 5% of N (or if $n/N \leq 0.05$), then $\frac{N-n}{N-1} \approx 1$ and thus this term can be omitted in Equation (7.2).

7.3 COMMENTS ON THE SAMPLING DISTRIBUTION OF THE MEAN

We next turn to a few observations pertaining to the sampling distribution of the mean. The basis for the following set of comments is Table 7.2:

1. Each of the sample means is approximately equal to the population mean ($\mu = 3$).
2. The sample means cluster much more closely about the population mean than do the original values of X, that is, $V(\bar{X}) = 0.33 < V(X) = 2$.
3. If larger samples were taken, their means would cluster even more closely about the population mean than before. (To see this, let $n = 4$. Then

$$V(\bar{X}) = \frac{\sigma^2}{n}\frac{N-n}{N-1} = \frac{2}{4}\frac{5-4}{5-1} = 0.125.$$

Clearly, $V(\bar{X})$ has decreased in value from 0.33.) In this regard, we may conclude that the larger the sample, the closer the sample mean is likely to be to the population mean. This property of the sample mean is referred to as consistency, that is, $\bar{X}$ is a *consistent* estimator of μ if for large samples the probability is close to 1 that $\bar{X}$ will be near μ or

$$P(\bar{X} \to \mu) \to 1 \; as \; n \to \infty.$$

So as n increases without bound, the sampling distribution of the mean becomes more and more concentrated about μ.

Next comes some important terminology.

4. The *standard error of the mean* will be calculated as $\sqrt{V(\bar{X})}$ and will be denoted as $\sigma_{\bar{X}}$. Clearly, the standard error of the mean is simply the standard deviation of the distribution of sample means or of $\bar{X}$. Since $V(\bar{X})$ serves as a measure of the average squared sampling error arising when the mean of a single sample is used to estimate the population mean, $\sigma_{\bar{X}} = \sqrt{V(\bar{X})}$ is expressed in terms of the original units of X. So given a value for $\sigma_{\bar{X}}$, we may conclude that, on the average, the $\bar{X}_j$s lie approximately $\sigma_{\bar{X}}$ units away from μ.

(It is important to note that the term "standard error" always refers to the "standard deviation of the sampling distribution of some statistic.")

What factors affect the magnitude of $\sigma_{\bar{X}}$? From Equations (7.1) and (7.2), let us first find:

(Infinite population case)

$$\sigma_{\bar{X}} = \frac{\sigma}{\sqrt{n}}, \tag{7.3}$$

(Finite population case)

$$\sigma_{\bar{X}} = \frac{\sigma}{\sqrt{n}}\sqrt{\frac{N-n}{N-1}}. \tag{7.4}$$

Then a glance at these two expressions reveals that $\sigma_{\bar{X}}$ varies directly with σ; it varies inversely with $\sqrt{n}$, that is, relative to this latter point,

$$\sigma_{\bar{X}} = \frac{\sigma}{\sqrt{n}} \rightarrow 0 \text{ as } n \rightarrow \infty$$

$$\sigma_{\bar{X}} = \frac{\sigma}{\sqrt{n}}\sqrt{\frac{N-n}{N-1}} \rightarrow 0 \; as \; n \rightarrow N.$$

In sum, the sampling distribution of $\bar{X}$ is a probability distribution that describes sampling variation—the sample-to-sample variation in $\bar{X}$ under simple random sampling. In fact, this variation reflects "only" random sampling error since $\bar{X}$ deviates from μ by a random amount. It thus constitutes "pure sampling error" since no other source of error or variation in the sample data affects the sampling distribution of the mean. Hence we are ignoring any type of nonsampling error (e.g., non-response, under- representation, and so on) and only picking up sample-to-sample variation due to the process of random selection. A measure of this pure sampling variation is the sample-to-sample standard deviation of $\bar{X}$ or the standard error of the mean $\sigma_{\bar{X}}$. So how close $\bar{X}$ is to μ in repeated sampling is dictated by the spread or dispersion of the distribution of sample means, and this is characterized by $\sigma_{\bar{X}}$.

Example 7.3

We previously found that for the Example 7.2 data set, $V(\bar{X}) = 0.33$. Hence $\sigma_{\bar{X}} = \sqrt{0.33} = 0.574$. Also, we previously defined the sampling error of $\bar{X}$ as $\bar{X} - \mu$. Given the sampling distribution of the mean appearing in Table 7.2, let us specify the sampling error for each value of $\bar{X}$ along with their associated probabilities (Table 7.3). Then, as expected,

$$E(\bar{X} - \mu) = (-1)\frac{1}{10} + \left(-\frac{2}{3}\right)\frac{1}{10} + \left(-\frac{1}{3}\right)\frac{2}{10} + \cdots + (1)\frac{1}{10} = 0.$$

TABLE 7.3 Sampling Error $\bar{X} - \mu$

$\bar{X} - \mu$	$P(\bar{X} - \mu = \bar{X}_j - \mu)$
-1	1/10
$-\frac{2}{3}$	1/10
$-\frac{1}{3}$	2/10
0	2/10
$\frac{1}{3}$	2/10
$\frac{2}{3}$	1/10
1	1/10
	1

that is, on the average, the sampling error is zero. (This is consistent with our previous notion that $E(\bar{X}) = \mu$ or that $\bar{X}$ is an unbiased estimator of μ.) Also,

$$\begin{aligned}
V(\bar{X} - \mu) &= E\left[(\bar{X} - \mu)^2\right] - [E(\bar{X} - \mu)]^2 \\
&= E\left[(\bar{X} - \mu)^2\right] - 0 \\
&= 1\left(\frac{1}{10}\right) + \frac{4}{9}\left(\frac{1}{10}\right) + \frac{1}{9}\left(\frac{2}{10}\right) + \cdots + 1\left(\frac{1}{10}\right) = 0.33.
\end{aligned}$$

Hence $V(\bar{X}) = V(\bar{X} - \mu)$ is a measure of the average squared sampling error arising when $\bar{X}$ estimates μ; and $\sigma_{\bar{X}} = \sqrt{V(\bar{X})} = 0.574$, that is, on the average, the individual $\bar{X}_j$s are approximately 0.574 units away from μ.

7.4 A CENTRAL LIMIT THEOREM

Up to this point in our discussion of the sampling distribution of the mean, we have not made any assumption about the form (shape) of the parent population. For convenience, let us assume that we are sampling from an infinite population. In this regard:

1. Suppose that our population variable X is $N(\mu, \sigma)$. Then $\bar{X}$ will be $N(\mu, \sigma/\sqrt{n})$ and thus the *standardized sample mean*

$$\bar{Z} = \frac{\bar{X} - \mu}{\sigma/\sqrt{n}} \text{ is } N(0, 1). \tag{7.5}$$

   Hence we can use the standard normal area table to calculate probabilities involving $\bar{X}$. For instance, given that X is $N(20,15)$ and $n = 100$, find $P(18 \leq \bar{X} \leq 25)$. To this end, we have

$$\begin{aligned}
&P\left(\frac{18 - 20}{15/\sqrt{100}} \leq \frac{\bar{X} - \mu}{\sigma/\sqrt{n}} \leq \frac{25 - 20}{15/\sqrt{100}}\right) \\
&= P(-1.33 \leq \bar{Z} \leq 3.33) = 0.4082 + 0.5 = 0.9082
\end{aligned}$$

   (See Fig. 7.4).
2. Suppose X is "not" $N(\mu, \sigma)$. Can we still use the $N(0, 1)$ area table to calculate probabilities involving $\bar{X}$? Interestingly enough, the answer is "yes" if n is large enough. Large enough" means $n > 30$. The "yes" answer is provided courtesy of the *Central limit theorem:* If random samples are selected from a population with mean μ and standard deviation σ, then, as n increases without bound,

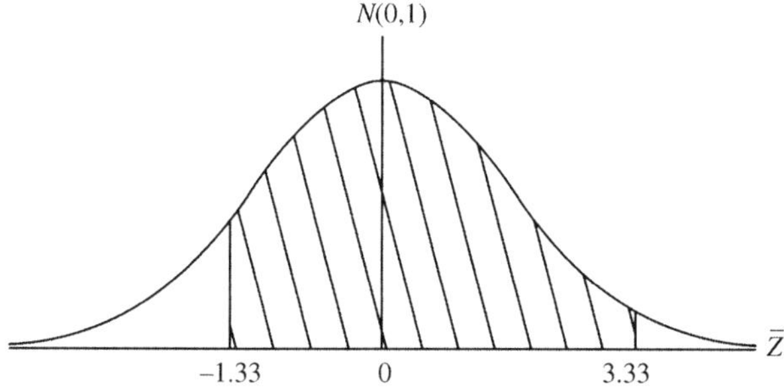

FIGURE 7.4 Area under $N(0, 1)$ from -1.33 to 3.33.

$$\bar{Z} = \frac{\bar{X} - \mu}{\sigma/\sqrt{n}} \rightarrow N(0, 1),$$

"regardless of the form of the parent population" (or $\bar{X} \rightarrow N(\mu, \sigma/\sqrt{n})$).

Have we seen the central limit theorem in operation? To answer this question, we need only examine the probability mass function for $\bar{X}$. Given Table 7.2, the said function can be graphed as Fig. 7.5. It is not hard to visualize that this pattern of points renders a reasonably good approximation to a normal curve, even for this highly abbreviated data set. To summarize:

(a) If X is $N(\mu, \sigma)$, then $\bar{X}$ is $N(\mu, \sigma/\sqrt{n})$ and thus $\bar{Z} = (\bar{X} - \mu)/(\sigma/\sqrt{n})$ is $N(0, 1)$.

(b) If X is not $N(\mu, \sigma)$, then $\bar{X}$ is approximately $N(\mu, \sigma/\sqrt{n})$ for large n and thus $\bar{Z} = (\bar{X} - \mu)/(\sigma/\sqrt{n})$ is approximately $N(0, 1)$.

As a practical matter, how is the sampling distribution of the mean used? As we shall soon see, it can be used to obtain an error bound on the sample mean as an estimate of the population mean. In addition, under random sampling, only

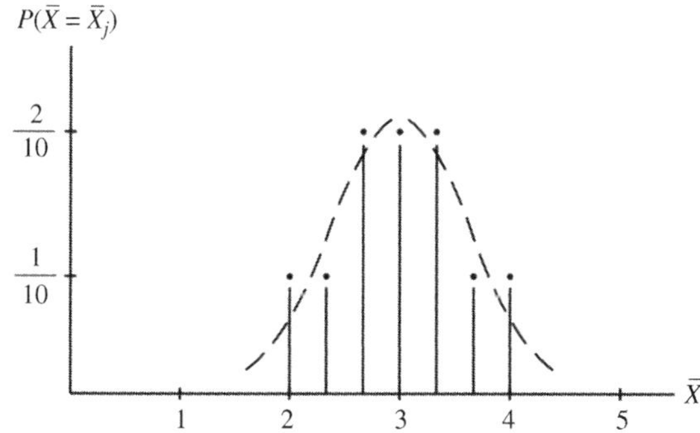

FIGURE 7.5 Probability mass function for $\bar{X}$.

rarely will $\bar{X}$ equal μ exactly. Hence the sampling distribution of the mean can be used to provide probability information about the sampling error $\bar{X} - \mu$ incurred when $\bar{X}$ is used to estimate μ. In the next chapter, these two issues are addressed in turn.

EXERCISES

1 Suppose a population of size $N = 5$ consists of the X values: 2, 4, 6, 8, 10. If we take all possible samples of size $n = 2$ without replacement, how many samples will we obtain? Find the sampling distribution of the mean. Graph its probability mass function. Next, find the following:

a. μ and σ

b. $E(\bar{X})$ and $V(\bar{X})$

c. The standard error of the mean

d. Verify that $E(\bar{X} - \mu) = 0$

2 Let the population variable X be $N(50, 10)$. What is the probability that the mean of a sample of size $n = 25$ will differ from μ by less than four units? Resolve this problem for $n = 50$ and for $n = 100$. What can we conclude?

3 Let the population variable X be $N(100, 10)$. What is the probability that the mean of a sample of size $n = 25$ will be within ± 5 units of μ? Resolve this problem for $\sigma = 16$. What can we conclude?

4 Suppose a population variable X consists of the values 1, 2, 3, 4, 5, 6. For samples of size $n = 3$ taken without replacement, determine the sampling distribution of the mean. Find $E(\bar{X})$ and $V(\bar{X})$. What is the standard error of the mean?

5 Given a sample of size $n = 64$ taken from a continuous population distribution with mean 60 and standard deviation 26, find the mean and variance of the sample mean.

6 Suppose a random sample of size $n = 10$ is taken from a population that is $N(25, 8)$. Describe the sampling distribution of $\bar{X}$.

7 Suppose a random sample of size $n = 40$ is obtained from a population with a mean of 60 and a standard deviation of 8. Must the population variable X be normally distributed in order for the sampling distribution of the mean $(\bar{X})$ to be approximately normal? Why? Describe the sampling distribution of the mean.

8 Suppose a random sample of size 15 is taken from a population with $\mu = 64$ and $\sigma = 16$. What must be true about the population distribution in order for the normal probability model to be applicable in determining probabilities regarding $\bar{X}$?

Assuming this requirement to hold, describe the sampling distribution of the mean. Also, find the following:

a. $P(\bar{X} > 66)$
b. $P(\bar{X} \leq 50)$
c. $P(60 \leq \bar{X} \leq 70)$
d. $P(\bar{X} = 57)$

9 The EPA has found that the miles per gallon (mpg) for a new hybrid vehicle is N (32, 4). Describe the sampling distribution of the mean. Additionally, find the following:

a. The probability that a randomly selected vehicle will get at least 34 mpg.
b. The probability that 10 randomly selected vehicles will have an average mpg in excess of 33.
c. The probability that 15 randomly selected vehicles will have an average mpg less than 30.

10 Suppose the population variable X is $N(3, 0.2)$ and $n = 25$. How large an interval must be chosen so that the probability is 0.95 that the sample mean $\bar{X}$ lies within $\pm$a units of the population mean μ?

11 Suppose the population variable X is $N(46, 13)$. How large a sample must be taken so that the probability is 0.95 that $\bar{X}$ will lie within ± 5 units of μ?

12 The following $n = 6$ data values: 33.35, 34.16, 33.90, 34.55, 33.29, and 34.41 represent the times (in minutes) for a craftsman to put the finishing touches on a table lamp. Is there any evidence to support the contention that the variable "finishing time" is normally distributed?

APPENDIX 7.A USING A TABLE OF RANDOM NUMBERS

Suppose our objective is to obtain a simple random sample of size n from a population of size N. One way to accomplish this task is to write the name of, say, each person in the population on a slip of paper (or if the elements of the population are not people, we can label each element using some identifier) and then "randomly" draw n slips of paper out of a container. While this process might be legitimate for small populations, it would certainly be unworkable for large ones. An alternative to this procedure is to employ a table of random numbers (Table 7.A.1).

Let the population consist of $N = 10$ items and suppose we want to select a random sample of size $n = 4$. In this instance, we need to use the single digit code $0, 1, 2, \ldots, 9$. Given these 10 digits, we now go to the population and arbitrarily tag each of its members with one of them. (Note that this procedure presupposes that we have at our

TABLE 7.A.1 Random Numbers

Row	Column Number									
Number	01–05	06–10	11–15	16–20	21–25	26–30	31–35	36–40	41–45	46–50
01	89,392	23,212	74,483	36,590	25,956	36,544	68,518	40,805	09,980	00467
02	61,458	17,639	96,252	95,649	73,727	33,912	72,896	66,218	52,341	97,141
03	11,452	74,197	81,962	48,443	90,360	26,480	73,231	37,740	26,628	44,690
04	27,575	04,429	31,308	02,241	01,698	19,191	18,948	78,871	36,030	23,980
05	36,829	59,109	88,976	46,845	28,329	47,460	88,944	08,264	00843	84,592
06	81,902	93,458	42,161	26,099	09419	89,073	82,849	09,160	61,845	40,906
07	59,761	55,212	33,360	68,751	86,737	79,743	85,262	31,887	37,879	17,525
08	46,827	25,906	64,708	20,307	78,423	15,910	86,548	08,763	47,050	18,513
09	24,040	66,449	32,353	83,668	13,874	86,741	81,312	54,185	78,824	00718
10	98,144	96,372	50,277	15,571	82,261	66,628	31,457	00377	63,423	55,141
11	14,228	17,930	30,118	00438	49,666	65,189	62,869	31,304	17,117	71,489
12	55,366	51,057	90,065	14,791	62,426	02,957	85,518	28,822	30,588	32,798
13	96,101	30,646	35,526	90,389	73,634	79,304	96,635	06,626	94,683	16,696
14	38,152	55,474	30,153	26,525	83,647	31,988	82,182	98,377	33,802	80,471
15	85,007	18,416	24,661	95,581	45,868	15,662	28,906	36,392	07,617	50,248
16	85,544	15,890	80,011	18,160	34,468	84,106	40,603	01,315	74,664	20,553
17	10,446	20,699	98,370	17,684	16,932	80,449	92,654	02,084	19,985	59,321
18	67,237	45,509	17,638	65,115	29,757	80,705	82,686	48,565	72,612	61,760
19	23,026	89,817	05,403	82,209	30,573	47,501	00135	33,955	50,250	72,592
20	67,411	58,542	18,678	46,491	13,219	84,084	27,783	34,508	55,158	78,742
21	48,663	91,245	85,828	14,346	09,172	30,168	90,229	04,734	59,193	22,178
22	54,164	58,492	22,421	74,103	47,070	25,306	76,468	26,384	58,151	06,646
23	32,639	32,363	05,597	24,200	13,363	38,005	94,342	28,728	35,806	06,912
24	29,334	27,001	87,637	87,308	58,731	00256	45,834	15,398	46,557	41,135
25	02,488	33,062	28,834	07,351	19,731	92,420	60,952	61,280	50,001	67,658

disposal a sampling frame or list of items in the population.) We next consult our random numbers table (remember that under simple random sampling we always sample without replacement) and choose an arbitrary starting point. For instance, let us start at row 03 and column 25. We thus have our first random digit, namely "0". Moving down this column from the value 0, we next encounter "8" (our second random digit) and then "9" (the third random digit). Continuing to move down this column we get "9" again. But since we are sampling without replacement, we skip the second "9" and select "7" as our fourth random digit. Hence our set of four random digits consists of 0, 8, 9, and 7. We now go to the population and determine which of its members were tagged with these digits. These population members thus constitute our simple random sample size $n = 4$.

What if the population consists of $N = 100$ items and we want to select a simple random sample of size $n = 10$? Now, we employ a two-digit code consisting of the 100 numbers 00, 01, 02, . . . , 99. We next arbitrarily tag each member of the population with one of these two-digit numbers. Again, our starting point in Table 7.A.1 is arbitrary. Suppose we start with row 05 and columns 21 and 22. Our first random number is thus "28." Moving down these columns from 28 we see that our complete sequence of $n = 10$ random numbers is

$$28, 09, 86, 78, 13, 82, 49, 62, 73, 83.$$

We now go back to the population and select those members that have been tagged with these 10 two-digit random numbers. They constitute our simple random sample of size $n = 10$.

If our population had $N = 1000$ items and we wanted to obtain a simple random sample of size $n = 25$, then our requisite three-digit code would involve the 1000 numbers $000, 001, 002, \ldots, 999$. Next, the population members are arbitrarily tagged with these three-digit numbers and a trip to Table 7.A.1 has us choose an arbitrary starting point (e.g., our first random number could be chosen from row 01 and columns 3, 4, and 5 or "392"). Moving down these columns away from 392 enables us to select 24 additional three-digit random numbers so that our entire set of 25 three-digit random numbers consists of the following:

392	902	228	544	663
458	761	366	446	164
452	827	101	237	639
575	010	152	026	334
829	144	007	411	488

We now revisit the population and determine which of its members received these three-digit numbers. These members thus make up our random sample of $n = 25$ items.

Given that a *bona fide* table of random numbers is unavailable, what is the next best thing to use? How about a telephone directory? (Not the Yellow Pages.) The last four digits of any telephone number are not assigned on a systematic basis. Hence one need only arbitrarily pick a page and a column within it, find a starting point, and then move down the column(s) to extract a set of random numbers.

Suppose that one does not have a well-defined sampling frame or list of items within the population. Then a sampling process called *systematic random sampling* will work nicely. Specifically, to obtain a *systematic random sample*, select every kth member from the population (k is termed the *sampling cycle*), with the first member selected by choosing a random number between 1 and k. More formally, the process of extracting a systematic random sample of size n from a population of known size N is as follows:

1. Round the quantity N/n down to the nearest integer. This gives the sampling cycle k.

2. Randomly select a number p between 1 and k.
3. Hence the pth item in the population is selected along with every kth member thereafter. Our systematic random sample thus consists of the following population elements:

$$p, p+k, p+2k, \ldots, p+(n-1)k.$$

4. This sampling process commences until we have selected n items from the population.

If N is not known, then k cannot be calculated—k must be determined judiciously so that a representative sample is obtained, that is, k should not be so large that n is unattainable; or k should not be so small that only a narrow slice of the population or a biased sample results. It is important to note that for systematic random sampling to be effective, the items in the population should be arranged in a random fashion to begin with. For instance, weekend shoppers exiting a suburban mall could be considered a random population as also the alphabetical listing of students enrolled at a local community college.

Example 7.A.1

Suppose that we want to obtain a simple random sample of size $n=35$ from a population whose size is estimated to be $N=600$. Under systematic random sampling, $N/n=600/35=17.4\approx 17=k$. Let us randomly select a number between 1 and 17. Let us say it is "10." Then we select the 10th member of the population and every 17th member thereafter, until a sample of 35 items is achieved. Thus our sample of size $n=35$ consists of the elements positioned or ordered as follows:

$$10, 27, 44, 61, 78, \ldots, 588$$

where $588=p+(n-1)\,k=10+34\,(17)$.

Given that we now know how to acquire a simple random sample, we may go a step further and check to see if any such sample data set can be treated "as if" it was drawn from a normal population. This is the task of Appendix 7.B.

APPENDIX 7.B ASSESSING NORMALITY VIA THE NORMAL PROBABILITY PLOT

How can we determine if a simple random sample taken from some unknown population is normally distributed (or approximately so)? One way to answer this question is to see if the sample data values plot out as a frequency distribution that is bell-shaped and symmetrical. As one might expect, for large data sets, a distinct normal pattern may emerge that would lead one to conclude that the underlying population is indeed approximately normally distributed. However, for small data sets, simple graphics might not work very well; the shape of the population might not be readily discernable from only a small subset of observations taken from it.

A convenient tool for assessing the normality of a data set is the *normal probability plot*—a graph involving the observed data values and their *normal scores* in the sample or "expected" Z-scores. (A more formal test of normality appears in Appendix 10.C.) The steps involved in obtaining a normal probability plot are:

1. Arrange the sample data values X_i, $i = 1, \ldots, n$, in an increasing (ordered) sequence.
2. Determine the probability level

$$p_j = \frac{j - 0.375}{n + 0.25}, \tag{7.B.1}$$

   where j serves to index the position of the data point X_i in the ordered sample data set.
3. Find the expected Z-score or normal score for p_j, Z_{p_j}, using the $N(0, 1)$ area table.
   Given that the population variable X is $N(\mu, \sigma)$, we may view the expected Z-score for p_j, Z_{p_j}, as the value of Z that cuts off $100\,p_j\%$ of the area in the left-hand tail of the $N(0, 1)$ distribution, that is, $P(Z \leq Z_{p_j}) = p_j$ (Fig. 7.B.1). Thus p_j is the expected area to the left of the Z_{p_j} (and thus to the left of X_j) if the data set is drawn from a normal population—it is the expected proportion of data points that are less than or equal to X_j. So if the underlying population is normal, we should be able to predict the area under $N(0, 1)$ to the left of X_j.
4. Plot the ordered observed sample data values X_j (placed on the horizontal axis) against the normal scores Z_{p_j} (plotted on the vertical axis).

Now, since $Z = (X - \mu)/\sigma$, it follows that $X = \mu + \sigma Z$ so that $X_j = \mu + \sigma Z_{p_j}$. Hence, if X is $N(\mu, \sigma)$, the X_js should be linearly related to the Z_{p_j} values, that is, "the normal probability plot of the observed data points against their expected Z-scores should be

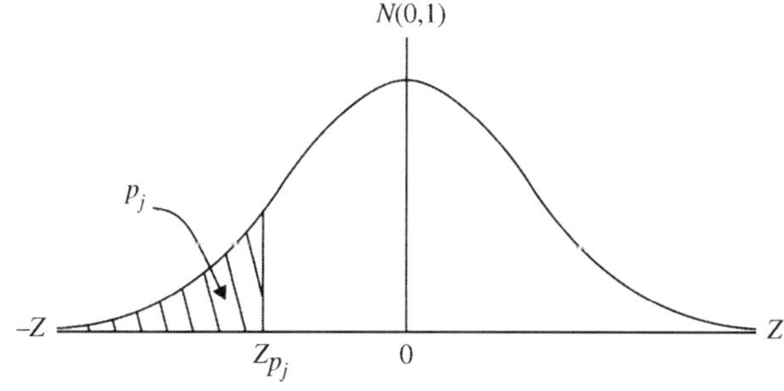

FIGURE 7.B.1 Expected Z-score for p_j.

TABLE 7.B.1 Sample Observations on the Population Variable *X*

17.34	21.50	20.00	18.97	23.11
21.98	22.00	19.01	14.77	28.62

approximately linear;" the points (X_j, Z_{p_j}) should lie approximately on a line with vertical intercept μ and slope σ. (The choice of p_j according to Equation (7.B.1) renders an unbiased estimator of σ (Blom, 1958). More on this issue in Chapter 12.)

Example 7.B.1

Given the sample data set depicted in Table 7.B.1, is there any compelling evidence to conclude that the parent population is normally distributed?

Table 7.B.2 organizes the work for carrying out the preceding four-step procedure:

Column 1: Values of index j

Column 2: X_js arranged in an increasing sequence

Column 3: p_js are determined "under the assumption of normality," for example,

$p_1 = \frac{1-0.375}{10+0.25} = 0.060$ [the area under $N(0, 1)$ to the left of 14.77 is 0.060 if X is $N(\mu, \sigma)$]

$p_2 = \frac{2-0.375}{10+0.25} = 0.159$ [the area under $N(0, 1)$ to the left of 17.34 is 0.159 if X is $N(\mu, \sigma)$], and so on.

Column 4: We use the $N(0, 1)$ area table to find the set of expected Z-scores or normal scores, for example,

$P(Z \leq Z_{p_1}) = p_1$ or $P(Z \leq -1.56) = 0.060$

$P(Z \leq Z_{p_2}) = p_2$ or $P(Z \leq -1.00) = 0.159$; and so on.

TABLE 7.B.2 Obtaining a Normal Probability Plot

(1) Index j	(2) Ordered X_is become the X_js	(3) p_j	(4) Expected Z_{p_j}
1	14.77	0.060	−1.56
2	17.34	0.159	−1.00
3	18.97	0.256	−0.65
4	19.01	0.353	−0.38
5	20.00	0.451	−0.13
6	21.50	0.549	0.13
7	21.98	0.646	0.38
8	22.00	0.744	0.65
9	23.11	0.841	1.00
10	28.62	0.0.940	1.56

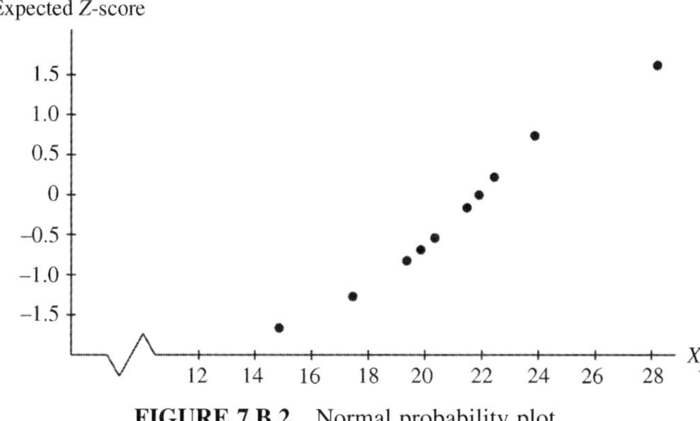

FIGURE 7.B.2 Normal probability plot.

The normal probability plot involving the X_j values (the ordered X_is) versus the expected Z-scores appears as Fig. 7.B.2. Although this plot does display some departure from perfect linearity, it does provide a reasonable approximation to a straight line. Hence we may conclude that our sample data set can be viewed "as if" it was extracted from a normal population.

APPENDIX 7.C RANDOMNESS, RISK, AND UNCERTAINTY[1]

7.C.1 Introduction to Randomness

What constitutes "randomness?" Basically, the term *random* is synonymous with "unpredictable." For example, if I flip a fair coin repeatedly, then the outcome on any flip (heads or tails) is random—it cannot be accurately predicted from trial to trial with any regularity (although there is a 50/50 chance of correctly guessing the outcome on any given toss of the coin).

This is in contrast to a situation in which one starts with a particular rule for generating outcomes with a fixed initial condition and then proceeds to produce a string of outcomes (numbers). Even though a fairly complicated pattern may emerge from the process and the pattern is indistinguishable from one produced by a purely random process (e.g., a coin toss), the bottom line is that this is not a truly random process. The process is purely deterministic and orderly since, from the known rule, the outcome at any future step is completely predictable. The values generated in this fashion are called *pseudo-random numbers* and will be discussed in somewhat greater detail below.

[1] The material presented herein draws heavily from: Tsonis (2008), Beltrami (1997), Rao (1999), Knight (1921), Hull and Dobell (1962), and Anderson and Brown (2005).

What if we observe a string of outcomes (ostensibly random) but we do not have the rule that produced it? If we can encode the pattern of outcomes and ultimately find the rule, then here too we can predict the system's behavior over time and thus pure randomness is not present. Note that in order to predict a future outcome, all steps leading to that outcome have to be explicitly calculated. But if no specific rule(s) can be discovered, then, in this instance, the string of outcomes is truly unpredictable and thus can be considered as random, that is, we can treat the string of values "as if" it were generated by a purely random process.

As indicated above, randomness reflects the unpredictability of an event; an outcome or event occurs without discernible design or cause. Randomness is a part of everyday life; we are typically concerned about the weather, worried about the morning bus being late because of traffic problems, anxious about how one's 401k plan will fare this month, and so on. Randomness can be a source of amusement (will I win tonight's lottery drawing or this week's office sports pool?) as well as trepidation (will the "big quake" hit the San Francisco area during my upcoming visit there?). And it can possibly serve as a safeguard against boredom (one may muse: "Maybe I'll meet some interesting people today in my job as receptionist?").

At first blush, randomness may be thought to possess the qualities of confusion, disorder, something jumbled or scrambled, and disorganization. This view of randomness is, however, much too naïve. Randomness is actually steeped in "rules." While the rule may be obvious (flipping a coin to get heads or tails) or hidden from us (the physical world is typically known imperfectly), the resulting outcomes are uncertain and thus unpredictable—we cannot perceive a cause or detect any pattern of behavior or semblance of order. This is what is meant when someone says: "The incident happened by blind chance." Here some outcome occurs at the intersection of an unlimited number of supposedly unrelated events, with each of these events occurring because of happenstance. Such contingent circumstances are thought to reflect unintentional accidents or are viewed as "an act of God" (some predetermined plan is being carried out via "divine intervention").

Many attempts have been made to thwart the vicissitudes of uncertainty. The ancients attempted to divine and even undermine the "will of the Gods" by engaging in games of chance. They believed that gaming or the simulation of the Gods' uncertain behavior would reveal the outcomes that "higher powers" had in store for them. Thus "fate" and "destiny," as directed by "divine intent," might be modified or even obstructed.

A more sophisticated attempt to harness uncertainty was made with the advent of probability theory. Now, one could achieve a fuller understanding of the patterns of outcomes due to chance by expressing in mathematical form the likelihood of occurrence of events. Probability theory served as the vehicle for separating in a precise fashion probable from improbable outcomes. Thus the disorder of belief in fateful happenings was replaced by a methodical way of assessing the odds of chance outcomes.

Armed with these new analytical tools, randomness was eventually regarded in a more pragmatic light. Certainly, misfortune and disruptions are still with us (e.g., natural disasters, economic dislocations, acts of terror and mindless violence, physical and mental health problems, and so on). However, randomness soon came to be viewed as a crucial or even healthy medium for the development and/or sustainability of life-enhancing processes. In this regard, uncertainty can be thought of as necessary for the spawning of innovation and diversity; it is a mechanism that sets in motion evolutionary adaptations and sparks the processes that are essential for the renewal of life. A confluence of chance factors may ensure the viability of different life forms and the resiliency of entire ecosystems. That is, chance disruptions (mutations or accidents of nature) in the physical world provide the raw material for natural selection in the long run. Environmental conditions favor or inhibit the development or survival of specific individuals or species. Thus it is happenstance that can alter the path of evolution; some species adapt and thrive while others disappear.

We know that if an individual number is random, then its value is unpredictable. Additionally, any such number must be drawn from a set of equiprobable values (e.g., it is obtained from, say, a uniform probability distribution). But when considering a sequence of random numbers, each element drawn from the sequence must be statistically independent of the others.

While randomness is inherent in the physical world (think of collisions between subatomic particles), it is also the case that uncertainty has served as the basis for developing theories of societal behavior. Although uncertainties exist at the individual level, a modicum of stability is present in the average behavior of a collection of individuals. This conclusion, reflecting order in disorder, emerges due to the Law of Large Numbers. That is, the degree of uncertainty in the average performance of individuals in a system decreases as the number of individuals increase so that the system as a whole exhibits a virtually deterministic form of behavior.

For instance, suppose a large number of individuals are asked to estimate, by sight, the length of an object. Certainly, some will overestimate the length while others will underestimate it. And some of these individuals will be close to the true length while others will be a bit farther from it. Each estimate can be treated as a random number—we cannot predict with any degree of certainty the value that any given person will state. Moreover, each individual is assumed to guess the length of the object independently of the others. So given that positive and negative deviations from the true length are equally likely and no one value can be expected to predominate, we expect that the actual measurements are normally distributed about the true length of the object or the measurement errors are normally distributed about zero. Thus small errors are likely to occur more frequently than large ones and very large errors are highly unlikely.

So while we cannot predict the actual type of measurement error that a single individual will make, we can predict, with a reasonable degree of certainty, how the errors resulting from measurements made by a large number of individuals behave: more people will make small errors than large ones; the larger the error, the fewer the number of people making it; approximately, the same number of individuals will overestimate as will underestimate the actual length; and the average error, taken over all individuals, will be zero.

7.C.2 Types of Randomness

In what follows, we shall recognize three varieties of randomness (this convenient taxonomy is due to Tsonis (2008)):

Type I Randomness: Due to lack of knowledge of how a system operates, that is, its rules of operation are unknown.

Type II Randomness: We do not read a system's signals fully and/or accurately or have infinite computing power.

Type III Randomness: Occurs because of the complex interactions of many environmental factors or systems.

We now examine each of these varieties in turn. (As our discussion commences, it is important to keep in mind that order and predictability results from specific rules of operation whereas randomness and unpredictability emerges because the rules are nonexistent or unresolvable.)

7.C.2.1 Type I Randomness

Due to our inability to determine the rules by which a system or the physical world operates. Either we cannot uncover or discover the rules or (trivially) no rules exist.

Consider the following aperiodic patternless sequence of numbers that can be generated by the function or rule

$$R_1: 2^n + n, n = 0, 1, 2, \ldots$$

As n increases in value from zero, we obtain the sequence

$$S_1:\ 1, 3, 6, 11, 20, 37, \ldots$$

Are the elements of this sequence random numbers? Obviously not since one can readily predict any digit in S_1 via R_1.

Let us reverse this process. That is, suppose we are first given S_1 and asked to find the rule (R_1) that generated it. After a bit of trial and error or numerical gymnastics we possibly could arrive at the rule R_1: $2^n + n$, $n = 0, 1, 2, \ldots$. Going from the rule to the sequence and vice versa is called *reversibility*, and it is the reversibility of the rule R_1 and the sequence S_1 that precludes S_1 from being a random sequence of numbers.

However, since S_1 is a sequence of even and odd integers, let us derive from it a new sequence S_2 in the following fashion. We introduce a second rule R_2: record the letter A if an even number occurs in S_1; and record the letter B when an odd number appears in S_1. This scheme thus renders a second (parallel) sequence.

$$S_2:\ B, B, A, B, A, B, \ldots$$

Now, suppose we were provided with S_2 but not the rule R_2 that generated it. It is quite possible that we could conclude that the A's and B's represent even and odd numbers, respectively, so that reversibility between S_2 and R_2 holds. However, reversibility does not hold between R_2 and S_1—for each A (respectively, B), there is an infinite number of even numbers (respectively, odd numbers) to choose from. And since R_2 introduces an element of uncertainty into the system that prevents us from uncovering R_1, we see that S_2 can be treated as a random sequence.

The upshot of this discussion is that if we observe an aperodic patternless sequence that has been generated by some rule that we are not able to uncover, then clearly we do not have the ability to predict future elements of the sequence and thus the sequence in question can be treated as random. For all intents and purposes, the sequence under discussion is no different from a purely random sequence if reversibility between the sequence and some rule that produced it does not hold. So if we experience irreversible processes that preclude us from discovering the rules of a system's operation, and we don't simply find ourselves in some circumstance in which rules do not exist, then the condition of unpredictability and thus randomness prevails.

7.C.2.2 Type II Randomness

Due to lack of infinite precision and computing power. Suppose we have a nonlinear albeit deterministic model (or rule) described by an iterative process in which the value X_{n+1} is determined from the previous value X_n, $n=0, 1, 2, \ldots$.

For $n=0$, X_o is our initial condition. Let $X_o=0.06517$. Now, let us change the initial condition slightly, for example, $X_o=0.06517$ is replaced by $X_o=0.06500$. Surprisingly, the two paths of the system (one with the initial condition 0.06517 and the other with the initial condition 0.06500) may be radically different (due to the nonlinearity of the process) and thus may not converge (if they converge at all) to the same value. The key notion here is that the system is highly sensitive to the initial conditions chosen. In fact, even for miniscule perturbations or errors of measurement of the initial conditions, the end results become highly unpredictable and thus random.

Clearly lack of precision, as reflected by using convenient or inexact approximations to initial conditions, influences the evolution of a nonlinear system or model, thus leading to an outcome that is much different from, say, previous ones. To avoid this circumstance and to make "on-target" predictions, infinite precision and infinite computing power is needed. Since we lack such resources, the exact state of the system after a few iterations will be unknown and consequently unpredictable, thus rendering all future states random or unpredictable as well. Thus randomness emerges even if the rules are known and/or the system is reversible. In fact, this state of affairs, in which sensitivity to initial conditions is important, is termed *chaotic* (more on that later).

7.C.2.3 Type III Randomness

Due to the continuing impact of the environment. Suppose we have a fixed period of time (in hours) in which to drive at the posted speed limit. How far will we travel in a given direction? Here time $\times$ speed $=$ distance traveled. It should be evident that "distance" is a random number since the traffic patterns, slow-downs/stops, possible accident delays, and so on are unpredictable. So, even though "time" and "speed" are specified and thus deterministic, "distance" is not fully knowable and thus is considered random.

What is the mechanism by which randomness is introduced into the system? Here randomness is injected into the operation of the system via the interaction of external environmental factors or players whose actions are beyond our control, for example, stop signs, other drivers, traffic lights, the right-of-way of emergency vehicles, and so on, obviously adversely impact distance traveled. Taken collectively, these environmental factors interact in a complex fashion to produce a process of operation that is essentially a random process.

Interestingly enough, as the preceding example illustrates, Type III Randomness can emerge even if we start with no randomness at all. A system that exhibits regular behavior can eventually display complicated behavior when it is subject to an external force that might also be very regular. Hence the operation of very simple rules of behavior, when exposed to other such rules, can, in combination, precipitate unpredictability. The collective behavior of interacting nonrandom environmental factors can produce randomness.

7.C.3 Pseudo-Random Numbers

Natural random numbers occur in an "innate" way, that is, they are generated by one of nature's physical processes such as: tracking successive male (we record a 1) and female (we record a 0) births at a hospital; tracing the path of a neutron or some other particle that is subject to random collisions; arrivals at a toll booth; counting the particles emitted by the decay of a radioactive element; atmospheric noise, and so on. In these examples, randomness is "inherent" and any calculation of these numbers involves a simulation of the underlying physical process involved.

Artificially generated random numbers are obtained by using a chance mechanism such as: flipping a fair coin; rolling a fair pair of dice; drawing tokens numbered 0, 1, . . . , 9 from a bag, one by one, under replacement; using a suitable probability distribution, and so on. Whether a random number has been produced naturally or artificially is inconsequential—each is thought to be a *true* or *pure random number*.

It is more or less the current belief in the mathematics community that, in order to obtain a valid sequence of random numbers, one should not rely on natural procedures but, instead, on suitably deterministic ones. Random numbers generated in this fashion (typically with the use of a computer) are termed *pseudo-random numbers*. Sequences of such numbers, which are obtained via a software function, are called pseudo-random because they are actually deterministic in nature, that is, given a specific source function and a fixed seed (starting) value, the same sequence of numbers can be generated time after time. If the pseudo-random number source function is well-designed, then the resulting sequence of numbers will appear to be truly random (provided that the sequence does not repeat too often). In fact, the sequence of pseudo-random numbers will be able to pass tests for randomness even though it was developed from a completely deterministic process.

As indicated above, pseudo-random numbers are produced by computer via so-called pseudo-random number generators (PRNGs)—algorithms that use mathematical formulae to develop a sequence of numbers that appear to be random. PRNGs are as follows:

a. Efficient: The algorithm can produce many numbers in a short time.
b. Deterministic: A given sequence of numbers can be reproduced if the seed value is known.
c. Periodic: The sequence will eventually repeat itself (after a long running time).

While PRNGs are not suitable for data encryption or gambling purposes, they are useful for extracting "random samples" (the entries in a table of random numbers are actually pseudo-random numbers) and for discovering chance mechanisms in the physical world and for explaining the occurrence of natural events.

7.C.4 Chaotic Behavior

It was mentioned above that the term "Type II Randomness" was used to describe the behavior of so-called chaotic systems—nonlinear dynamic systems that are highly sensitive to initial conditions. For chaotic systems, small differences in initial conditions (ostensibly due to rounding errors) yield widely divergent outcomes or terminal states. And this occurs even though these systems are completely deterministic (no random components are involved), with their future paths fully dependent on the starting point chosen. Moreover, system outcomes are generally predictable only in the very short run since the initial conditions provide but scant information about the system's path. In general, any deterministic system that has limited predictability due to its dependence on initial conditions is deemed *chaotic*.

Many natural systems exhibit chaotic behavior: the growth of certain species in ecosystems; earthquakes due to shifts in the earth's crust; the magnetic fields of planets orbiting the sun; and, of course, weather patterns. Additionally, medical researchers are attempting to predict the occurrence of seemingly random epileptic seizures by monitoring a patient's behavior at the outset (the initial conditions) of these seizures.

For a chaotic system, a miniscule perturbation in its starting point can lead to a significant difference in its future trajectory and thus in its future behavior. This result was aptly pointed out by Edward Lorenz (1972) in a paper entitled *Predictability: Does the Flap of a Butterfly's Wings in Brazil Set off a Tornado in Texas?* Lorenz contended that the flapping of a butterfly's wings represents an arbitrarily small change in the initial conditions of the system, which subsequently precipitates a chain of events leading to an outcome of considerable consequence. Had the butterfly not flapped its wings, the system's trajectory and terminal state would have been markedly different. So if we start with only a modicum of information about the system (the initial conditions), then, as time passes, the system becomes less and less predictable—only short-run prognostications are relevant. The circumstance just described has come to be known as the *Butterfly Effect*—the difference in the starting points of two trajectories is so small as to be akin to a butterfly flapping its wings. So if a dynamic system is deterministic with no random elements, it does not necessarily mean that the system is predictable; its future behavior may be hostage to its initial conditions. And if sensitivity to initial conditions actually holds, then, as stated earlier, chaotic behavior is said to obtain. This is why constant weather updates are so important; when meteorological models are run, only short-run forecasts are made on the basis of current conditions. And it is the current conditions that serve as the initial conditions for the next run of the model and thus the next short-run prediction.

What does chaotic behavior over time look like? Let us note first that no observed real-world time series is made up of only pure unadulterated signal; it typically contains a random error or noise component. So even if a system is largely deterministic, some element of randomness will certainly be present. This said, a deterministic system always has an error term that is either small and remains stable and regular or chaotic in that it increases exponentially over time. Thus chaotic systems are fairly orderly and display a pattern. And it is the search for order that has led researchers to create models that simulate the processes that a system will experience over time.

7.C.5 Risk and Uncertainty

This section considers how the concepts of risk and uncertainty relate to the notion of randomness. Probably the seminal distinction between risk and uncertainty was made by Frank Knight (1921). Although the two terms "risk" and "uncertainty" tend to be used interchangeably, Knight posited that *risk* is present when future events occur with knowable probabilities, while *uncertainty* is present when future events have unknowable probabilities. Thus risk can be quantified (either by using historical data or by observation/experimentation); uncertainty is not measurable.

In a "risky situation," decisions can be made on the basis of the information contained in an *ex ante* (before the fact) known probability distribution. And if the possible outcomes of a choice are specified, then we can compute a probability-weighted average or expectation (e.g., an expected utility or cost) as the basis for decision making. In a "climate of uncertainty," the probability distribution of a random outcome is unknown—uncertainty cannot be quantified. (Before the housing crisis, most realtors operated under the assumption that house prices would rise indefinitely because there was no shortage of buyers. Since the probable path in house prices is not knowable, decision making in the housing market occurred largely under conditions of uncertainty.) In short:

Risk: Involves randomness with knowable probabilities;

Uncertainty: Involves randomness with unknowable probabilities.

How should we view the notion of risk? Most risk usually results from instability in the expected outcome—a risky situation has the potential for pronounced "ups and downs;" it displays considerable variation in the possible outcomes of a situation. For instance, holding a stock that pays \$100 per year in dividends and whose price only rarely changes is not very risky. But keeping a stock whose price moves up and down in an erratic fashion over time is risky.

One measure of risk is the standard deviation of the possible outcomes of some decision situation. In this regard, it should be intuitively clear that the greater the magnitude of the standard deviation, the greater the risk incurred. For example, consider the cost of repairing two portable or laptop computers (called simply A and B). Let the expected or average value of the repair-cost distribution for each be \$250. Suppose the standard deviation for the A distribution is \$100 while the standard deviation for the B distribution is \$40. If the cost of repairs is normally distributed for each laptop model, what is the probability that the repairs for each computer will cost more than \$300? For computer A,

$$P(X_A > 300) = P\left(Z_A > \frac{300 - 250}{100} = 0.5\right) = 0.3085,$$

and for computer B,

$$P(X_B > 300) = P\left(Z_B > \frac{300 - 250}{40} = 1.25\right) = 0.1056.$$

Hence the probability that the repairs will cost more than \$300 is about 31% for the type A computer; it is about 11% for the type B computer. Since type A has greater absolute variation associated with its distribution relative to that of B, it follows that, in terms of repair cost variability, the type A computer is a much more risky proposition.

Since decision making in the face of risk involves a situation in which the probability of an outcome can be determined, it follows that the outcome can be insured against (e.g., automobile accidents, house fires, and so on). However, given that uncertainty arises from circumstances involving outcomes with unknown probabilities, such outcomes cannot be insured against.

How does one insure against risk? We may view *insurance* as an agreement or contract whereby, for a stipulated payment (the *premium*), an insurer agrees to pay to the policyholder a defined amount (the *benefit*) upon the occurrence of an unlikely but high-cost event. Thus a policyholder exchanges an unknown loss for the payment of a known premium. The risk of any unanticipated loss is transferred from the policyholder to the insurer.

How does an insurance company deal with the risk it assumes? Insurance companies use risk pooling to moderate the impact of the risk they take on. *Risk pooling* is a way of spreading risk over similarly situated individuals. By insuring many customers, losses are shared by averaging them together. Risk pooling works because of the Law of Large Numbers. When applied to insurance activities, the law states that when a large number of individuals face a low-probability outcome, the proportion actually experiencing the outcome will be close to the expected proportion. Thus the larger the policy pool, the more predictable are the losses incurred. (Think of averaging together more and more numbers in some limited range. Clearly, the average will become more and more stable in that unusually high and low values will tend to cancel each other out. For instance, on the first roll of a fair pair of dice the sum of the faces showing can be any integer value between 2 and 12 inclusive. As you roll the pair of dice again and again, the average of the sums obtained will get closer and closer to 7. And if you roll the pair of dice, say, 10,000 times, then the average will be very stable around the expected value or long-run average of 7.) Hence insurance companies charge a premium based on the average or expected losses for the pool (and the potential for variation about the average) plus a certain markup to cover administrative costs in order to have reserves sufficient to cover all of the projected claims submitted by the pool. When they pool many individuals, the average loss is very stable so that they assume minimal risk themselves.

For risk pooling to be successful, the individual risks within the pool must be mutually independent. Independent risks tend to move up and down at different times and not in concert. When risks fluctuate in this fashion, they tend to cancel each other out. (It is far less risky to provide flight insurance for 100 individuals traveling on 100 different flights than for 100 individuals traveling on the same flight.)

Example 7.C.1

An *insurance loss distribution* is a combination of two other distributions—the *frequency of loss distribution* (involving the random variable X_F, which represents the number of losses that will occur in a given time period) and the *severity of loss distribution* (involving the random variable X_S, which depicts the amount of the loss—its severity—given that a loss has actually occurred. For instance, suppose an insured motorist has the following frequency of loss distribution (Table 7.C.1) and accompanying severity of loss distribution (Table 7.C.2).

Combining these two (independent) distributions results in the loss (due to accident) distribution for random variable X (Table 7.C.3). (Remember that for independent events A and B, $P(A \cap B) = P(A) \cdot P(B)$.)

TABLE 7.C.1 Frequency of Loss Distribution

Number of losses X_F	Probability
0	0.90
1	0.10
>1	0.00
	1.00

TABLE 7.C.2 Severity of Loss Distribution

Severity of losses X_S ($)	Probability
1,000	0.60
5,000	0.35
15,000	0.05
	1.00

TABLE 7.C.3 Loss Distribution

Loss incurred X ($)	Probability
0	0.900
1,000	0.060 [=0.10 (0.60)]
5,000	0.035 [=0.10 (0.35)]
15,000	0.005 [=0.10 (0.05)]
	1.000

Given this loss distribution, the motorist's expected loss is thus

$$\begin{aligned} E(X) &= \sum X_i P(X = X_i) \\ &= 0(0.90) + 1,000(0.06) + 5,000(0.035) + 15,000(0.005) \\ &= \$310. \end{aligned}$$

So on average, the motorist spends $310 on repairs due to accidents in a given time period.

To assess the riskiness of the motorist's loss distribution (the potential variation between possible losses and the expected loss), we need to determine the standard deviation of X.

We first find

$$\begin{aligned} V(X) &= E(X^2) - E(X)^2 = \sum X_i^2 P(X = X_i) - E(X)^2 \\ &= (0)^2(0.90) + (1,000)^2(0.06) + (5,000)^2(0.035) \\ &\quad + \left(15{,}000^2\right)(0.005) - (310)^2 \\ &= 1,963{,}900. \end{aligned}$$

Then the standard deviation of X is $\sqrt{V(X)} - \$1,401.39$, our measure of the average variability in losses about the expected loss.

With a standard deviation as large as the one found in the preceding example problem (there exists considerable variation in possible losses), it would be prudent for the motorist to purchase an insurance policy so as to transfer some of the risk inherent in the loss distribution to an insurance company. Let us next look to the formation of an insurance pool.

Suppose now that n motorists, each with the same loss distribution, decide to purchase the same type of insurance policy from a given insurance company. Does the insurance company assume the aggregate risk of the n individual independent policyholders? Not at all. The aggregate level of risk incurred by the insurance company is smaller than the sum of the risks of all n individual policyholders. To rationalize this conclusion, let us consider the following theorem.

Theorem 7.C.1. Let $X_1, X_2, \ldots, X_n$ be a set of n independent random variables such that each X_i has the same mean $E(X_i) = \mu$ and the same variance $V(X_i) = \sigma^2, i = 1, \ldots, n$. Let $S_n = X_1 + X_2 + \ldots, + X_n$. Then

$$E(S_n) = E(X_1) + E(X_2) + \ldots + E(X_n) = n\mu$$

$$V(S_n) = V(X_1) + V(X_2) + \ldots + V(X_n) = n\sigma^2$$

and thus the standard deviation of S_n is

$$\sqrt{V(S_n)} = \sqrt{n}\,\sigma,$$

where σ is the standard deviation for each X_i.

If we think of $\sqrt{V(S_n)}$ as the aggregate risk taken on by the insurance company and $\sum \sqrt{V(X_i)}$ as the sum of the risks associated with the n independent policyholders, then it follows that

$$\sqrt{V(S_n)} = \sqrt{n}\,\sigma < \sum \sqrt{V(X_i} = n\sigma.$$

Hence the risk assumed by the insurance company (the standard deviation of the risk pool) is less than the sum of the risks associated with the n independent individual policyholders. So if, for instance, $n = 100$ and each of the independent individual policyholders has $E(X) = \$310 = \mu$ and $\sqrt{V(X)} = \$1{,}401.39 = \sigma$ (see Example 7.C.1), then

$$\sqrt{V(S_n)} = \sqrt{100}(1401.39) = \$14{,}013.90 < 100(1401.39) = \$140{,}139.00.$$

It is also instructive to consider the relative variation between the insurance company's loss distribution and the loss distribution for an individual policyholder. We know from our previous discussion of variability (see Section 3.7) that when comparing the dispersions of two (or more) distributions, the coefficient of variation is a useful analytical device. Remember that the coefficient of variation (V) is defined as the standard deviation divided by the mean. For the insurance company, the coefficient of variation is $V_I = \sqrt{V(S_n)}/E(S_n) = \sqrt{n}\sigma/n\mu = \sigma/\sqrt{n}\mu$; and for the individual policyholder, the coefficient of variation is $V_P = \sqrt{V(X_i)}/E(X_i) = \sigma/\mu$. Clearly, $V_I < V_P$. So, given n independent policyholders, it is evident that $V_I = \sigma/\sqrt{n}\mu \rightarrow 0$ *as* $n \rightarrow \infty$, that is, the insurance company's relative risk, as indexed by the coefficient of variation, approaches zero as the size of the risk pool increases without bound. Clearly, large pools of independent and similarly situated customers are preferred by the company to small ones. In terms of the results obtained in Example 7.C.1,

$$V_I = \frac{\sqrt{V(S_n)}}{E(S_n)} \times 100 = \frac{\sigma}{\sqrt{n}\mu} \times 100 = \frac{1403.39}{\sqrt{100}(300)} \times 100 = 45.2\%$$

$$V_P = \frac{\sqrt{V(X_i)}}{E(X_i)} \times 100 = \frac{\sigma}{\mu} \times 100 = \frac{1403.39}{310} \times 100 = 452.1\%.$$

So for the insurance company, the standard deviation is 45.2% of the mean; for an individual customer, the standard deviation is 452.1% of the mean. Thus there is much less risk associated with the insurance company's loss distribution relative to that for a given policyholder.

8

CONFIDENCE INTERVAL ESTIMATION OF μ

8.1 THE ERROR BOUND ON $\bar{X}$ AS AN ESTIMATOR OF μ

In the discussion that follows, we shall assume that the population variable X is $N(\mu, \sigma)$ or n is sufficiently large. (Why?) Then a *95% error bound on $\bar{X}$ as an estimate of μ* is

$$\pm 1.96\sigma_{\bar{X}} \text{ or } \pm 1.96\frac{\sigma}{\sqrt{n}}. \tag{8.1}$$

Once this error bound or maximum tolerable error level is calculated, we can conclude that: we are 95% confident that $\bar{X}$ will not differ from μ by more than $\pm 1.96\sigma_{\bar{X}}$. Where did the 1.96 come from? As one might have guessed, it comes from the $N(0, 1)$ area table (see Fig. 8.1). Using this process, the reader can easily verify that *90% and 99% error bounds on $\bar{X}$ as an estimate of μ* are $\pm 1.645\sigma_{\bar{X}}$ and $\pm 2.58\sigma_{\bar{X}}$, respectively.

Example 8.1

A sample of size $n = 36$ is taken from a $N(\mu, 10)$ population. Place a 95% error bound on $\bar{X}$ as an estimate of μ. Using Equation (8.1), we have $\pm 1.96(10/\sqrt{36})$ or ± 3.26. Hence may be 95% confident that $\bar{X}$ will not differ from μ by more than plus or minus 3.26 units.

Statistical Inference: A Short Course, First Edition. Michael J. Panik.

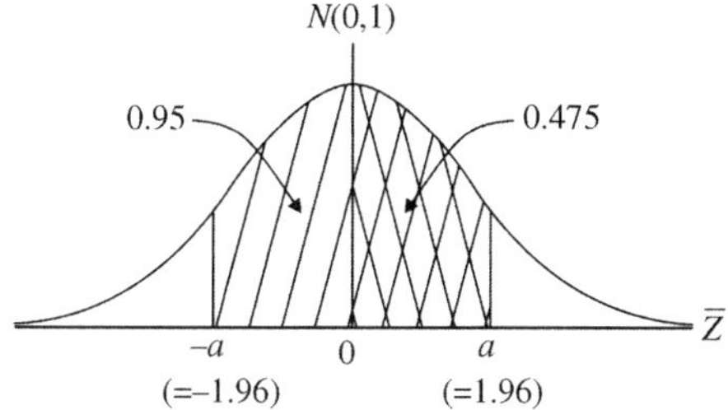

FIGURE 8.1 (1) Total area from $-a$ to a is 0.95. Hence the area from 0 to a is 0.475; (2) Find 0.475 in the body of the $N(0, 1)$ area table and read it in reverse to get a $\bar{Z}$ value; (3) Here $\bar{Z} = 1.96$.

Next, suppose X is $N(\mu, \sigma)$ with σ known. What is the probability that $\bar{X}$ from a sample of size n will be within $\pm$ a units of μ? (See Fig. 8.2 for the interval in question.) Here we are interested in finding $P(\mu - a \leq \bar{X} \leq \mu + a)$. But this is equivalent to the probability that the sampling error $\bar{X} - \mu$ lies between $\pm a$ or $P(-a \leq \bar{X} - \mu \leq a)$. Then transforming to a probability statement involving $\bar{Z}$ enables us to find

$$P\left(\frac{-a}{\sigma/\sqrt{n}} \leq \frac{\bar{X}-\mu}{\sigma/\sqrt{n}} \leq \frac{a}{\sigma/\sqrt{n}}\right) = P\left(\frac{-a}{\sigma/\sqrt{n}} \leq \bar{Z} \leq \frac{a}{\sigma/\sqrt{n}}\right). \tag{8.2}$$

Given Equation (8.2), a trip to the $N(0, 1)$ area table will subsequently provide the desired result.

Example 8.2

Given that X is $N(\mu, 10)$, what is the probability that $\bar{X}$ determined from a sample of size $n = 25$ will be within ± 4 units of μ? Here we want to find

$$\begin{aligned}
P(\mu - 4 \leq \bar{X} \leq \mu + 4) &= P(-4 \leq \bar{X} - \mu \leq 4) \\
&= P\left(\frac{-4}{10\sqrt{25}} \leq \frac{\bar{X}-\mu}{\sigma/\sqrt{n}} \leq \frac{4}{10/\sqrt{25}}\right) \\
&= P(-2 \leq \bar{Z} \leq 2) = 2(0.4772) = 0.9544.
\end{aligned}$$

If the population variable X is not normally distributed, can we still perform this calculation? (Explain your answer.)

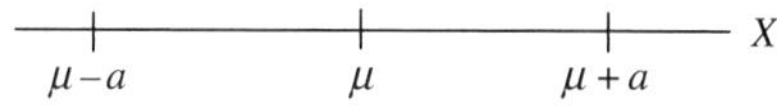

FIGURE 8.2 The event $\mu - a \leq \bar{X} \leq \mu + a$.

8.2 A CONFIDENCE INTERVAL FOR THE POPULATION MEAN μ (σ KNOWN)

In attempting to estimate μ from a simple random sample of size n, we previously employed the sample mean $\bar{X}$ as a point estimator for μ, where a *point estimator* reports a single numerical value as the estimate of μ. Now, let us report a whole range of possible values rather than a single point estimate. This range of values is called an *interval estimate* or *confidence interval* for μ, that is, it is a range of values that enables us to state just how confident we are that the reported interval contains μ. What is the role of a confidence interval for μ? It indicates how precisely μ has been estimated from the sample; the narrower the interval, the more precise the estimate.

As we shall now see, we can view a confidence interval for μ as a "generalization of the error bound concept." In this regard, we shall eventually determine our confidence limits surrounding μ as

$$\bar{X} \pm \text{error bound}, \tag{8.3}$$

where now the term $\pm$ error bound is taken to be our *degree of precision*.

Let us see how all this works. To construct a confidence interval for μ, we need to find two quantities L_1 and L_2 (both function of the sample values) such that, before the random sample is drawn,

$$P_{CI}(L_1 \leq \mu \leq L_2) = 1 - \alpha, \tag{8.4}$$

where L_1 and L_2 are, respectively, lower and upper confidence limits for μ and $1 - \alpha$ is the *confidence probability*—it is the probability that we will obtain a sample such that the interval (L_1, L_2), once calculated, will contain μ (Fig. 8.3). Here $1 - \alpha$ is chosen or pegged in advance of any sampling and typically equals 0.95 (or 0.90 or 0.99). Once $1 - \alpha$ is specified and the confidence interval (L_1, L_2) for μ is determined, our conclusion regarding μ can be framed as: we may be $100(1 - \alpha)\%$ confident that the interval from L_1 to L_2 contains μ.

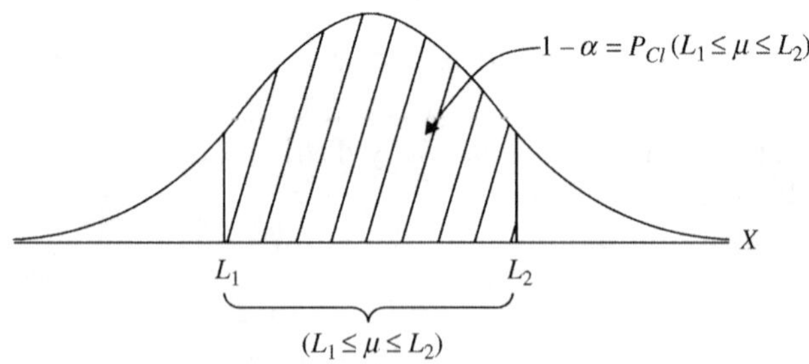

FIGURE 8.3 A $100(1 - \alpha)\%$ confidence interval for μ.

To actually find L_1 and L_2 we need to make an assumption and to introduce some specialized notation/concepts. First, let us assume that the population variable X is $N(\mu, \sigma)$ with σ known. Then $\bar{X}$ is $N(\mu, \sigma/\sqrt{n})$ and thus $\bar{Z} = (\bar{X} - \mu)/(\sigma/\sqrt{n})$ is standard normal or $N(0, 1)$. Second, let us define lower and upper percentage points for the standard normal distribution:

a. Consider the probability statement $P(\bar{Z} \geq \bar{Z}_\alpha) = \alpha$. Clearly, this expression defines an area under the right-hand tail of the $N(0, 1)$ distribution (Fig. 8.4). Here α is chosen first, that is, any α implies a $\bar{Z}_\alpha$ value. Hence $\bar{Z}_\alpha$, which depends on the specified α, is the *point* on the positive $\bar{Z}$-axis such that the probability of a larger value is α—it is termed an *upper α percentage point* for $N(0,1)$.

   For instance, given $\alpha = 0.05$, let us find $\bar{Z}_\alpha = \bar{Z}_{0.05}$ (Fig. 8.5). It should be apparent that a *lower α percentage point* (denoted $-\bar{Z}_\alpha$) for $N(0, 1)$ can be defined using the probability statement $P(\bar{Z} \leq -\bar{Z}_\alpha) = \alpha$.

   Next, since we are interested in determining a confidence "interval," let us find the probability of $\bar{Z}$ falling into some interval.

b. Consider the probability statement

$$P(-\bar{Z}_{\alpha/2} \leq \bar{Z} \leq \bar{Z}_{\alpha/2}) = 1 - \alpha \tag{8.5}$$

   (Fig. 8.6). Here $\bar{Z}_{\alpha/2}$ is an *upper $\alpha/2$ percentage point* for the $N(0, 1)$ distribution—it is the point on the positive $\bar{Z}$-axis such that the probability of a larger value is $\alpha/2$, that is, $\bar{Z}_{\alpha/2}$ cuts off an area equal to $\alpha/2$ in the right-hand tail of $N(0, 1)$. (The *lower $\alpha/2$ percentage point* $-\bar{Z}_{\alpha/2}$ is defined in an analogous fashion.)

   For instance, if $1 - \alpha = 0.95$, then $\alpha = 0.05$ and thus $\alpha/2 = 0.025$. Then $\bar{Z}_{0.025}$ can be determined from Fig. 8.7.

We are now in a position to find a $100(1 - \alpha)\%$ confidence interval for μ. Let us start with Equation (8.5) and substitute the definition of $\bar{Z}$ into this expression. We consequently obtain:

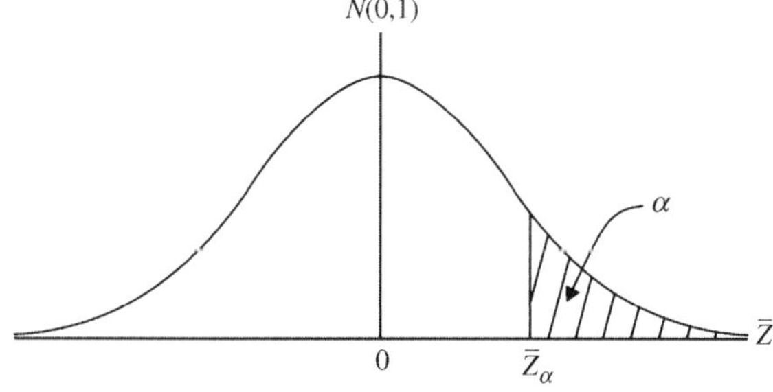

FIGURE 8.4 $\bar{Z}_\alpha$ is an upper percentage point for $N(0, 1)$.

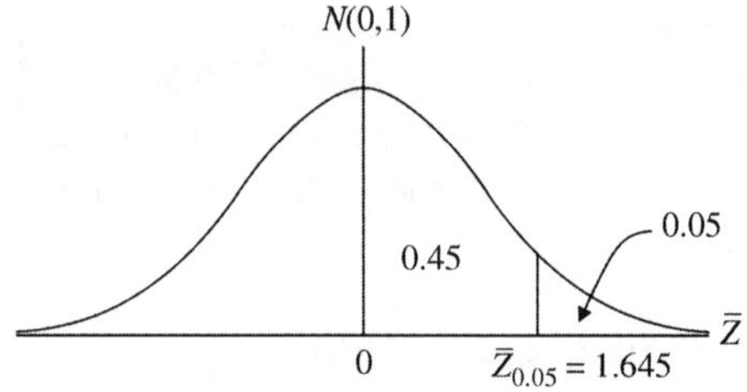

FIGURE 8.5 (1) $\alpha = 0.05$ is the area under the right-hand tail of $N(0, 1)$. Hence the remaining area to the right of 0 is 0.45; (2) Find 0.45 in the body of the $N(0, 1)$ table and read the table in reverse to find $\bar{Z}_{0.05} = 1.645$.

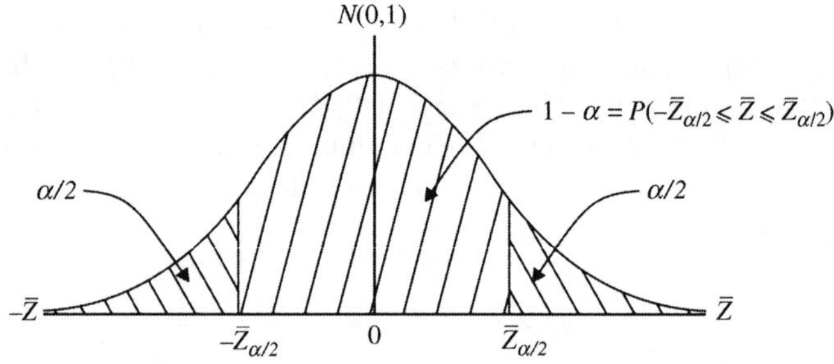

FIGURE 8.6 $\bar{Z}_{\alpha/2}$ is an upper $\alpha/2$ percentage point for $N(0, 1)$; $-\bar{Z}_{\alpha/2}$ is a lower $\alpha/2$ percentage point for $N(0, 1)$.

$P\left(-\bar{Z}_{\alpha/2} \leq \frac{\bar{X}-\mu}{\sigma/\sqrt{n}} \leq \bar{Z}_{\alpha/2}\right) = 1 - \alpha$ [Multiply each term in parentheses by $\sigma/\sqrt{n}$.]

$P\left(-\bar{Z}_{\alpha/2}\frac{\sigma}{\sqrt{n}} \leq \bar{X} - \mu \leq \bar{Z}_{\alpha/2}\frac{\sigma}{\sqrt{n}}\right) = 1 - \alpha$ [Subtract $\bar{X}$ from each term in parentheses.]

$P\left(-\bar{X} - \bar{Z}_{\alpha/2}\frac{\sigma}{\sqrt{n}} \leq -\mu \leq -\bar{X} + \bar{Z}_{\alpha/2}\frac{\sigma}{\sqrt{n}}\right) = 1 - \alpha$ [Multiply each term in parentheses by –1. What happens to the sense of the inequalities?]

$$P\Bigg(\underbrace{\bar{X} - \bar{Z}_{\alpha/2}\frac{\sigma}{\sqrt{n}}}_{L_1} \leq \mu \leq \underbrace{\bar{X} + \bar{Z}_{\alpha/2}\frac{\sigma}{\sqrt{n}}}_{L_2}\Bigg) = 1 - \alpha. \tag{8.6}$$

(Note that this last probability statement mirrors Equation (8.4).) Hence a $100(1 - \alpha)\%$ *confidence interval for* μ is

$$\bar{X} \pm \bar{Z}_{\alpha/2}\frac{\sigma}{\sqrt{n}}. \tag{8.7}$$

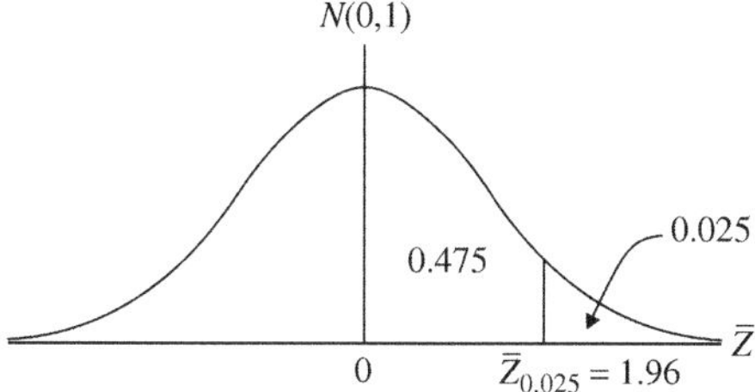

FIGURE 8.7 (1) $\alpha/2 = 0.025$ is the area under the right-hand tail of $N(0, 1)$. Hence the remaining area to the right of 0 is 0.475; (2) Find 0.475 in the body of the $N(0, 1)$ table and read it in reverse to find $\bar{Z}_{0.025} = 1.96$.

Note that when minus holds, we get the lower confidence limit L_1; and when plus holds, we get the upper confidence limit L_2.

Let us now focus on the term $1 - \alpha$:

a. In Equation (8.6), $1 - a$ is the "confidence probability" because $\bar{X}$ is a random variable. However, once the sample realization of $\bar{X}$ is obtained, $\bar{X}$ is no longer random and thus: (i) Equation (8.6) is no longer a probability statement—it is now called a *confidence statement*; and (ii) $1 - a$ is no longer a probability—it is simply called our *100(1 – a)% confidence coefficient.*
b. The confidence coefficient $1 - a$ is alternatively termed the *level of reliability,* with $\bar{Z}_{\alpha/2}$ serving as the *100(1 – a)% reliability coefficient.* We may view the notion of reliability as a long-run concept that emerges under repeated sampling from the same population. For instance, let $1 - a = 0.95$. Then the level of reliability (95%) is the proportion of time the confidence interval contains μ, that is, if we take many random samples of size n from the population and if we construct an interval such as Equation (8.7) for each of them, then, in the long run, 95% of all these intervals would contain μ and 5% of them would not.
c. What is it that a confidence interval *does not* tell us? It does not give us the *probability* that μ falls within a specific interval. This is because μ is a fixed constant and not a random variable (remember that only random variables have probabilities associated with them.)

So what exactly is it that we are confident about? We are confident in the *methodology* employed to obtain the confidence limits L_1 and L_2. That is, for $1 - \alpha = 0.95$, saying that we are 95% confident that μ lies within the interval (L_1, L_2) means that we have used a technique that brackets μ for 95% of all possible samples of size n. Hence 95% is our *success rate*—our method of estimation would give us the correct result 95% of the time if applied repeatedly. Hence the notion of *repeated sampling* is key to understanding a confidence coefficient.

In this regard, a confidence interval determined from a particular sample may or may not contain μ. That is, if we extract a simple random sample from a population and compute the confidence limits according to Equation (8.7), then we do not actually know whether the 95% confidence interval obtained from our sample is one of the 95% that contains μ, or one of the 5% that does not. However, if we repeat this process of confidence interval estimation a large number of times, then, in the long run, 95% of all the intervals calculated would house μ, and 5% of them would not.

If one is wedded to the concept of probability, then all we can say is: under repeated sampling from the same population, the probability is 0.95 that we have used a technique that produces an interval containing μ.

Example 8.3

Given that a population variable X is $N(\mu, 10)$, find a 95% confidence interval for μ given that $n = 25$ and $\bar{X} = 15$. Since $1 - \alpha = 0.95$, $\alpha = 0.05$ and thus $\alpha/2 = 0.025$. Hence $\bar{Z}_{0.025} = 1.96$ (see Fig. 8.7). Then from Equation (8.7), we obtain

$$L_1 = \bar{X} - \bar{Z}_{\alpha/2}\frac{\sigma}{\sqrt{n}} = 15 - 1.96\left(\frac{10}{5}\right) = 11.08$$

$$L_2 = \bar{X} + \bar{Z}_{\alpha/2}\frac{\sigma}{\sqrt{n}} = 15 + 1.96\left(\frac{10}{5}\right) = 18.92,$$

or $(L_1, L_2) = (11.08, 18.92)$. Thus we may be 95% confident that the interval from 11.08 to 18.92 contains μ.

How precisely have we estimated μ? We are within $\pm\bar{Z}_{\alpha/2}\frac{\sigma}{\sqrt{n}} = \pm 1.96\left(\frac{10}{5}\right) = \pm 3.92$ units of μ with 95% reliability. (Note that we did not ask: "How 'accurately' has μ been estimated?" The *accuracy* of our estimate is $\bar{X} - \mu$. But since μ is unknown, the degree of accuracy cannot be determined—we can only offer a level of precision of our estimate of μ.)

There is actually a third way to explain our result. We can interpret our confidence interval findings in terms of the error bound concept, that is, we may be 95% confident that $\bar{X}$ will not differ from μ by more than ± 3.92 units.

Example 8.4

Suppose X is $N(\mu, 8)$. Find a 99% confidence interval for μ when $n = 16$ and $\bar{X}$ was found to be 52. Since $1 - \alpha = 0.99$, $\alpha = 0.01$ and thus $\alpha/2 = 0.005$ (See Fig. 8.8).

Then from 8.7,

$$L_1 = 52 - 2.58(8/4) = 46.84$$

$$L_2 = 52 + 2.58(8/4) = 57.16$$

or $(L_1, L_2) = (46.84, 57.16)$. Hence we may be 99% confident that the interval from 46.84 to 57.16 contains μ; or, in terms of precision, we are within $\pm 2.88(8/4) = \pm 5.16$ units of μ with 99% reliability; or, using the error bound notion, we are 99% confident that $\bar{X}$will not differ from μ by more than ± 5.16 units.

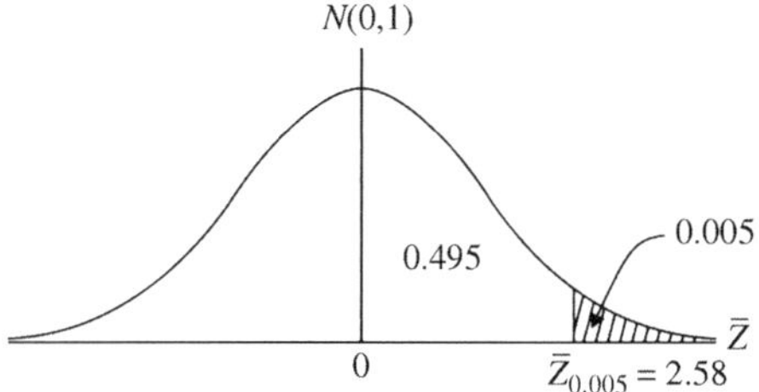

FIGURE 8.8 (1) $\alpha/2 = 0.005$ is the area under the right-hand tail of the $N(0, 1)$ curve. Thus the remaining area to the right of zero is 0.495; (2) Find 0.495 in the body of the $N(0, 1)$ table and read it in reverse to find $\bar{Z}_{0.005} = 2.58$.

What influences the width of the confidence interval (L_1, L_2)? Let us denote the width of (L_1, L_2) as

$$w = L_2 - L_1 = 2\bar{Z}_{\alpha/2}\frac{\sigma}{\sqrt{n}}. \tag{8.8}$$

In general, w varies "directly" with $1-\alpha$ (since the magnitude of $1-\alpha$ determines $\bar{Z}_{\alpha/2}$) and σ; it varies "inversely" with $\sqrt{n}$. Hence the larger is $1-\alpha$ and σ, the wider is the confidence interval for μ; and the larger is $\sqrt{n}$, the narrower is the confidence interval for μ.

8.3 A SAMPLE SIZE REQUIREMENTS FORMULA

Let us arrange Equation (8.8) so as to obtain

$$n = \left(\frac{\bar{Z}_{\alpha/2}\sigma}{w/2}\right)^2. \tag{8.9}$$

What is the significance of this expression? Since our goal is to estimate the population mean μ, we should not simply start the estimation process by collecting some arbitrary number of observations on X and then determine either a point or interval estimate of μ. Sampling is not costless—there may be a sizable expenditure, stated in terms of dollars or effort or time, involved in collecting data. In this regard, even before any data collection is undertaken, we should have in mind certain target levels of precision and reliability that we would like to attain. This is where Equation (8.9) comes into play. It is called a *sample size requirements formula* because of the *a priori* requirements of precision and reliability imposed on the sampling process. Hence Equation (8.9) gives us the sample size required for a degree of precision of $\pm w/2$ with $100(1-\alpha)\%$ reliability.

Example 8.5

Let $\sigma = 10$. How large of a sample should be taken so that the 95% confidence interval for μ will not be greater than 6 units in width or $w \leq 6$? Here $1-\alpha = 0.95$, $\alpha = 0.05$, and thus $\alpha/2 = 0.025$. Then $\bar{Z}_{0.025} = 1.96$ and $w/2 = 3$. From Equation (8.9),

$$n = \left(\frac{1.96x10}{3}\right)^2 = 42.68 \approx 43$$

(Note that we should "round up" since n will seldom be an integer.) Hence a sample of size 43 will meet our requirements, that is, 43 is the sample size required for a degree of precision of ± 3 units with 95% reliability.

Suppose we wanted to be "twice as precise" in estimating μ, that is, now $w/2 = 1.5$. How would n have to be adjusted? As a "rule of thumb," being twice as precise requires that we "quadruple" the sample size. This is easily verified since now

$$n = \left(\frac{1.96 \times 10}{1.5}\right)^2 = 170.74 \approx 171,$$

where $171 = 4 \times 42.68$. In general,

$$n' = \left(\frac{\bar{Z}_{\alpha/2}\sigma}{\frac{1}{2}(w/2)}\right)^2 = 4\left(\frac{\bar{Z}_{\alpha/2}\sigma}{(w/2)}\right)^2 = 4\text{n}.$$

8.4 A CONFIDENCE INTERVAL FOR THE POPULATION MEAN μ (σ UNKNOWN)

In the discussion that follows, we shall assume that we are sampling from a normal population. We determined above that a $100(1-\alpha)\%$ confidence interval for the population mean μ is $\bar{X} \pm \bar{Z}_{\alpha/2}\frac{\sigma}{\sqrt{n}}$ when σ is known. But if σ is unknown, it must be estimated by

$$s = \sqrt{\frac{\sum (X_i - \bar{X})^2}{n-1}} = \sqrt{\frac{\sum X_i^2}{n-1} - \frac{(\sum X_i)^2}{n(n-1)}}^{1}.$$

Additionally, if n is "small" ($n \leq 30$), then the statistic

$$t = \frac{\bar{X} - \mu}{s/\sqrt{n}} \tag{8.10}$$

follows a *t distribution with* $n-1$ *degrees of freedom*. So once s replaces σ and n is small, the resulting statistic [Eq. (8.10)] is no longer $N(0, 1)$.

[1] We divide $\sum (X_i - \bar{X})^2$ by $n-1$ to obtain s^2 as an unbiased estimator for σ^2, that is, $E(s^2) = \sigma^2$ (under repeated sampling from the same population, s^2 is, on the average, equal to σ^2). Had we divided $\sum (X_i - \bar{X})^2$ by n so as to obtain s_1^2, then s_1^2 would be a biased estimator for σ^2, with the amount of bias equal to $-\sigma^2/n$, that is, $E(s_1^2) = \sigma^2 - \sigma^2/n$. This said, it can be shown that s^2/n is an unbiased estimator for the variance of $\bar{X}$ (i.e., for $\sigma^2{}_{\bar{X}}$).

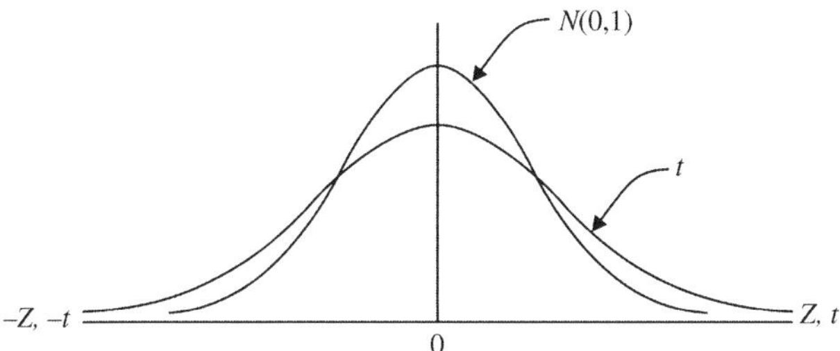

FIGURE 8.9 The t distribution compared with the $N(0,1)$ distribution.

Let us examine the properties of the sampling distribution of the t statistic [Eq. (8.10)]:

1. It is symmetrical about a mean of zero and asymptotic to the horizontal axis.
2. There is a different t distribution for each sample size, that is, t varies with $n - 1$. This dependency on $n - 1$ is termed *degrees of freedom*.
3. $V(t) > 1$ (Fig. 8.9) and $V(t) \to 1$ as $n \to \infty$, that is, $t \to N(0,1)$ as $n \to \infty$.

Hence the t distribution resembles the standard normal distribution; it is a bit more variable than the standard normal distribution (since $V(t) > 1$ for finite n) but is virtually indistinguishable from the standard normal curve for large n. In fact, as $n \to \infty$, the t distribution collapses onto the $N(0,1)$ distribution.

Let us consider the percentage points of the t distribution. Given the probability statement $P(t \geq t_{\alpha,n-1}) = \alpha$, it should be evident that $t_{\alpha,n-1}$, the *upper percentage point for the t distribution*, is the point on the positive t-axis such that the probability of a larger value is α, where α constitutes the area in the right-hand tail of the t distribution (Fig. 8.10a). Similarly, given $P(t \leq -t_{\alpha,n-1}) = \alpha$, the *lower percentage point for the t distribution*, $-t_{\alpha,n-1}$, is the point on the negative t-axis such that the probability of a smaller value is α. Now, α is an area under the left-hand tail of the t distribution (Fig. 8.10b). And for the probability statement

$$P(-t_{\alpha/2,n-1} \leq t \leq t_{\alpha/2,n-1}) = 1 - \alpha, \tag{8.10}$$

(now α is divided between the two tails of the t distribution), $t_{\alpha/2,n-1}$ and $-t_{\alpha/2,n-1}$ are, respectively, the *upper and lower $\alpha/2$ percentage points of the t distribution* (Fig. 8.10c).

For instance, from Table A.2:

$t_{\alpha,n-1} = t_{0.05,6} = 1.943$; $-t_{\alpha,n-1} = -t_{0.01,20} = -2.528$;

$t_{\alpha/2,n-1} = t_{0.10,7} = 1.415$; $t_{\alpha/2,n-1} = t_{0.005,10} = 3.169$; and $t_{\alpha/2,n-1} = t_{0.025,8} = 2.306$.

Given that a $100(1 - \alpha)\%$ confidence interval for the population mean μ has the general form $P_{CI}(L_1 \leq \mu \leq L_2) = 1 - \alpha$, we may employ Equation (8.10) to obtain

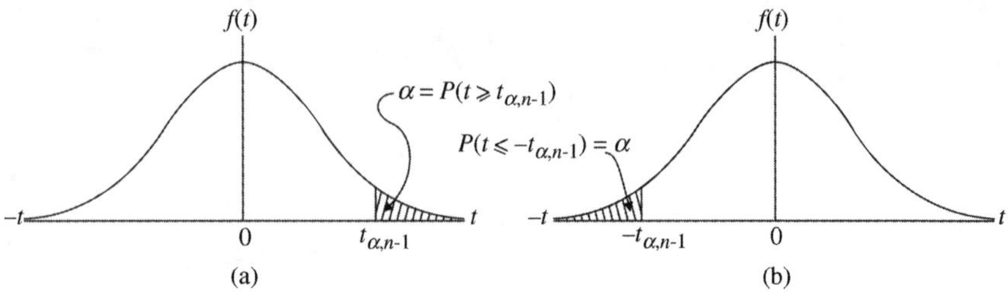

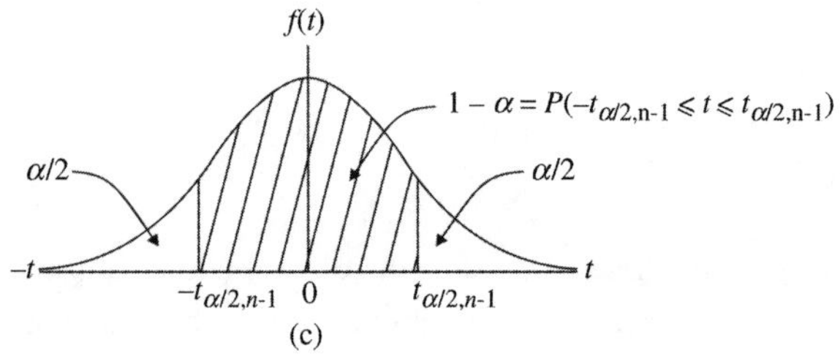

FIGURE 8.10 Percentage points of the t distribution.

$$P\left(-t_{\alpha/2,n-1} \leq \frac{\bar{X}-\mu}{s/\sqrt{n}} \leq t_{\alpha/2,n-1}\right) = 1-\alpha$$

or

$$P\left(\underbrace{\bar{X} - t_{\alpha/2,n-1}\frac{s}{\sqrt{n}}}_{L_1} \leq \mu \leq \underbrace{\bar{X} + t_{\alpha/2,n-1}\frac{s}{\sqrt{n}}}_{L_2}\right) = 1-\alpha. \tag{8.11}$$

Once the sample realizations of $\bar{X}$ and s are substituted into Equation (8.11), we obtain a $100(1-\alpha)\%$ confidence interval (L_1, L_2) for the population mean μ, that is, we may be $100(1-\alpha)\%$ confident that the interval from L_1 to L_2 contains μ. In sum, given than σ is unknown and n is small, a $100(1-\alpha)\%$ confidence interval for μ is

$$\bar{X} \pm t_{\alpha/2,n-1}\frac{s}{\sqrt{n}}. \tag{8.12}$$

Example 8.6

Suppose that from a random sample of size $n=25$ it is determined that $\bar{X}=16$ and $s=8$. Find a 95% confidence interval for μ. Since $1-\alpha=0.95$, $\alpha=0.05$, and thus $\alpha/2=0.025$. Then $t_{\alpha/2,n-1} = t_{0.025,24} = 2.064$ (Table A.2). A substitution of our sample information into Equation (8.12) gives us

$$L_1 = \bar{X} - t_{\alpha/2,n-1}\frac{s}{\sqrt{n}} = 16 - 2.064\left(\frac{8}{\sqrt{25}}\right) = 12.69$$

$$L_2 = \bar{X} + t_{\alpha/2,n-1}\frac{s}{\sqrt{n}} = 16 + 20.64\left(\frac{8}{\sqrt{25}}\right) = 19.30.$$

Hence we may be 95% confident that the interval from 12.69 to 19.30 [or $(L_1, L_2) = (12.69, 19.30)$] contains μ. How precisely have we estimated μ? We are within $\pm 20.64\left(\frac{8}{5}\right) = \pm 3.30$ units of μ with 95% reliability. This result may be stated in terms of the error bound concept: we are 95% confident that $\bar{X}$ will not differ from μ by more than ± 3.30 units.

Our approach to the confidence interval estimation of μ can now be summarized in Fig. 8.11.

Example 8.7

Last week's profit/sale values (in hundreds of \$) selected from a random sample of invoices were: 2.1, 3.0, 1.2, 6.2, 4.5, and 6.1. Find a 99% confidence interval for average profit/sale. Here $n = 6$, $\sum X_i = 22.1$, $\sum X_i^2 = 99.55$, $\bar{X} = 3.683$, $s = 1.905$, and, for $1 - \alpha = 0.99$, $t_{0.005,5} = 4.032$. Then from Equation (8.12), we can readily determine

$$3.683 \pm 4.032\left(\frac{1.905}{\sqrt{6}}\right) \text{ or } 3.683 \pm 3.14.$$

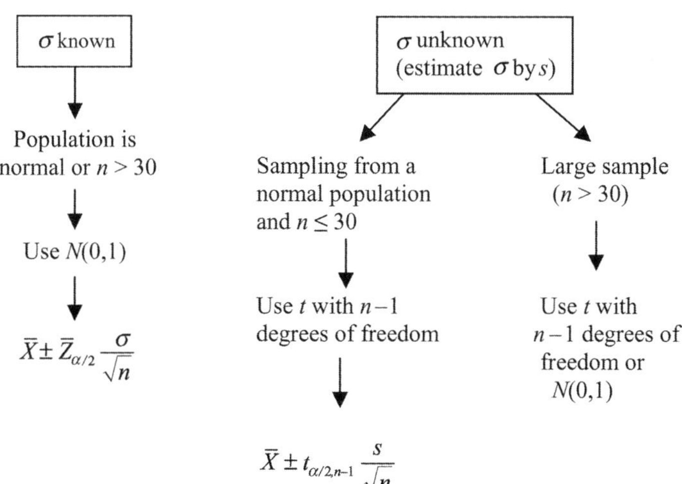

FIGURE 8.11 Finding a $100(1-\alpha)\%$ confidence interval for μ when sampling from a $N(\mu, \sigma)$ population.

so that $(L_1, L_2) = (0.543, 6.823)$, that is, we may be 99% confident that the population average profit/sale value lies between \$54.30 and \$682.50. What assumption are we implicitly making about the population of profit/sale figures?

A key assumption underlying the use of the t distribution to determine a confidence interval for μ when σ is unknown is that we are sampling from a normal population. In this circumstance the t statistic is symmetrical. However, as a practical matter, no actual population can be expected to be "exactly" normal. In fact, departures from normality typically occur because the population distribution is skewed or has heavy tails. (*Heavy tailed distributions* are fairly symmetrical but display outliers or an inordinate amount of data in the tails when compared to a normal distribution.) Given this caveat, what is the implication of the lack of normality for confidence interval estimation using the t statistic? When sampling from a population distribution that is highly skewed or has heavy tails and n is small, the sampling distribution of the t statistic is not symmetrical but is highly skewed. Hence using the t distribution to determine a confidence interval for μ is inappropriate. So in the case of heavy-tailed distributions or distributions exhibiting substantial skewness, we should not rely on Equation (8.12) to estimate the population mean. Appendix 8.A provides us with an alternative estimation procedure.

In view of the results presented in Appendix 8.A, we may modify Fig. 8.11 as follows (Fig. 8.12):

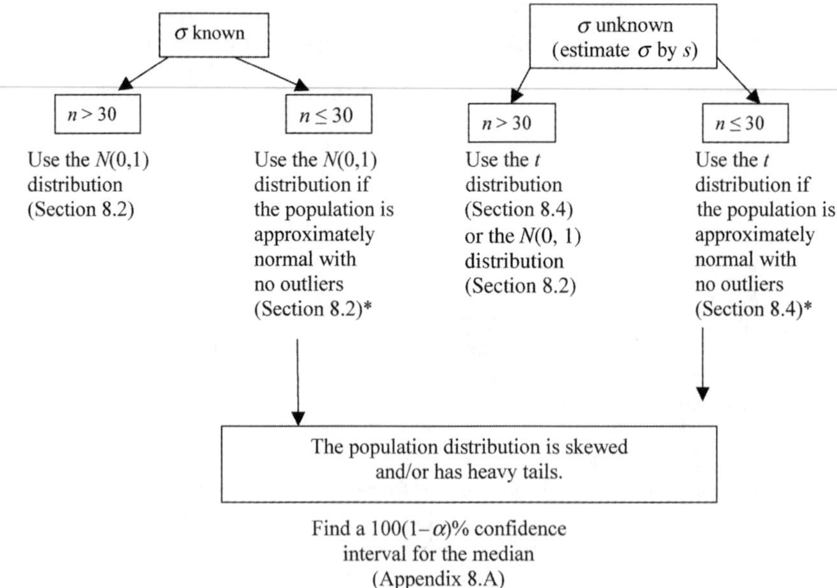

FIGURE 8.12 Finding a $100(1-\alpha)\%$ confidence interval for μ.

EXERCISES

1 Given a sample of size $n = 35$ from a $N(\mu,10)$ population, the sample mean was found to be $\bar{X} = 15$. Find the following:

a. A 95% confidence interval for μ

b. A 90% confidence interval for μ

c. A 99% confidence interval for μ

2 Given samples from a normal population with known variance, find the following:

a. The confidence coefficient if $n = 16$, $\sigma^2 = 64$, and the total width of the confidence interval for the mean is 3.29 units.

b. The sample size when $\sigma^2 = 100$ and the 95% confidence interval for the mean is 17.5–22.5.

c. The known variance when $n = 100$ and the 99% confidence interval for the mean is 23.26 units in width.

3 Assuming random samples from known populations having known standard deviations, find confidence intervals for the population means having the specified degree of confidence:

a. $n = 9$, $\bar{X} = 20$, $\sigma = 3$, $1 - \alpha = 0.95$

b. $n = 16$, $\bar{X} = 52$, $\sigma = 8$, $1 - \alpha = 0.99$

c. $n = 25$, $\bar{X} = 120$, $\sigma = 20$, $1 - \alpha = 0.90$

4 A recent poll asked college juniors how many times they actually sat down and wrote a letter home during the previous academic year. How many students are needed to estimate the average number of letters written last year to within one letter with 95% reliability? (A pilot study revealed that $\sigma = 15.4$ letters.)

5 A recent poll asked high school students how much time they spend each week texting. How many students are required to estimate average texting time per week to within 0.5 hr with 90% reliability if a pilot study indicated that $\sigma = 6.5$ hr?

6 Find the t-value such that:

a. The area in the right tail is 0.10 with 25 degrees of freedom.

b. The area in the right tail is 0.05 with 30 degrees of freedom.

c. The area to the left of the t-value is 0.01 with 18 degrees of freedom.

d. The area to the left of the t-value is 0.05 with 6 degrees of freedom.

e. The area in the right tail is 0.02 with 19 degrees of freedom.
f. The area in the right tail is 0.10 with 32 degrees of freedom.

7 Find the critical value of t that corresponds to the following

a. 90% confidence with 20 degrees of freedom.
b. 95% confidence with 16 degrees of freedom.

8 Find the value of t such that:

a. The probability of a larger value is 0.005 when degrees of freedom is 27.
b. The probability of a smaller value is 0.975 when degrees of freedom is 14.
c. The probability of a larger value (sign ignored) is 0.90 when degree of freedom is 40.

9 Assuming unknown population variances, find 95% confidence intervals for the means based on the following sets of sample values:

a. $n=9,\ \sum X_i = 36,\ \sum (X_i - \bar{X})^2 = 288$
b. $n=16,\ \sum X_i = 64,\ \sum (X_i - \bar{X})^2 = 180$
c. $n=16,\ \sum X_i = 320, \sum X_i^2 = 6{,}640$
d. $n=25,\ \sum X_i = 500, \sum X_i^2 = 12{,}400$

10 Assuming unknown population variances, find 99% confidence limits for the means based on the following data sets:

a. 2, 10, 4, 8, 4, 14, 8, 12, 10
b. 1, 3, 5, 6, 6, 3

11 A sample of nine plots of land had a mean yield of 110 g and an estimated standard deviation of 12 g. Find a 98% confidence interval for the population mean yield.

12 The mean weight gain of 15 head of cattle fed a given growth hormone during a 30-day trial period was 40 lbs. The coefficient of variation was found to be 10%. Find a 95% confidence interval for the true mean weight gain of the cattle.

13 A group of 25 children were given a standard IQ test. They were subsequently introduced into a special program designed to increase one's IQ and then tested at the end of the program. The difference between the second and first scores had a mean equal to 4. The estimated standard deviation was 5. Find a 95%

confidence interval for the true mean difference. Do you think the program was successful in increasing IQs?

14 Profit per sale, tabulated last week from a random sample of invoices for a used car dealer, was (in thousands of \$): 2.1, 3.0, 1.2, 6.2,3.7, 4.5, 5.1. Find a 90% confidence interval for average profit per sale.

15 Given the following random sample of $n = 25$, find a 95% confidence interval for the population median. (For this data set, does the median provide a better measure of central tendency than the mean?) X: 3, 1, 97, 6, 7, 13, 15, 4, 87, 2, 20, 6, 6, 3, 10, 18, 35, 42, 17, 3, 50, 19, 1, 17, 115.

APPENDIX 8.A A CONFIDENCE INTERVAL FOR THE POPULATION MEDIAN *MED*

When a population distribution is highly skewed or has heavy tails, the median is a better measure of central location than the mean. Moreover, we noted at the end of Section 8.4 that using the t statistic when sampling from such populations is not appropriate and can lead to misleading interval estimates of μ, especially for small samples. Given that the sample median *med* serves as an estimator for the population median *Med*, let us construct a $100(1-\alpha)\%$ confidence interval for *Med* that is centered on *med*. Before doing so, however, we need to examine the notion of an order statistic. This concept will enable us to develop a convenient notational device for specifying a confidence interval for the population median.

Specifically, the *order statistics* associated with the random sample X_1, $X_2, \ldots, X_n$ are the sample values arranged in an increasing sequence and denoted as follows:

$$X_{(1)}, X_{(2)}, \ldots, X_{(n)} \tag{8.A.1}$$

where, for $1 \leq i \leq n$,

$$\begin{aligned} X_{(1)} &= \text{Min}(X_i) \\ X_{(2)} &= \text{Next largest } X_i \\ &\vdots \\ X_{(n)} &= \text{Max } X_i. \end{aligned}$$

For instance, given the seven sample observations on X: 6, 2, 4, 8, 19, 3, 10, the set of order statistics are as follows:

$$X_{(1)} = 2(= \text{Min } X_i),\ \ X_{(2)} = 3,\ \ X_{(3)} = 4,\ \ X_{(4)} = 6,\ \ X_{(5)} = 8,$$

$$X_{(6)} = 10,\ \ X_{(7)} = 19(= \text{Max } X_i).$$

Stated in terms of these order statistics, we may depict the sample median as follows:

$$\text{med} = \begin{cases} X_{((n+1)/2)} & \text{if } n \text{ is odd} \\ \left(X_{(n/2)} + X_{(n/2+1)}\right)/2 & \text{if } n \text{ is even.} \end{cases} \tag{8.A.2}$$

Given the preceding set of $n = 7$ observations, find the sample median. Since n is odd, Equation (8.A.2) renders the following:

$$\text{med} = X_{(8/2)} = X_{(4)} = 6.$$

If we had the sample data set X: 3, 6, 4, 1, 9, 10, then, since $n = 6$ is even, Equation (8.A.2) dictates the following:

$$X_{(1)} = 1,\ \ X_{(2)} = 3,\ \ X_{(3)} = 4,\ \ X_{(4)} = 6,\ \ X_{(5)} = 9,\ \ X_{(6)} = 10$$

$$med = \frac{X_{(3)} + X_{(4)}}{2} = \frac{4+6}{2} = 5$$

(we average the values of $X_{(3)}$ and $X_{(4)}$).

Suppose we have a random sample consisting of the n values $X_1, X_2, \ldots, X_n$. To determine a confidence interval for the population median, let us first form the set of order statistics $X_{(1)}, X_{(2)}, \ldots, X_{(n)}$. Then applying a normal approximation to the binomial distribution with $p = 0.5$ enables us to form, for $n > 10$, an approximate $100(1-\alpha)\%$ confidence interval for the population median as

$$L_1 = X_{(k)} \le Med \le X_{(n-k+1)} = L_2, \tag{8.A.3}$$

where

$$k = \frac{n - Z_{\alpha/2}\sqrt{n} - 1}{2} \tag{8.A.4}$$

and L_1 and L_2 denote the lower and upper confidence limits, respectively. Once k is determined, we "round out" to $X_{(k)}$ and $X_{(n-k+1)}$. Note that since the binomial distribution is discrete, the exact confidence level will be a bit larger than $100(1-\alpha)\%$. So once L_1 and L_2 are determined, we may conclude that we are "at least" $100(1-\alpha)\%$ confident that the population median lies within the limits L_1 and L_2.

Example 8.A.1

Suppose that a random sample of size $n = 20$ on a variable X results in the following set of order statistics:

$$2, 3, 5, 5, 7, 8, 9, 9, 10, 15, 19, 24, 25, 27, 29, 36, 45, 81, 91, 100.$$

Here $\bar{X} = 540/20 = 27$ while, from Equation (8.A.2),

$$med = \frac{X_{(10)} + X_{(11)}}{2} = \frac{15 + 19}{2} = 17.$$

Clearly, the value of $\bar{X}$ has been inflated by some large (extreme) values since 13 of the data points lie below $\bar{X}$ and only 6 lie above it. Clearly, the median in this instance is a much better indicator of centrality than the mean. For $1 - \alpha = 0.95$, Equation (8.A.4) yields

$$k = \frac{20 - 1.96\sqrt{20} - 1}{2} = 10.23461/2 = 5.12,$$

so that, via Equation (8.A.3) (after rounding out),

$$X_{(k)} = X_{(5)} = 7$$
$$X_{(n-k+1)} = X_{(20-5.12+1)} = X_{(16)} = 36.$$

Hence we are at least 95% confident that the population median lies between $L_1 = 7$ and $L_2 = 36$.

9

THE SAMPLING DISTRIBUTION OF A PROPORTION AND ITS CONFIDENCE INTERVAL ESTIMATION

9.1 THE SAMPLING DISTRIBUTION OF A PROPORTION

Our goal in this section is to estimate the parameter p of a binomial population. Remember that a binomial population is one whose elements belong to either of two classes—success or failure. Previously, the binomial random variable X was defined as the number of successes obtained in n independent trials of a simple alternative experiment and p denoted the probability of a success. Now, X will be taken to be the number of successes observed in a simple random sample of size n and p will represent the proportion of successes in the population. (Obviously, $1-p$ is the population proportion of failures.)

The problem that we now face is the estimation of p. Also, since we are basing our estimate of p on sample information, we need to obtain some idea of the error involved in estimating p. In order to adequately address these issues, we need to examine the characteristics of the sampling distribution of the sample proportion. To this end, let us extract a random sample of size n from a binomial population and find the *observed relative frequency of a success*

$$\hat{p} = \frac{X}{n} = \frac{\text{Observed number of success}}{n}. \tag{9.1}$$

Statistical Inference: A Short Course, First Edition. Michael J. Panik.

As we shall see shortly, $\hat{p}$ serves as our "best" estimate for p. Now, we know that $\hat{p}$ is a random variable that varies under random sampling, depending upon which sample is chosen. And since $\hat{p}$ is a random variable, it has a probability distribution called the *sampling distribution of the sample proportion* $\hat{\mathbf{p}}$—a distribution showing the probabilities (relative frequencies) of getting different sample proportions ($\hat{p}_j$'s) from random samples of size n taken from a binomial population.

Why study the sampling distribution of $\hat{p}$? How shall we view its role? It shows how proportions vary, due to chance, under random sampling. So by studying the sampling distribution of $\hat{p}$, we can learn something about the error arising when the sample proportion determined from a given sample is used to estimate p. To assess the characteristics of interest of the sampling distribution of $\hat{p}$, we need to start by examining its mean and variance. In what follows, we shall assume that n is large.

Let us first consider the mean of the random variable $\hat{p}$. It can be shown that

$$E(\hat{p}) = p, \tag{9.2}$$

that is, the average value of $\hat{p}$ taken over all possible random samples of size n, equals p. Hence $\hat{\mathbf{p}}$ *is said to be an unbiased estimator for p* (since "on the average", $\hat{p}$ is on target). In this regard, the sampling distribution of $\hat{p}$ is centered on p (Fig. 9.1). Additionally, it can be demonstrated that the *variance of $\hat{p}$ (with p known)* is

$$V(\hat{p}) = \sigma_{\hat{p}}^2 = \frac{p(1-p)}{n}. \tag{9.3}$$

Then the *standard error of* $\hat{\mathbf{p}}$ *(with p known)* is

$$\sigma_{\hat{p}} = \sqrt{\frac{p(1-p)}{n}}. \tag{9.4}$$

Hence $\sigma_{\hat{p}}$ is the standard deviation of the sampling distribution of p (with p known), that is, on the average, the individual $\hat{p}_j$'s are approximately $\sigma_{\hat{p}}$ units away from p.

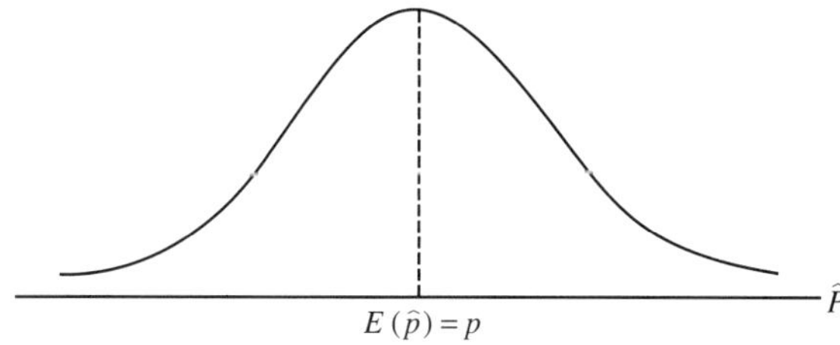

FIGURE 9.1 $\hat{p}$ is an unbiased estimator for p.

Since p is typically unknown, it will be estimated by $\hat{p}$. Then an *unbiased estimator for* $\sigma_{\hat{p}}^2$ is

$$s_{\hat{p}}^2 = \frac{\hat{p}(1-\hat{p})}{n}, \tag{9.5}$$

which is termed the *estimated variance of* $\hat{\mathbf{p}}$; it is the estimated variance of the sampling distribution of $\hat{p}$. Given that $\hat{p}-p$ represents the *sampling error* connected with using $\hat{p}$ to estimate p, we can view $s_{\hat{p}}^2$ as the average squared sampling error arising when $\hat{p}$ is used to estimate p. Then

$$s_{\hat{p}} = \sqrt{\frac{\hat{p}(1-\hat{p})}{n}} \tag{9.6}$$

is *the estimated standard error of* $\hat{p}$.

What about the form of the sampling distribution of $\hat{p}$? Since $\hat{p}$ is actually the "average number of successes" in a sample of size n, we can gain some insight into the form of the sampling distribution of $\hat{p}$ by invoking the Central Limit Theorem: for sufficiently large n, the *standardized observed relative frequency of a success,*

$$U = \frac{\hat{p}-p}{\sigma_{\hat{p}}} = \frac{\hat{p}-p}{\sqrt{\dfrac{p(1-p)}{n}}} \tag{9.7}$$

is approximately $N(0,\ 1)$. In fact, as noted earlier, the approximation works well provided $np(1-p) > 10$. Hence we can use the standard normal area table to calculate probabilities involving $\hat{p}$. For instance, suppose a random sample of size $n = 100$ is taken from an infinite (binomial) population with $p = 0.72$. Then from the above discussion, we know that

$$E(\hat{p}) = p = 0.72, \ \ \sigma_{\hat{p}} = \sqrt{\frac{0.72(1-0.72)}{100}} = 0.045.$$

What is the probability of finding at least $X = 76$ successes in the sample? To answer this, we first determine that $np(1-p) = 20.16 > 10$. Hence we can legitimately calculate

$$\begin{aligned} P(X \geq 76) &= P\left(\frac{X}{n} \geq \frac{76}{100}\right) = P(\hat{p} \geq 0.76) \\ &= P\left(\frac{\hat{p}-p}{\sigma_{\hat{p}}} \geq \frac{0.76-0.72}{0.045} = 0.88\right) = P(U \geq 0.88) \\ &= 0.5 - 0.3106 = 0.1894 \end{aligned}$$

via the $N(0,\ 1)$ area table.

As a practical matter, how is the sampling distribution of the proportion $\hat{p}$ applied? As will be demonstrated below, it can be used to obtain an error bound on the sample proportion as an estimate of the population proportion. Additionally, under random sampling, only rarely will $\hat{p}$ equal p exactly. Hence the sampling distribution of $\hat{p}$ can be used to provide us with probability information about the sampling error $\hat{p} - p$ incurred when $\hat{p}$ is used to estimate p.

9.2 THE ERROR BOUND ON $\hat{p}$ AS AN ESTIMATOR FOR p

In what follows, we shall assume that n is small relative to N or $n < 0.05N$. (Why?) Also, for n sufficiently large, the standard normal probabilities will render a satisfactory approximation to probabilities involving $\hat{p}$ for any p not too close to 0 or 1 or, as noted above, for $np(1-p) > 10$.

Given that $\hat{p}$ serves as an estimator for p, a *95% error bound on $\hat{p}$ as an estimator for p* is

$$\pm 1.96 s_{\hat{p}} \text{ or } \pm 1.96\sqrt{\hat{p}(1-\hat{p})/n}. \tag{9.8}$$

Once this error bound or maximum tolerable error value is determined, our conclusion is: we are 95% confident that $\hat{p}$ will not differ from p by more than $\pm 1.96\ s_{\hat{p}}$. Since U [see Equation (9.7)] is standard normal, the 1.96 value is obtained from the $N(0, 1)$ area table. The reader can easily demonstrate that 90% and 99% error bounds on $\hat{p}$ as an estimator for p are $\pm 1.645 s_{\hat{p}}$ and $\pm 2.58\ s_{\hat{p}}$, respectively.

Example 9.1

In an interview of $n = 1500$ individuals, $X = 527$ answered "yes" to the question: "Are you satisfied with the overall performance of your congressman?" What is the best estimate of the proportion p of persons in the population who would offer an affirmative response? What error is involved in estimating p? From $\hat{p} = X/n = 527/1500 = 0.3513$, we can obtain the following:

$$s_{\hat{p}} = \sqrt{\frac{0.3513(0.6487)}{1500}} = 0.0116$$

and thus, via Equation (9.8),

$$95\% \text{ error bound}: \pm 1.96\ (0.0116) = \pm 0.0227,$$

that is, we may be 95% confident that $\hat{p}$ will not differ from p by more than $\pm 2.27\%$.

Next, suppose X is the observed number of successes in a sample of size n taken from a binomial population. What is the probability that $\hat{p} = X/n$ will fall within $\pm a$ units of p?

Here we are interested in finding $P(p-a \leq \hat{p} \leq p+a)$. But this is equivalent to the probability that the sampling error $\hat{p}-p$ lies between $\pm a$ or $P(-a \leq \hat{p}-p \leq a)$. Then transforming to a probability statement involving U enables us to find

$$P\left(\frac{-a}{s_{\hat{p}}} \leq \frac{\hat{p}-p}{s_{\hat{p}}} \leq \frac{a}{s_{\hat{p}}}\right) = P\left(\frac{-a}{s_{\hat{p}}} \leq U \leq \frac{a}{s_{\hat{p}}}\right). \tag{9.9}$$

Example 9.2

What is the probability that $\hat{p}=0.47$ determined from a sample of size $n=250$ taken from a binomial population will be within $\pm 3\%$ of p? Since

$$s_{\hat{p}} = \sqrt{\frac{0.47(0.53)}{250}} = 0.032,$$

$$P(p-0.03 \leq \hat{p} \leq p+0.03) = P(p-0.03 \leq \hat{p} \leq p+0.03)$$

$$= P\left(\frac{-0.03}{0.032} \leq \frac{\hat{p}-p}{s_{\hat{p}}} \leq \frac{0.03}{0.032}\right) = P(-0.94 \leq U \leq 0.94)$$

$$= 2(0.3264) = 0.6528.$$

9.3 A CONFIDENCE INTERVAL FOR THE POPULATION PROPORTION (OF SUCCESSES) p

As an alternative to simply using $\hat{p}$ (determined from a simple random sample of size n taken from a binomial population) as a point estimator for p, let us determine a whole range of possible values that p could assume. This range of values is an *interval estimate* or *confidence interval for p*, that is, it is a range of values that allows us to state just how confident we are that the reported interval contains p. Once this interval is obtained, it enables us to state just how precisely p has been estimated from the sample; the narrower the interval, the more precise the estimate.

As will be indicated below, a confidence interval "generalizes the error bound concept." In this regard, the confidence limits bounding p will be expressed as

$$\hat{p} \pm \text{error bound}, \tag{9.10}$$

where the term $\pm$ error bound serves as our *degree of precision.*

To construct a confidence interval for p, we need to find two quantities L_1 and L_2 such that, before any sampling is undertaken,

$$P_{CI}(L_1 \leq p \leq L_2) = 1-\alpha, \tag{9.11}$$

where L_1 and L_2 are lower and upper confidence limits for p, respectively, and $1-\alpha$ is the *confidence probability*, which is chosen in advance. Once $1-\alpha$ is specified and L_1

and L_2 are calculated, our conclusion concerning p is framed as: we may be $100(1-\alpha)\%$ confident that the interval from L_1 to L_2 contains p.

To determine the $100(1-\alpha)\%$ confidence interval (L_1, L_2) for p, let us state, on the basis of Equation (9.7), that for sufficiently large n, the quantity

$$U = \frac{\hat{p} - p}{s_{\hat{p}}} = \frac{\hat{p} - p}{\sqrt{\frac{\hat{p}(1-\hat{p})}{n}}} \tag{9.12}$$

is approximately $N(0, 1)$. Then

$$P\left(-U_{\alpha/2} \leq U \leq U_{\alpha/2}\right) = 1 - \alpha, \tag{9.13}$$

where $-U_{\alpha/2}$ and $U_{\alpha/2}$ are, respectively, *lower and upper α/2 percentage points* for the $N(0, 1)$ distribution (Fig. 9.2).

Given Equation (9.13), let us insert the definition of U [see Equation (9.12)] into this equation so as to obtain the following:

$$P\left(-U_{\alpha/2} \leq \frac{\hat{p} - p}{s\hat{p}} \leq U_{\alpha/2}\right) = 1 - \alpha$$ [Multiply each term in parentheses by $s_{\hat{p}}$.]

$$P\left(-U_{\alpha/2}s_{\hat{p}} \leq \hat{p} - p \leq U_{\alpha/2}s_{\hat{p}}\right) = 1 - \alpha$$ [Subtract $\hat{p}$ from each term in parentheses.]

$$P(-\hat{p} - U_{\alpha/2}s_{\hat{p}} \leq -p \leq -\hat{p} + U_{\alpha/2}s_{\hat{p}}) = 1 - \alpha$$ [Multiply each term in parentheses by –1.]

$$P(\underbrace{\hat{p} - U_{\alpha/2}s_{\hat{p}}}_{L_1} \leq p \leq \underbrace{\hat{p} + U_{\alpha/2}\, s_{\hat{p}}}_{L_2}) = 1 - \alpha \tag{9.14}$$

where $s_{\hat{p}}$ is determined from Equation (9.6). Clearly, this probability statement mirrors Equation (9.11). Hence a *100(1 – α)% confidence interval for p* is

$$\hat{p} \pm U_{\alpha/2}s_{\hat{p}}. \tag{9.15}$$

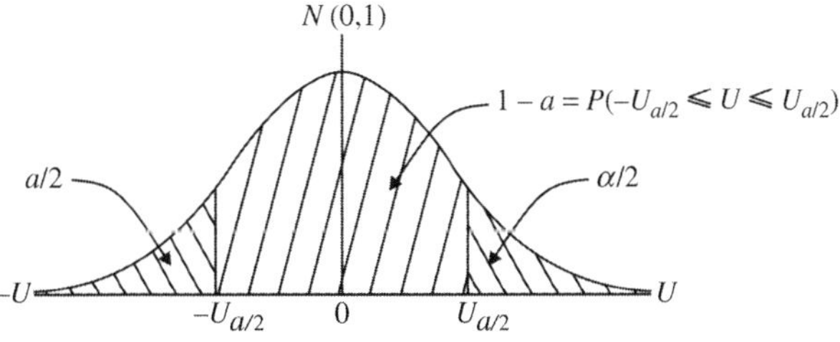

FIGURE 9.2 $U_{\alpha/2}$ is an upper $\alpha/2$ percentage point for $N(0, 1)$; $-U_{\alpha/2}$ is a lower $\alpha/2$ percentage point for $N(0, 1)$.

So when minus holds, we get the lower confidence limit L_1; and when plus holds, we obtain the upper confidence limit L_2.

In Equation (9.14) $1-\alpha$ is a confidence probability since $\hat{p}$ is a random variable. But once the sample realization of $\hat{p}$ is obtained ($\hat{p}$ is no longer random), $1-\alpha$ is no longer a probability—it is the *100(1 – α)% confidence coefficient or level of reliability,* with $U_{\alpha/2}$ serving as the *100(1 – α)% reliability coefficient.*

Example 9.3

Suppose that in a sample size of $n=100$ it is found that $X=64$ items display a particular characteristic. Find a 95% confidence interval for p, the proportion of the population having this characteristic. Here $\hat{p}=X/n=0.64$, $s_{\hat{p}}=\sqrt{\hat{p}\,(1-\hat{p})/n}=0.048$, and $1-\alpha=0.95$. Then $\alpha/2=0.025$ and, since U is $N(0, 1)$, $U_{\alpha/2}=U_{0.025}=1.96$ (a review of Fig. 8.7 may be helpful). Then from Equation (9.15),

$$L_1=\hat{p}-U_{\alpha/2}\,s_{\hat{p}}=0.64-1.96(0.048)=0.546$$

$$L_2=\hat{p}+U_{\alpha/2}\,s_{\hat{p}}=0.64+1.96(0.048)=0.734.$$

Hence we may be 95% confident that the interval from 0.546 to 0.734 contains p.

How precisely have we estimated p? We are within $\pm U_{\alpha/2}\,s_{\hat{p}}=\pm 1.96\,(0.048)=\pm 0.094$ or $\pm 9.4\%$ of p with 95% reliability. Alternatively, we can invoke the error bound concept and conclude that we are 95% confident that $\hat{p}$ will not differ from p by more than $\pm 9.4\%$.

What influences the width of the confidence interval for p? From Equation (9.15),

$$w=L_2-L_1=2U_{\alpha/2}\sqrt{\frac{\hat{p}(1-\hat{p})}{n}}. \tag{9.16}$$

Thus w varies "directly" with $1-\alpha$ (since the size of $1-\alpha$ determines $U_{\alpha/2}$); it varies "inversely" with $\sqrt{n}$. So the larger is $1-\alpha$, the wider is (L_1, L_2); and the larger is $\sqrt{n}$, the narrower is (L_1, L_2).

9.4 A SAMPLE SIZE REQUIREMENTS FORMULA

Given Equation (9.16), let us solve for n or

$$n=\frac{(U_{\alpha/2})^2\hat{p}(1-\hat{p})}{(w/2)^2}. \tag{9.17}$$

Let us refer to this expression as the *sample size requirements formula*. What are the requirements? As noted above when employing Equation (8.9), they are "precision" and "reliability." Hence Equation (9.17) provides us with the sample size required for a degree of precision of $\pm w/2$ with $100(1-\alpha)\%$ reliability. (Note that $w/2$ should always be expressed as a decimal.)

Example 9.4

Suppose we would like to estimate the proportion of individuals with a particular characteristic to within ±3% of p (the population proportion of individuals displaying this characteristic) with 95% reliability. How large of a sample should be taken? Here $U_{\alpha/2} = U_{0.025} = 1.96$ and $\pm w/2 = \pm 0.03$. Then from Equation (9.17),

$$n = \frac{(1.96)^2 \hat{p}(1-\hat{p})}{(0.03)^2}.$$

Since we do not as yet have our sample, we cannot determine $\hat{p}$. (It seems we have found ourselves "up a tree without a creek.") There are three possible ways to resolve this difficulty:

1. Use a prior estimate of p; or
2. Conduct a small-scale pilot study so as to obtain a preliminary estimate of p; or, if neither of these options are feasible, then
3. Set $\hat{p} = \frac{1}{2}$. ($\hat{p} = \frac{1}{2}$ is the value of $\hat{p}$ at which $s_{\hat{p}}$ is at a maximum. Hence we will obtain a sample size a bit larger than we actually need, but, "that's life.")

Under option 3,

$$n = \frac{(1.96)^2 \frac{1}{2} \cdot \frac{1}{2}}{(0.03)^2} = 1067,$$

a generous sample size. Here 1067 represents the sample size required for a degree of precision of ±0.03 with 95% reliability.

EXERCISES

1 Given the following results obtained from random samples from infinite binomial populations, find the best estimate for $p(\hat{p})$ and the standard error of $\hat{p}$:

a. $n = 25$, $X = 10$
b. $n = 64$, $X = 32$
c. $n = 100$, $X = 64$
d. $n = 400$, $X = 140$

2 Out of a sample of 1000 individuals, 672 gave the answer "yes" to a particular question. What is the best estimate of p, the proportion of those who would respond "yes?" What is the standard error of $\hat{p}$?

3 Assuming a large population, describe the sampling distribution of $\hat{p}$ for each of the following cases:

a. $n = 300, \quad p = 0.07$

b. $n = 500, \quad p = 0.04$

c. $n = 1000, \quad p = 0.10$

d. $n = 100, \quad p = 0.75$

4 Suppose a random sample of size $n = 75$ is extracted from an infinite (binomial) population and the proportion of items in the population with a given property is $p = 0.8$. Describe the sampling distribution of $\hat{p}$. Also, find:

a. The probability of obtaining at least $X = 65$ items with the property.

b. The probability of obtaining fewer than $X = 25$ items with the property.

c. The probability of obtaining between 35 and 60 items with the property.

d. The probability that less than 24% of the items have the property.

e. The probability that greater than 63% of the items have the property.

5 According to a recent department store poll, 7% of its customers do not have a charge card and instead pay by check or cash. Suppose a random sample of 500 customers is obtained. Describe the sampling distribution of $\hat{p}$. Then find the following:

a. The probability that more than 10% of the customers do not have a charge card.

b. The probability that fewer than 50 of the customers do not have a charge card.

6 We have been assuming that our random samples of size n were taken from an infinite (binomial) population or, equivalently, that $n/N \leq 0.05$, where N is the population size. If $n/N > 0.05$, then we need to modify $\sigma_{\hat{p}}$ by introducing the finite population correction $\sqrt{(N - n)/(N - 1)}$ so as to obtain

$$\sigma_{\hat{p}} = \sqrt{\frac{p(1-p)}{n}}\sqrt{\frac{N-n}{N-1}}.$$

If in the preceding problem (Problem 9.5) the sample of 500 is taken from the store's customer list that contains 5,000 names, describe the sampling distribution of $\hat{p}$. Use this new information to rework parts (a) and (b) above.

7 On the opening day of fishing season, 1000 fish were caught in Lake Gearse by rod and reel. It was found that 290 of them were smallmouth bass. Find a 95% confidence interval for the percentage of smallmouth bass among fish in the lake.

8 In a sample of 100 metal castings, 20 were found to have defects resulting from the uneven heating of the mold. Determine a 99% confidence interval for the true proportion of all castings that have defects.

9 In a clinical trial of 850 subjects who received a 10 mg dose of a new wonder drug for fighting diabetes, 48 reported dizziness as a side effect. Find a 95% confidence interval for the true proportion of the drug recipients who reported dizziness.

10 A researcher wishes to estimate the proportion of veterans who are jobless after 3 yr of being discharged from the armed services. What sample size n should be used if he/she wishes the estimate to be within $\pm 3\%$ of the true proportion with 95% reliability and:

a. A prior estimate of $\hat{p} = 0.65$ is available from a pilot study.
b. No prior estimate of $\hat{p}$ is available.

11 Last year, 39% of the individuals polled by a market research firm answered "yes" to a particular question. The margin of error was ± 0.02 and the estimate was made with 99% reliability. How many people were polled?

APPENDIX 9.A RATIO ESTIMATION

A procedure that is akin to that employed to estimate a population proportion p is ratio estimation. It can be used to determine a population ratio proper or to estimate a population mean. *Ratio estimation* involves taking measurements on two characteristics of interest of a sampling unit—one corresponding to the numerator and the other corresponding to the denominator of the ratio. In this regard, let a sample of size n consist of the n ordered pairs of observations $(X_i, Y_i,)$, $i = 1, \ldots, n$, where the sample values $Y_1, \ldots, Y_n$ refer to some measurable characteristic of the sampling unit that appears in the numerator of the ratio, and the sample values $X_1, \ldots, X_n$ pertain to some measurable characteristic of the sampling unit that appears in the denominator of the ratio. In what follows, the X_i's will serve as subsidiary information used for ratio estimation purposes.

At this point, some notation is in order. Let the population ratio appear as follows:

$$R = \frac{\tau_Y}{\tau_X} = \frac{\sum_{j=1}^{N} Y_j}{\sum_{j=1}^{N} X_j} = \frac{N\mu_Y}{N\mu_X} = \frac{\mu_Y}{\mu_X}, \tag{9.A.1}$$

where the population totals for the Y- and X-characteristics are denoted as τ_Y and τ_X, respectively, and μ_Y and μ_X are the respective means for the said characteristics. Then an estimator for R is the *sample ratio*

$$\hat{R} = \frac{Y}{X} = \frac{\sum_{j=1}^{N} Y_i}{\sum_{j=1}^{N} X_i} = \frac{n\bar{Y}}{n\bar{X}} = \frac{\bar{Y}}{\bar{X}}, \tag{9.A.2}$$

where $\bar{Y}$ and $\bar{X}$ are estimators for μ_Y and μ_X, respectively.

How does the sample ratio in Equation (9.A.2) differ from the "ratio" used earlier to estimate a population proportion p? As indicated above, an estimator for p is $\hat{p} = X/n = \sum_{i=1}^{n} X_i/n$, where X_i is taken to be a Bernoulli random variable, that is, $X_i = 0$ or 1, where a "1" is recorded if the sampling unit has some particular characteristic of interest and a "0" is recorded if it does not. The following example will clarify the essential difference between $\hat{R}$ and $\hat{p}$.

Example 9.A.1

In a large apartment complex of $N = 100$ units, some of the apartments have one bedroom while others have two bedrooms. A random sample of size $n = 10$ apartments yielded the following results (Table 9.A.1). Then an estimate of the population proportion of units having one bedroom is $\hat{p} = X/10 = 6/10$. Moreover, if a second random sample of $n = 10$ apartment units was taken, a new estimate of the proportion of apartments with only one bedroom might be, say, 5/10. Notice that in each of these ratios the denominator $n = 10$ is a constant while "only the numerator is the realization of a random variable X."

However, what if we were interested in estimating the number of occupants per bedroom? In this instance, the quantities

$$Y = \sum_{i=1}^{10} Y_i = \text{Total number of occupants}$$

$$X = \sum_{i=1}^{10} X_i = \text{Total number of bedrooms}$$

have realization 21 and 14, respectively. Then the number of occupants per bedroom would be $Y/X = 21/14 = 1.5$ (i.e., for every 10 bedrooms, there are 15 occupants). Here Y/X will be termed a *ratio estimate* of the population number of occupants per bedroom. Note that Y/X is different from $\hat{p} = X/n$; in the former estimator "both the numerator and denominator are random variables." Hence a ratio estimator has a

TABLE 9.A.1 Apartment Characteristics

Apartment Number	Number of Bedrooms (X-Characteristic)	Number of Occupants (Y-Characteristics)
07	1	2
43	1	1
61	2	4
31	1	1
57	2	3
09	1	2
97	1	2
93	1	1
72	2	2
25	2	3
	14	21

denominator that varies from sample to sample and thus a ratio estimator is not a proportion in a strict or narrow sense of the word. In fact, to determine whether a sample proportion or a ratio estimator should be used to determine some unknown population value, one need only ponder the following question: "If different random samples of size n were taken from the same population, would the denominator of Y/X be the same for each sample? If the answer is "no," then ratio estimation involving Y/X is applicable. But if the answer is "yes," then determining a simple sample proportion X/n is warranted.

It is well known that $\hat{R}$ is a biased but consistent estimator of the population ratio R, that is, the bias may be ignored for moderately large samples ($n > 30$). Moreover, the bias is zero when all of the sample points (X_i, Y_i), $i = 1, \ldots, n$, fall on a straight line passing through the origin.

Next, it can be demonstrated that the variance of $\hat{R}$ may be expressed as follows:

$$V(\hat{R}) = E(\hat{R} - E(\hat{R}))^2 = \frac{1}{\mu_X^2}\left(\frac{N-n}{nN}\right)\frac{\sum_{i=1}^{n}(Y_i - RX_i)^2}{n-1}. \tag{9.A.3}$$

Then the estimated standard error of $\hat{R}$ is

$$s_{\hat{R}} = \left[\frac{1}{\bar{X}^2}\left(\frac{N-n}{nN}\right)\frac{\sum_{i=1}^{n}(Y_i - \hat{R}X_i)^2}{n-1}\right]^{1/2} \tag{9.A.4}$$

where, for computational expedience, we have

$$\sum_{i=1}^{n}(Y_i - \hat{R}X_i)^2 = \sum_{i=1}^{n}Y_i^2 - 2\hat{R}\sum_{i=1}^{n}X_iY_i + \hat{R}^2\sum_{i=1}^{n}X_i^2. \tag{9.A.5}$$

If N is unknown, we may assume that $1 - \frac{n}{N}$ is approximately 1. Then, under this assumption, Equation (9.A.4) becomes

$$s'_{\hat{R}} = \left[\frac{1}{n\bar{X}^2}\frac{\sum_{i=1}^{n}(Y_i - \hat{R}X_i)^2}{n-1}\right]^{1/2}. \tag{9.A.4.1}$$

To determine how precisely $\hat{R}$ has been estimated from the sample data, we note first that, for large n, the sampling distribution of the sample ratio $\hat{R}$ is approximately normal. Hence a *100(1 – α)% confidence interval for R* is

$$\hat{R} \pm Z_{\alpha/2}s_{\hat{R}}. \tag{9.A.6}$$

Example 9.A.2

Using the sample data appearing in Table 9.A.1, let us construct a 95% confidence interval for the population ratio R depicting the number of occupants per bedroom.

TABLE 9.A.2 Estimation of Occupants per Bedroom

Number of Bedrooms (X_i)	Number of Occupants (Y_i)	X_i^2	Y_i^2	X_iY_i
1	2	1	4	2
1	1	1	1	1
2	4	4	16	8
1	1	1	1	1
2	3	4	9	6
1	2	1	4	2
1	2	1	4	2
1	1	1	1	1
2	2	4	4	4
2	3	4	9	6
14	21	22	53	33

Transforming Table 9.A.1 into a work table (Table 9.A.2) yields, from Equations (9.A.2) and (9.A.5),

$$\hat{R} = \frac{Y}{X} = \frac{21}{14} = 1.5$$

$$\begin{aligned}\sum\nolimits_{i=1}^{10}(Y_i - 1.5X_i)^2 &= 53 - 2(1.5)(33) + (1.5)^2(22) \\ &= 3.5.\end{aligned}$$

Then from Equation (9.A.4),

$$S_{\hat{R}} = \left[\frac{1}{1.96}\left(\frac{90}{1000}\right)\frac{3.5}{9}\right]^{1/2} = 0.133.$$

For $\alpha = 0.05$, Equation (9.A.6) renders $1.5 \pm 1.96\ (0.133)$ or $(L_1, L_2) = (1.24, 1.76)$, that is, we may be 95% confident that the true or population number of occupants per bedroom lies between 1.24 and 1.76 (or the number of occupants per 100 bedrooms lies between 124 and 176).

Next, we may also utilize Equations (9.A.1) and (9.A.2) to formulate a ratio estimator for the population mean $\mu_Y = (\tau_X/\tau_Y)\mu_X$ as

$$M_Y = \left(\frac{\bar{Y}}{\bar{X}}\right)\mu_X, \tag{9.A.7}$$

where it is assumed that μ_X is known. (Although the simple random sample estimator $\bar{Y}$ is unbiased for μ_Y, the reduced variance of the ratio estimator in Equation (9.A.7) usually compensates for the bias associated with ratio estimation of the mean.)

Moreover, the variance of M_Y is

$$V(M_Y) = \mu_X^2 V(\hat{R}) = \left(\frac{N-n}{nN}\right)\frac{\sum_{i=1}^{n}(Y_i - \hat{R}X_i)^2}{n-1} \tag{9.A.8}$$

and thus the estimated standard error of M_Y is

$$s_{M_y} = \left[\left(\frac{N-n}{nN}\right)\frac{\sum_{i=1}^{n}(Y_i - \hat{R}X_i)^2}{n-1}\right]^{1/2}. \tag{9.A.9}$$

And if we again assume a large sample n, then a *100(1 – α)% confidence interval for* μ_Y is given by

$$M_Y \pm Z_{\alpha/2} s_{M_Y}.$$

In sum, a ratio estimator is typically selected when:

a. The denominator in $\hat{R}$ is not n but is actually a random variable.
b. There is a strong association between a Y-characteristic and some supplemental or subsidiary X-characteristic in the population.

10

TESTING STATISTICAL HYPOTHESES

The predominant operational theme addressed in this book is that of statistical inference–drawing conclusions about a population via random sampling coupled with the use of probability theory to assess the reliability of those conclusions. There are two sides to the "inferential coin," so to speak. These are estimation and testing. As indicated in Chapters 8 and 9, a confidence interval is used to estimate an unknown parameter using sample data. As we shall now see, a hypothesis test (or significance test) is carried out to determine how strong the sample evidence is regarding some claim or assertion about a parameter.

We previously found that confidence interval estimation was based upon the application of the sampling distribution of a statistic and its attendant probability reckoning. The same is true of hypothesis testing. That is, hypothesis testing also utilizes the concept of probability, obtained from the sampling distribution of a statistic, to determine what would happen if we applied our test procedure many times in succession in the long run. It is the notion of probability that enables us to specify, in a quantitative fashion, how credible our test results are.

10.1 WHAT IS A STATISTICAL HYPOTHESIS?

In general, a *hypothesis* is essentially an unproved theory or assertion; it is tentatively accepted as an explanation of certain facts or observations. A *statistical hypothesis* is a testable hypothesis—it specifies a value for some parameter (call it θ)

Statistical Inference: A Short Course, First Edition. Michael J. Panik.

of a theoretical population distribution. For instance, θ might be a mean or variance or proportion of successes in a population. (This is in contrast to what is called a *maintained hypothesis*—an assumption that we are willing to believe in and which is thought to hold at least approximately, for example, the population is normally distributed.)

To develop a test procedure, we must actually specify a "pair of hypotheses" in order to admit alternative possibilities for a population. The first hypothesis in the pair is the *null hypothesis* (denoted H_0)—it is the hypothesis to be tested and either rejected or not rejected. The null hypothesis is "always assumed to be true." (Incidentally, the word "null" is to be interpreted in the context of "no difference" between the true value of θ and its hypothesized value.) The null hypothesis may be either *simple* (θ is hypothesized to equal a single numerical value) or *compound* (θ is hypothesized to fall within a range of values.) The second hypothesis in the pair is the *alternative hypothesis* (denoted as H_1)—it is the hypothesis that states what the population would look like if the null hypothesis were untrue.

For instance, three typical cases or pairs of hypotheses can be formulated:

Case 1	Case 2	Case 3
H_0: $\theta = \theta_o$ H_1: $\theta \neq \theta_o$	H_0: $\theta = \theta_o$ H_1: $\theta > \theta_o$	H_0: $\theta = \theta_o$ H_1: $\theta < \theta_o$

Here θ_o is termed the *null value* of θ (e.g., we may have H_0: $\theta = 3$ versus H_1: $\theta > 3$, where $\theta_o = 3$). Note that in each of these cases equality always belongs to the null hypothesis. This is because H_0 is always "assumed to be true."

10.2 ERRORS IN TESTING

It should be intuitively clear that since sample information is used to test hypotheses about a population parameter, an element of uncertainty is involved in assessing the test results. Hence it is possible to reach a "false conclusion." The two types of errors encountered are:

Type I Error (TIE): Rejecting H_0 when it is actually true.

Type II Error (TIIE): Not rejecting H_0 when it is actually false.

In fact, the complete set of possible actions given the status of the null hypothesis are presented in Table 10.1.

The risks associated with wrong decisions are the (conditional) probabilities of committing Type I and Type II Errors:

P (Type I Error) $= P$ (reject H_0/H_0 true) $= \alpha$.

P (Type II Error) $= P$ (do not reject H_0/H_0 false or H_1 true) $= \beta$.

TABLE 10.1 Types of Errors

		State of Nature	
		H_o true	H_o false (or H_1 true)
Action taken on the basis of sample information	Do not reject H_o	No error	Type II error
	Reject H_o	Type I error	No error

10.3 THE CONTEXTUAL FRAMEWORK OF HYPOTHESIS TESTING

In what sort of context does the notion of hypothesis testing arise? Suppose that you and a friend are having a drink during happy hour at the local watering hole and one of the patrons seated nearby places a pair of dice on the bar and asks your friend if he would like to play a game of chance. Your friend is curious and the stranger, who introduces himself as Lefty, explains the game: "if the sum of the faces showing on any roll is 7, then you must pay me \$5.00; but if any other sum of the faces obtains, then I will pay you \$2.00." Your friend, after being told by Lefty that the pair of dice is fair, decides to play the game. Lefty asks your friend to roll the dice. The game begins and your friend rolls eight consecutive 7s. "Wow!" exclaims your friend, "You sure are lucky." Lefty replies: "Keep rolling, a sum other than 7 is due to occur." You, however, are by now slightly suspicious of Lefty and his pet dice. You make the following quick calculation on a napkin:

> If the pair of dice is fair, the probability of a sum of 7 on any roll is 1/6. While rolling eight consecutive 7s seems farfetched, it might just be a fluke—Lefty might just be extremely lucky, or not. You set out to determine the likelihood of getting eight 7s in a row under the assumption that the pair of dice is fair. Since the rolls are independent, you calculate

$$\begin{aligned} P(\text{eight consecutive 7s}) &= P(7 \cap 7 \cap 7 \cap 7 \cap 7 \cap 7 \cap 7 \cap 7) \\ &= (1/6)^8 = 0.0000005934. \end{aligned}$$

Hence there is about 1 chance in 1,679,616 of rolling eight consecutive 7s. Since you are now not willing to attribute Lefty's good fortune to chance, you reject the notion that the pair of dice is fair. You could be wrong, but the odds are that you are not. So, two considerations present themselves:

1. Lefty is a very lucky guy, or
2. Lefty's pair of dice is not fair.

Should you inform your friend that something is amiss, or was your friend unfortunate enough to experience a run of unusually bad luck? This conundrum reflects the "essence of hypothesis testing:"

(i) You make an assumption about the state of nature. (The probability of rolling a 7 with a fair pair of dice is 1/6 or H_0: $P(7) = 1/6$. Your suspicion regarding Lefty's pair of dice is that $P(7) > 1/6$ or H_1: $P(7) > 1/6$.)

(ii) You gather sample evidence. (You feel that eight rolls of the pair of dice is adequate for the experiment at hand.)

(iii) You determine if the sample evidence (under the assumption about the state of nature) contradicts H_0. You feel that it does because the probability of getting eight 7s in a row "if H_0 were true" is "too small" to be attributed exclusively to sampling variation alone. Hence it is highly unlikely that H_0 is true. Given the above probability result, α, the probability of a TIE, would be miniscule.

Other instances of formulating null and alternative hypotheses along with their associated Type I and Type II Errors, are:

1. The manufacturer of a particular type of 6 v battery claims that it will last an average of 400 hr under normal conditions in a particular model of camp lantern. A Boy Scout troop leader wants to determine if the mean life of this battery is overstated by the manufacturer. Here,

$$H_0 : \mu = \mu_o = 400$$
$$H_1 : \mu < 400.$$

TIE: Occurs if H_0 is rejected when, in fact, H_0 is true. A TIE is made if the sample evidence leads the troop leader to believe that μ is less than 400 hr when, in actuality, μ is not less than 400.

TIIE: Occurs if H_0 is not rejected when, in fact, H_1 is true. A TIIE is made if the sample evidence prompts the troop leader to not reject H_0 when, in reality, $\mu < 400$ hr.

2. ACE Packing and Transit Co. boasts of only a 1.5% breakage rate on items shipped overnight. A dealer in glass art feels that this rate might be much too low. Here,

$$H_0 : p = p_o = 0.015$$
$$H_1 : p > 0.015.$$

TIE: Occurs if H_0 is rejected when it is actually true. A TIE is made when sample evidence prompts the dealer to believe $p > 0.015$ when, in fact, p does not exceed 0.015.

TIIE: Occurs if H_0 is not rejected when it is the case that H_1 is true. A TIIE is made if the sample result leads the dealer to not reject H_0 when, in actuality, $p > 0.015$.

3. According to a spokesman from the outreach program of a local church group, the mean value of clothing and other usable household items donated by

parishioners over the last few years has been fairly stable at around \$147.00 per year. The pastor wants to determine if the level of giving will be different for this year. Now,

$$H_0 : \mu = \mu_o = \$147.00$$

$$H_1 : \mu \neq \$147.00.$$

TIE: Occurs if H_0 is rejected when, in reality, it is true. A TIE is made if the sample data prompts the pastor to believe that $\mu \neq \$147.00$ when, in fact, $\mu = \$147.00$.

TIIE: Occurs if H_0 is not rejected when H_1 is true. A TIIE emerges if the sample evidence leads the pastor to not reject H_0 when it is the case that $\mu \neq \$147.00$.

Some general and more commonplace environments involving hypothesis testing now follows:

10.3.1 Types of Errors in a Legal Context

Suppose a felony trial is being conducted. Given our judicial system:

H_0: The defendant is innocent (assumed true).
H_1: The defendant is guilty.

The court seeks evidence for guilt. If the prosecutor cannot make a case, the jury renders a verdict of "not guilty." (Note that they do not declare the defendant "innocent." He/she may indeed be guilty, but the evidence against the defendant was not compelling. This is why, in hypothesis testing, we say that we either "reject" or "do not reject" H_0. We don't "accept" H_0. The null hypothesis may actually be untrue—it is just that the data could not discredit it.) What sorts of errors can occur?

TIE: Finding an innocent person guilty.
TIIE: Failing to convict a guilty person.

Type I Errors should be much less common than Type II Errors. The court will tolerate a high probability of a Type II Error in order to diminish the probability of a Type I Error. The legal system should avoid convicting an innocent person at all costs.

10.3.2 Types of Errors in a Medical Context

An individual is feeling poorly and goes to the local hospital emergency room. From the viewpoint of the medical personnel:

H_0: The individual is healthy (assumed true).
H_1: The individual is unhealthy.

The doctors seek evidence that the person is unhealthy. Either the individual tests positive for some malady, or does not. Under this process, the person is never declared "healthy;" the individual may actually be unhealthy, but there was insufficient clinical evidence to uncover a medical problem. In this venue the types of errors are:

TIE: A healthy person tests positive (a false positive).

TIIE: An unhealthy person tests negative (a false negative).

Here Type II Errors should be much less common than Type I Errors. Type II Errors can have dire consequences (treatment is withheld or delayed). Hence the medical community will tolerate a high probability of a Type I Error in order to diminish the probability of a Type II Error. Concluding that an unhealthy person is healthy is anathema.

10.3.3 Types of Errors in a Processing or Control Context

A production manager is monitoring a temperature gauge for a plating process. The temperature reading is supposed to vary only within a narrow range and the accuracy of the gauge is critical if the plating material is to adhere to the host item. From the viewpoint of the manager:

H_0: The process is in control or working as designed.

H_1: The process is not working as designed or is out of control.

The manager seeks evidence that the process is out of control. If it is, remedial action must be taken. However, no action is warranted if the magnitude of any temperature variation is deemed acceptable. In this environment, the errors that can be made are:

TIE: Shutting down or adjusting a process that is in control.

TIIE: Letting a process that is out of control continue operating.

The probability of a Type II Error should be kept much lower than that of a Type I Error since a production process delivering defective items could be much more costly to deal with (e.g., having to correct defects or scrap ruined items) than simply attempting to modify a process that is currently operating correctly.

10.3.4 Types of Errors in a Sports Context

Conflicts over judgment calls occur all too frequently at sporting events. In making an assessment of the efficacy of calls, we need to consider the possibility that:

H_0: The referee's or umpire's call is correct.

H_1: The referee's or umpire's call is erroneous.

If a league review of a questionable call is requested by a team, the league seeks evidence of an erroneous decision via the review of films, interviews, and so on. Any such review can result in the following:

TIE: Concluding that a correct call is erroneous.

TIIE: Failing to detect an erroneous call.

Here Type II Errors should be much less common than Type I Errors—failing to institute corrective action could have disastrous consequences for the sport. The league tolerates a high probability of a Type I Error in order to diminish the probability of a Type II Error. The league should avoid concluding that erroneous calls are correct if at all possible.

10.4 SELECTING A TEST STATISTIC

In order to conduct a hypothesis test about a population parameter θ, we need a *test statistic* (denoted by $\hat{\theta}$)—a random variable whose sampling distribution is known under the assumption that the null hypothesis H_0: $\theta = \theta_o$ is true. Why is a test statistic needed? Because rules pertaining to the rejection or non-rejection of the null hypothesis can be determined for a particular sample outcome by studying the range of $\hat{\theta}$ values. Specifically, the range of $\hat{\theta}$ values will be partitioned into the two disjoint or mutually exclusive subsets R and $\bar{R}$ (the complement of R), where,

R is the *critical region*: It is the region of rejection that contains the sample outcomes least favorable to H_0.

$\bar{R}$ is the *region of non-rejection*: It contains the sample outcomes most favorable to H_0.

Next, we need to determine the location and size of the critical region R. In general, the alternative hypothesis H_1 determines the location of R while α, the probability of a Type I Error, determines its size. In fact, α will be termed the *level of significance* of the test. Typically, $\alpha = 0.01$ or 0.05 or 0.10. So if, say, $\alpha = 0.05$, then we are conducing our hypothesis test at the $100\alpha\%$ or 5% level of significance. The *critical value* of $\hat{\theta}$ (denoted by $\hat{\theta}_c$) is the value of $\hat{\theta}$ that separates R and $\bar{R}$. As will be seen shortly, $\hat{\theta}_c$ comes from the $N(0,1)$ table or the t table (which is why it is often termed the *tabular value* of $\hat{\theta}$). So for a fixed sample size n and level of significance α, $\hat{\theta}_c$ depends on the form of H_1 and the distribution of the test statistic $\hat{\theta}$.

10.5 THE CLASSICAL APPROACH TO HYPOTHESIS TESTING

Armed with the concepts, we can now specify what is called the *classical approach to hypothesis testing*:

Set the level of significance α, the probability of incorrectly rejecting H_0 when it is actually true, equal to some small value (0.01 or 0.05 or 0.10). So if we set, say, $\alpha = 0.05$, then if we take many random samples of size n from the population and we repeat our test procedure for each of them, then, in the long run, we are willing to make a TIE 5% of the time. Thus 5% is the proportion of time that our test methodology renders the incorrect result of rejecting H_0 when it is actually true. Then, in accordance with H_1, choose the critical region or region of rejection R so that the probability of obtaining a value of $\hat{\theta}$ in R equals α when H_0 is rejected at the α level of significance.

The classical approach to hypothesis testing is executed by the following *stepwise procedure for hypothesis testing*:

1. Formulate H_0 (assumed true) and H_1 (locates R).
2. Specify α (determines the size of R).
3. Select a test statistic $\hat{\theta}$ ($\bar{Z}$ or t or U) whose sampling distribution is known (it is $N(0,1)$ or t distributed) under the assumption that H_0: $\theta = \theta_o$ is true.
4. Find R (this involves determining $\hat{\theta}_c$ from the $N(0,1)$ or t table).
5. Given H_0: $\theta = \theta_o$, compute the appropriate *standardized test statistic* (the *calculated value*) to be compared with $\hat{\theta}_c$, for example,

   (a) For H_0: $\mu = \mu_o$ (μ_o is the null value of μ), determine

   $$\bar{Z}_o = \frac{\bar{X} - \mu_o}{\frac{\sigma}{\sqrt{n}}} \text{ or } t_0 = \frac{\bar{X} - \mu_o}{\frac{s}{\sqrt{n}}}.$$

   (b) For H_0: $p = p_o$ (p_o is the null value of p), determine

   $$U_o = \frac{\hat{p} - p_o}{\sqrt{\frac{p_o(1-p_o)}{n}}}.$$

6. Apply the *classical decision rule*: reject H_0 if the calculated value is a member of R.

10.6 TYPES OF HYPOTHESIS TESTS

The type of test performed will be either a one-tail test (involving either the right or left tail of the sampling distribution of the $\hat{\theta}$ test statistic) or a two-tail test (involving both tails of the same). In general, the type of test reflects the location of R which, as we noted above, is determined by H_1.

(a) *One-tail test* (upper tail)

$$H_0 : \mu \leq \mu_o$$
$$H_1 : \mu > \mu_o$$

Suppose σ is known. Then our test statistic is $\bar{Z}$. (If σ is unknown and thus estimated by s, then we would use the t statistic for small n.) Since H_0 is

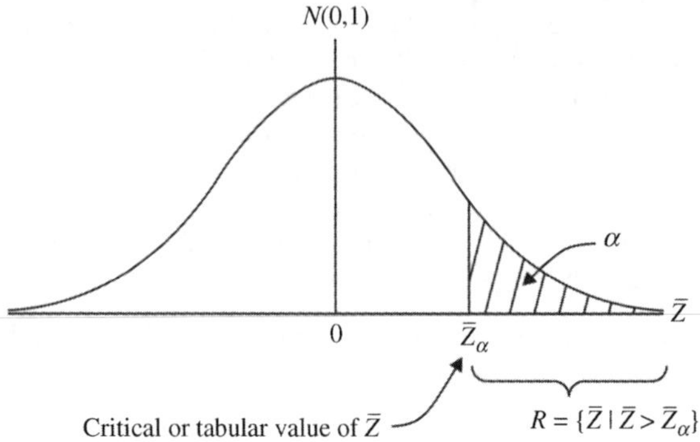

FIGURE 10.1 Upper-tail critical region.

supported by "small values" of $\bar{Z}_o$, R must be located under the right-hand tail of the $N(0,1)$ distribution (H_1 "points right"). We thus reject H_0 for large values of $\bar{Z}_o$(Fig. 10.1). So if $\bar{Z}_o > \bar{Z}_\alpha$(the calculated value of $\bar{Z}$ exceeds its tabular value), we fall within the critical region and thus we reject H_0 at the α level of significance.
If $\bar{Z}_o > \bar{Z}_\alpha$ or

$$\frac{\bar{X} - \mu_o}{\sigma/\sqrt{n}} > \bar{Z}_\alpha,$$

then $\bar{X} > \mu_o + (\frac{\sigma}{\sqrt{n}})\bar{Z}_\alpha$ and we reject H_0 since $\bar{X}$ exceeds μ_o by "too much," where our measure of "too much" is $(\frac{\sigma}{\sqrt{n}})\bar{Z}_\alpha$ (the upper error bound on $\bar{X}$ as an estimate of μ).

(b) *One-tail test* (lower tail)

$$H_0 : \mu \geq \mu_o$$
$$H_1 : \mu < \mu_o$$

Now H_0 is supported by "large values" of $\bar{Z}_o$ so that R must be located under the left-hand tail of the $N(0,1)$ distribution (H_1 "points left"). Hence we reject H_0 for small values of $\bar{Z}_o$ (Fig. 10.2). So if $\bar{Z}_o < -\bar{Z}_\alpha$ (the calculated value of $\bar{Z}$ falls short of its tabular value), we fall within the critical region and thus we reject H_0 at the α level of significance. With $\bar{Z}_o < -\bar{Z}_\alpha$ or

$$\frac{\bar{X} - \mu_o}{\frac{\sigma}{\sqrt{n}}} < -\bar{Z}_\alpha,$$

it follows that $\bar{X} < \mu_o - (\frac{\sigma}{\sqrt{n}})\bar{Z}_\alpha$ so that we reject H_0 since $\bar{X}$ falls short of μ_o by "too much," where our measure of "too much" is $-(\frac{\sigma}{\sqrt{n}})\bar{Z}_\alpha$ (the lower error bound on $\bar{X}$ as an estimate of μ).

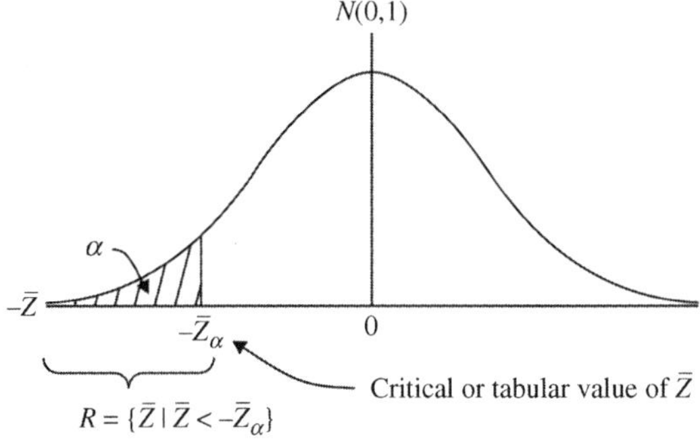

FIGURE 10.2 Lower-tail critical region.

(c) *Two-tail test*

$$H_0 : \mu = \mu_o$$
$$H_0 : \mu \neq \mu_o$$

In this instance, either large or small values of $\bar{Z}_o$ support H_1 so that R is located under each tail of the $N(0,1)$ distribution, with the area under each tail corresponding to $\alpha/2$ (α must be distributed between the two tails of the $N(0,1)$ curve). Hence we reject H_0 either for large or small values of $\bar{Z}_o$ (Fig. 10.3). So if either $\bar{Z}_o > \bar{Z}_{\alpha/2}$ or $\bar{Z}_o = -\bar{Z}_{\alpha/2}$ (both cases can be subsumed under $|\bar{Z}_o| > \bar{Z}_{\alpha/2}$), then we fall within the critical region and thus we reject H_0 at the α level of significance. With $|\bar{Z}_o| > \bar{Z}_{\alpha/2}$, we have

$$|\bar{Z}_o| = \left| \frac{X - \mu_o}{\frac{\sigma}{\sqrt{n}}} \right| > \bar{Z}_{\alpha/2}$$

so that $\bar{X} < \mu_o - (\frac{\sigma}{\sqrt{n}})\bar{Z}_{\alpha/2}$ or $\bar{X} > \mu_o + (\frac{\sigma}{\sqrt{n}})\bar{Z}_{\alpha/2}$. Hence we reject H_0 if either $\bar{X}$ falls short of μ_o by "too much" or $\bar{X}$ exceeds μ_o by "too much."

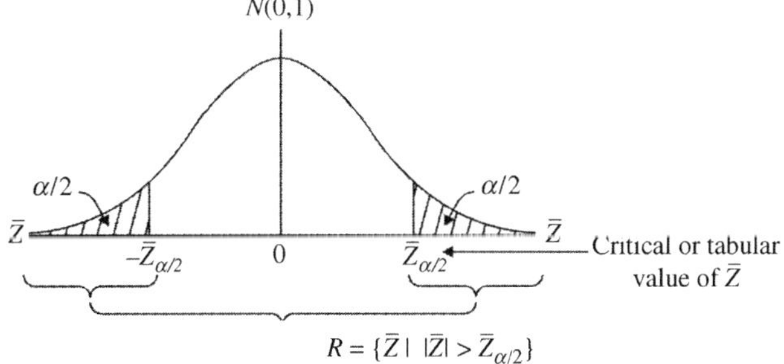

FIGURE 10.3 Two-tail critical region.

10.7 HYPOTHESIS TESTS FOR μ (σ KNOWN)

Example 10.1

A sample of size $n = 25$ is taken from a population that is $N(\mu,6)$ (μ is unknown but σ is known) and it is determined that $\bar{X} = 43.75$. Test H_0: $\mu = \mu_o = 40$ versus H_1: $\mu > 40$ at the $\alpha = 0.05$ level of significance. Following part (a) above,

$$\bar{Z}_o = \frac{\bar{X} - \mu_o}{\sigma/\sqrt{n}} = \frac{43.75 - 40}{6/\sqrt{25}} = 3.125.$$

Since $\bar{Z}_\alpha = \bar{Z}_{0.05} = 1.645$ (Fig. 10.4), $R = \{\bar{Z}|\bar{Z} > 1.645\}$. With $\bar{Z}_o = 3.125 > 1.645$, we are within R so that we reject H_0 at the 0.05 level of significance. Here $\bar{X} = 43.75$ exceeds $\mu_o = 40$ by 3.75 units—that is greater than $(\sigma/\sqrt{n})\bar{Z}_\alpha = 1.974$, the lower limit for "too much."

Example 10.2

A sample of size $n = 25$ is extracted from a $N(\mu,10)$ population. Given that $\bar{X} = 48.66$, test H_0: $\mu = \mu_o = 50$, against H_1: $\mu < 50$ at the $\alpha = 0.10$ level. Following part (b) above,

$$\bar{Z}_o = \frac{\bar{X} - \mu_o}{\sigma/\sqrt{n}} = \frac{48.66 - 50}{10/\sqrt{25}} = 0.67.$$

With $-\bar{Z}_\alpha = -\bar{Z}_{0.10} = -1.28$ (Fig. 10.5), it follows that $R = \{\bar{Z}|\bar{Z} < -1.28\}$. Since $\bar{Z}_o$ is not an element of R, we cannot reject H_0 at the 0.10 level of significance. Clearly, $\bar{X} = 48.66$ falls short of $\mu_o = 50$ by only -1.34 units—that is well above $-(\sigma/\sqrt{n})\bar{Z}_\alpha = -2.56$, the upper limit for "too much." Is there a moral to this story? Failure to reject H_0 means only that the sample data are consistent or in harmony with H_0; it is not that we have undisputable evidence that H_0 is true. In fact, H_0 may not be true—it is just that the data (and the chosen α) could not make a strong enough case against H_0.

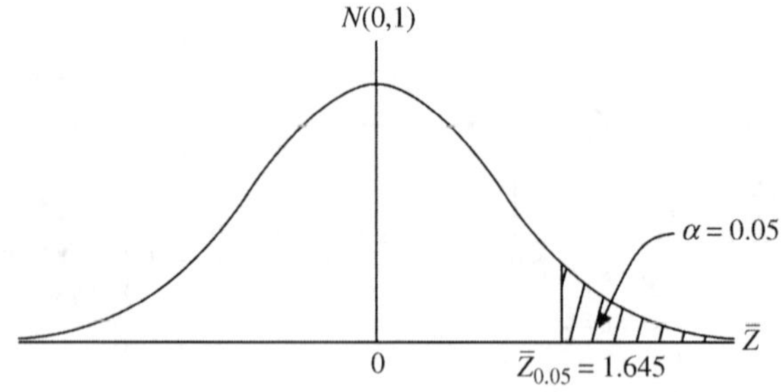

FIGURE 10.4 Right-tail test with $\alpha = 0.05$.

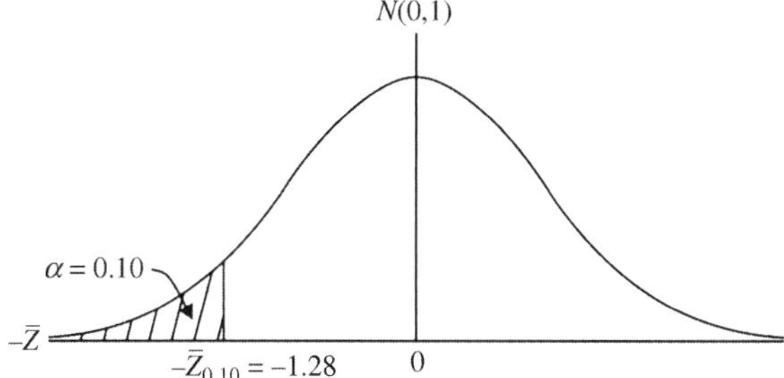

FIGURE 10.5 Left-tail test with $\alpha = 0.10$.

Example 10.3

Test H_0: $\mu = \mu_o = 100$ versus H_1: $\mu \neq 100$ at the $\alpha = 0.01$ level if it is known that the population is $N(\mu,7)$ and a sample of size $n = 64$ yields $\bar{X} = 104.17$. Following part (c) above,

$$|\bar{Z}_o| = \left|\frac{\bar{X} - \mu_o}{\frac{\sigma}{\sqrt{n}}}\right| = \left|\frac{104.17 - 100}{\frac{7}{\sqrt{64}}}\right| = 4.766.$$

Since $\bar{Z}_{\frac{\alpha}{2}} = \bar{Z}_{0.005} = 2.58$ (Fig. 10.6), $R = \{\bar{Z} \| \bar{Z}| > 2.58\}$. With $|\bar{Z}_o|$ a member of R, we can reject H_0 at the 0.01 level of significance.

10.8 HYPOTHESIS TESTS FOR μ (σ UNKNOWN AND n SMALL)

Given that σ is unknown and must be estimated by s and the sample size n is small, we must use the t distribution. To this end, let us consider the following tests.

Example 10.4

A manufacturer of security lamps claims that, on the average and under normal conditions, they will not draw more than 2.20 amps of current. A sample of 16 security lamps yielded $\bar{X} = 2.65$ amps with $s = 0.77$ amps. Should we reject the manufacturer's claim at the $\alpha = 0.05$ level? Here we seek to test H_0: $\mu = \mu_o = 2.20$, against H_1: $\mu > 2.20$. Then

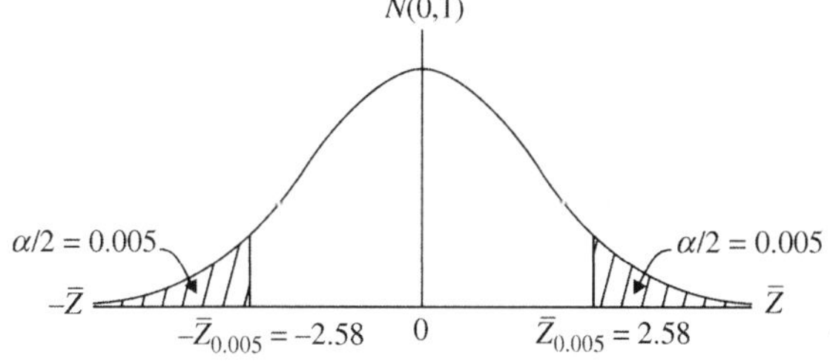

FIGURE 10.6 Two-tail test with $\alpha = 0.01$.

$$t_o = \frac{\bar{X} - \mu_o}{\frac{s}{\sqrt{n}}} = \frac{2.65 - 2.20}{\frac{0.77}{\sqrt{16}}} = 2.34.$$

With $t_{\alpha,\, n-1} = t_{0.05,15} = 1.753$, we see that $R = \{t | t > 1.753\}$. Since t_o falls within the critical region, we reject H_0 (the manufacturer's claim) at the 0.05 level of significance.

Have we proved the manufacturer wrong? The answer is "no." It is important to remember that in hypothesis testing we are never "proving" or "disproving" anything—we are simply making an "assumption" H_0 about μ, gathering some data, and then, in terms of a given probability level α, determining if we should reject that assumption on the basis of sample evidence. Nothing has been proven. In fact, for a different level of α, just the opposite action is called for, that is, with $\alpha = 0.01$, $t_{0.01,15} = 2.602$ so that Z_o is not within $R = \{t | t > 2.602\}$. Now, we cannot reject H_0 for $\alpha = 0.01$.

Example 10.5

A market research firm has a client whose sales last year averaged \$180.75 per customer. The client wants to know if average sales per customer have changed significantly for this year. The market research firm takes a random sample of $n = 25$ invoices and finds $\bar{X} = \$175.90$ and $s = \$9.82$. Let $\alpha = 0.05$. Here H_0: $\mu = \mu_o = \$180.75$ is tested against H_1: $\mu \neq \$185.75$ (since a "change" means that the average can either increase or decrease for this year) with

$$|t_o| = \left| \frac{\bar{X} - \mu_o}{\frac{s}{\sqrt{n}}} \right| = \left| \frac{175.90 - 180.75}{\frac{9.82}{\sqrt{25}}} \right| = 2.47.$$

Since $t_{\frac{\alpha}{2},\, n-1} = t_{0.025,24} = 2.064$, we have $R = \{t | |t| > 2.064\}$. Given that t_o is a member of R, we reject H_0 at the 0.05 level. At this level, there has been a statistically significant change in average sales.

What if the market research firm decides to perform a stricter test of H_0 versus H_1, say, at the $\alpha = 0.01$ level? Does the conclusion reached by the previous test stay the same? With $\alpha = 0.01$, $t_{\frac{\alpha}{2},\, n-1} = t_{0.05,24} = 2.797$ so that $R = \{t | |t| > 2.797\}$. Now t_o is not a member of R—we do not reject H_0. So at the $\alpha = 0.01$ level, there has not been a statistically significant change in sales.

Our approach to testing hypotheses about μ can now be summarized in Fig. 10.7.

Before we turn to hypothesis tests concerning the parameter p of a binomial population, let us consider a few salient features of hypothesis testing. These features can be collectively entitled "points to ponder:"

1. In all of our hypothesis testing above, we either decided to reject or not reject H_0. However, there is actually a third option; we can *reserve judgment*. That is, if our calculated value (Z_o or t_o) falls very close to the critical value of the test

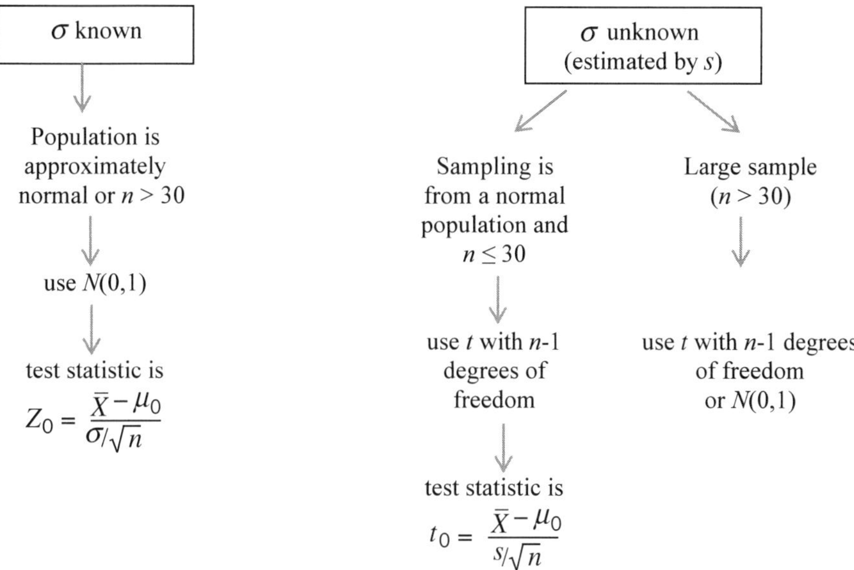

FIGURE 10.7 Conducting α-level hypothesis tests for μ when sampling from a $N(\mu, \sigma)$ population. (If the normality assumption is untenable, a nonparametric test is provided in Appendix 10.B).

statistic, then we might want to retest H_0 with, say, a larger sample, for example, if $\alpha = 0.05$ and $R = \{\bar{Z}||\bar{Z}| > 1.96\}$ and we find that $|\bar{Z}_o| = 2.00$ or $|\bar{Z}_o| = 1.94$, then we may want to reserve judgment. All this might be viewed as being "too close for comfort."

2. What is the connection between confidence interval estimation and hypothesis testing? The connection holds for a "two-tail test." Specifically, a 100(1 – α)% confidence interval for μ contains all μ_o's (null values for μ) that would "not be rejected" at the α level of significance. For instance, let us return to Example 10.5. For $\alpha = 0.05$, we rejected H_0: $\mu = \mu_o = \$180.75$. A 95% confidence interval for μ is therefore $\bar{X} \pm t_{\frac{\alpha}{2}, n-1} \frac{s}{\sqrt{n}}$ or $175.90 \pm 2.064\ (9.82/\sqrt{25})$ or $(L_1, L_2) = (171.846, 179.954)$. Hence we are 95% confident that μ lies between \$171.84 and \$179.95. Since $\mu_o = \$180.75$ is not a member of this interval, H_0 was rejected at the 0.05 level.

   For $\alpha = 0.01$, we could not reject H_0. A 99% confident interval for μ is $175.90 \pm 2.797\ (\frac{9.82}{\sqrt{25}})$ or now $(L_1, L_2) = (170.407, 181.393)$. Hence we are 99% confident that μ lies within these limits. Since $\mu_o = \$180.75$ lies within this interval, H_0 was not rejected at the 0.01 level.

3. What does it mean when we say that a particular result is statistically significant? If something is *statistically significant*, then there is only a very small probability that the observed sample result arose solely because of chance factors or is due solely to sampling error. Something *systematic* (i.e., behaving in a predictable fashion) is going on.

4. Quite often when the results of a statistical study are presented the researcher is asked to state his/her research hypothesis. Basically, the *research hypothesis* is the alternative hypothesis—it is what you want to demonstrate. (The null hypothesis is essentially a "throw away.") For instance, records indicate that a company's current gasoline engine gets 25 mpg on the average. The engineering unit has developed a newer and (ostensibly) more efficient or "greener" version of this engine. The unit performs some tests and finds that average miles per gallon increases to 28.9. To determine if "the newer version is better than the old version" (the research hypothesis), the engineers want to test H_0: $\mu = \mu_o = 25$ (the benchmark) against H_1: $\mu > 25$ at a particular level of significance. Hence rejecting H_0 supports H_1 (which is what they would like to demonstrate).

10.9 REPORTING THE RESULTS OF STATISTICAL HYPOTHESIS TESTS

How are the results of hypothesis tests typically presented? As we shall now see, so-called *p*-values are used. If we harken back to our classical hypothesis test procedure, one of the steps was to select a level of significance or α value. Hence α is our "chosen" level of significance. An alternative to this action is to let the data itself determine the "actual" level of significance of the test, that is, we shall determine the test's *p*-value—its observed or actual level of significance. Specifically, the *p*-value is the probability of obtaining a calculated value of the test statistic at least as large as the one observed if H_0 is true. In this regard, the larger the *p*-value, the more likely it is that the null hypothesis is true; and the smaller the *p*-value, the more likely it is that the null hypothesis is not true.

Thus a test of significance is based on the notion that: a sample outcome that would happen only infrequently if H_0 were true provides compelling evidence that H_0 is not true. And we assess the strength of the evidence by determining a probability (a *p*-value) that reflects how infrequently this outcome would occur if H_0 were true.

For instance, suppose we are testing H_0: $\mu = \mu_o$ versus H_1: $\mu > \mu_o$ and we get a calculated $\bar{Z}$ value of $\bar{Z} = 1.79$. What is the test's *p*-value? To answer this, we need to execute the following steps:

1. Find the area under the $N(0,1)$ curve from 0 to 1.79. From the $N(0,1)$ area table we get 0.4633 (Fig. 10.8).

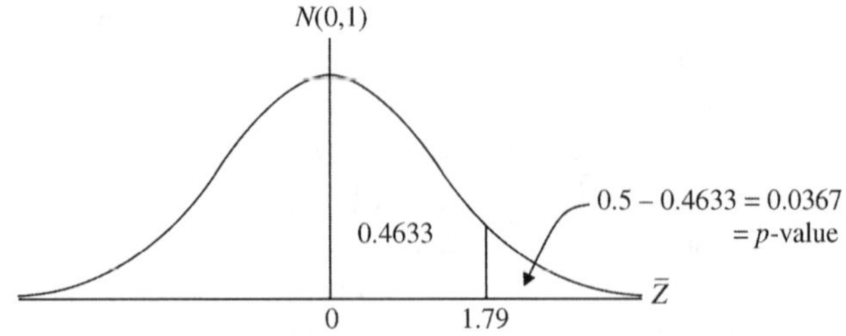

FIGURE 10.8 *p*-value for a one-tail test (upper tail).

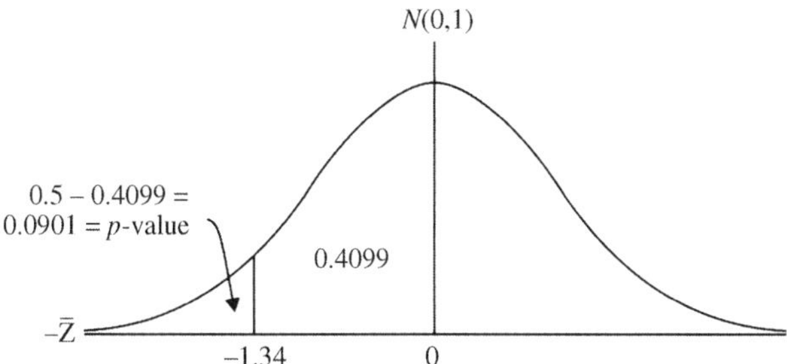

FIGURE 10.9 *p*-value for a one-tail test (lower tail).

2. Find the remaining area under the $N(0,1)$ curve from 1.79 to $+\infty$. This is $0.5 - 0.4633 = 0.0367$.

Hence there exists about a 3.7% chance of finding a calculated value of $\bar{Z}$ "at least as large" as 1.79 if H_0 is true.

Next, suppose we are testing H_0: $\mu = \mu_o$ against. H_1: $\mu < \mu_o$ and we obtain a calculated $\bar{Z}$ of $\bar{Z}_o = -1.34$. What is the *p*-value for this test? Let us find the area under the $N(0,1)$ curve from -1.34 to 0 (Fig. 10.9). This is 0.4099. Then the remaining area under the $N(0,1)$ curve from $-\infty$ to -1.34 is $0.5 - 0.4099 = 0.0901$. Hence we have about a 9% chance of finding a calculated value of $\bar{Z}$ "at least as small" as -1.34 if H_0 is true.

The preceding two cases considered one-tail tests. Let us now determine the *p*-value associated with a two-tail test. To this end, let us assume that we are testing H_0: $\mu = \mu_o$ against H_1: $\mu \neq \mu_o$ and the calculated $\bar{Z}$ is $|\bar{Z}_o| = 1.53$.

Now, the area under the $N(0,1)$ curve from 0 to 1.53 is 0.4370 (Fig. 10.10). Hence the remaining area under the $N(0,1)$ curve from 1.53 to $+\infty$ is 0.0630. But this latter area, under a two-tail test, is $0.0630 = p\text{-value}/2$ or the $p\text{-value} = 0.1260$. Note that, for a two-tail test, the *p*-value can alternatively be viewed as $P(\bar{Z} < -1.53) + P(\bar{Z} > 1.53) = 0.0630 + 0.0630 = 0.1260$.

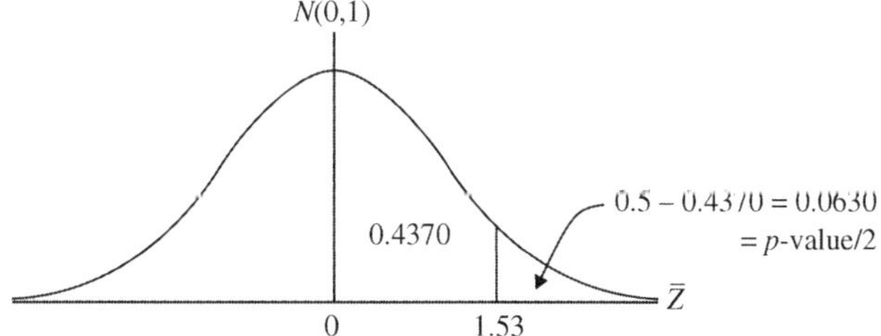

FIGURE 10.10 *p*-value for a two-tail test.

Finally, let us assume that we are testing H_0: $\mu = \mu_o$ against H_1: $\mu > \mu_o$ and we obtain a calculated $t = 1.86$ and $n = 20$. For $n - 1 = 19$ degrees of freedom, we see that, within the $d.f. = 19$ row of the t table,

$$1.729 < 1.86 < 2.093.$$

Hence

$$0.025 < p\text{-value} < 0.05,$$

as evidenced by the right-tail areas provided by the associated column headings.

How are p-values typically utilized in practice? We noted above that the smaller the p-value, the more likely it is that H_0 is not true. Hence "small" p-values lead to the rejection of H_0. Now, in conducting hypothesis tests, one typically has in mind some threshold or minimally acceptable level of α for rejection purposes, say, $\alpha = 0.05$. Hence any time a realized p-value < 0.05, H_0 is rejected.

Example 10.6

Suppose the population variable X is $N(\mu, 5.6)$ and a random sample of $n = 35$ yields $\bar{X} = 50$. Our objective is to test H_0: $\mu = \mu_o = 48$ against H_1: $\mu > 48$. Let our "comfort level" regarding α be 0.05. Since

$$\bar{Z}_o = \frac{\bar{X} - \mu_o}{\frac{\sigma}{\sqrt{n}}} = \frac{50 - 48}{\frac{5.6}{\sqrt{35}}} = 2.11,$$

the area under the $N(0,1)$ curve from 0 to 2.11 is 0.4826 (Fig. 10.11). Hence the p-value $= 0.5 - 0.4826 = 0.0174$. With the p-value less than our minimum tolerable α level of 0.05, we reject H_0—we have less than a 2% chance of finding a calculated value at least as large as 2.11 if H_0 is true.

A few salient points pertaining to the p-value for a hypothesis test are worth repeating/amplifying:

1. A p-value is computed under the assumption that H_0 is true. Hence it is the "maximum" chance of making a TIE if we indeed reject H_0.

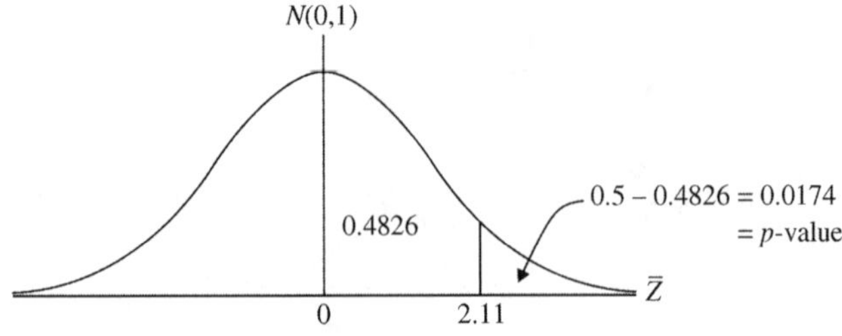

FIGURE 10.11 Right-tail p-value.

2. The role of the p-value is to quantify the degree of evidence against H_0 that is provided by the sample.
3. Hence the p-value tells us just how unlikely the sample outcome would be if H_0 were true. Thus highly unlikely outcomes (as evidenced by their small p-values) offer strong evidence that H_0 is not true.

   How small a p-value is needed to provide compelling evidence against H_0? The answer to this question depends on: (a) the plausibility of H_0; and (b) the consequences of rejecting H_0. In this regard, there is no unambiguous distinction between a test result that is significant or not—we only have increasingly strong evidence for the rejection of H_0 as the p-value decreases in size.
4. Suppose the p-value ≤ 0.05 (respectively ≤ 0.01). Then we have a test result that, in the long run, would happen no more than once per 20 (respectively, per 100) samples if H_0 were true. Thus the smaller the p-value, the more inconsistent H_0 is with the sample data.
5. The p-value is the probability that the test statistic would assume a realization as large, or larger, than that actually observed if H_0 is true. So under a test concerning, say, the population mean μ, we compare $\bar{X}$ with the null value μ_o. Then the p-value reveals how unlikely an $\bar{X}$ this extreme is if H_0: $\mu = \mu_o$ is true.

10.10 HYPOTHESIS TESTS FOR THE POPULATION PROPORTION (OF SUCCESSES) *p*

Suppose we extract a random sample of size n (>30) from a binomial population and we seek to test a hypothesis about p, the population proportion of successes. Under H_0: $p = p_o$, we know that, for a binomial random variable X,

$$E(X) = np_o$$

$$V(X) = np_o(1 - p_o).$$

Let us construct our *standardized test statistic* as

$$\begin{aligned} U_o &= \frac{X - E(X)}{\sqrt{V(X)}} = \frac{X - np_o}{\sqrt{np_o(1 - p_o)}} \\ &= \frac{\frac{X}{n} - p_o}{\sqrt{\frac{np_o(1 - p_o)}{n^2}}} = \frac{\hat{p} - p_o}{\sqrt{\frac{p_o(1 - p_o)}{n}}}, \end{aligned} \tag{10.1}$$

which is $N(0,1)$ distributed for large n. Here Equation (10.1) serves as the calculated value of our test statistic under H_0: $p = p_o$ (assumed true).[1]

Looking to the types of tests, we have the following:

(a) *One-tail test* (upper tail)

$$H_0 : p \leq p_o$$
$$H_1 : p > p_o$$

For α small, $R = \{U | U > U_\alpha\}$ (Fig. 10.12).

(b) *One-tail test* (lower tail)

$$H_0 : p \geq p_o$$
$$H_1 : p < p_o$$

Now, $R = \{U | U < U_\alpha\}$ (Fig. 10.13).

(c) *Two-tail test*

$$H_0 : p = p_o$$
$$H_1 : p \neq p_o$$

In this instance, $R = \{U || U | > U_{\alpha/2}\}$, where again α is small (Fig. 10.14).

Example 10.7

Last year, in a particular company about 35% of the assembly workers were in favor of union representation. Since then a campaign has been undertaken to get more workers to favor joining a union. However, management has been, according to some observers, overly generous with the worker's wage increases and fringe benefits. On balance, has there been a significant change in the proportion of assembly workers who favor unionization? Here we will test H_0: $p = 0.35$, against H_1: $p \neq 0.35$. Let $\alpha = 0.05$ so that $R = \{U || U | > 1.96\}$.

[1] Up to this point in our discussion of *standardized variables*, we have actually encountered three distinct $N(0,1)$ variables:

1. X is $N(\mu, \sigma) \rightarrow Z = \frac{X-\mu}{\sigma} \sim N(0, 1)$
2. $\bar{X} \rightarrow \bar{Z} = \frac{\bar{X}-\mu}{\frac{\sigma}{\sqrt{n}}} \sim N(0, 1)$ (for large n)
3. $\hat{p} \rightarrow U = \frac{\hat{p}-p}{\sqrt{\frac{p(1-p)}{n}}} \sim N(0, 1)$ (for large n)

where "$\rightarrow$" means "is transformed to" and "$\sim$" means "is distributed as." So while Z, $\bar{Z}$, and U are all "different" variables, they each follow the "same" $N(0,1)$ distribution.

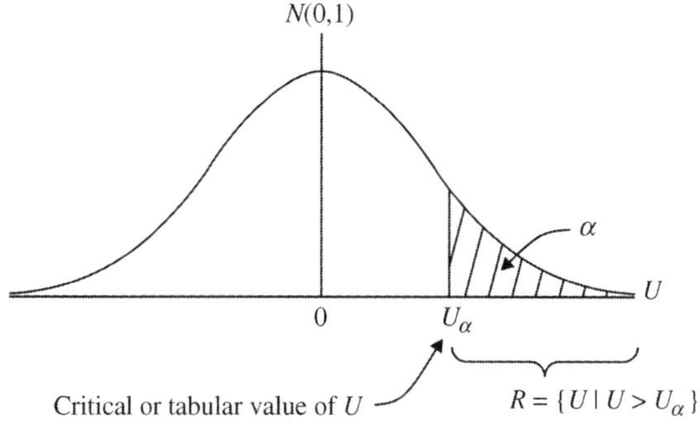

FIGURE 10.12 Upper-tail critical region.

A random sample of $n = 100$ workers yielded $X = 40$ workers who were in favor of joining a union. Then $\hat{p} = X/n = 0.40$ and

$$|U_o| = \left|\frac{\hat{p} - p_o}{\sqrt{\frac{p_o(1-p_o)}{n}}}\right| = \left|\frac{0.40 - 0.35}{\sqrt{\frac{0.35(0.65)}{100}}}\right| = 1.04.$$

Since 1.04 does not fall within the critical region, we cannot reject H_0 at the 0.05 level; at this level, there has not been a statistically significant change in the proportion of workers favoring unionization.

In terms of the tests p-value, the area under the $N(0,1)$ curve from 0 to 1.04 is 0.3508 (Fig. 10.15).

Then p-value/2 $= 0.5 - 0.3508 = 0.1492$ and thus the p-value $= 0.2984 > 0.05$ (our "comfort level of α"). Hence there is about a 30% chance of finding a calculated $|U_o|$ at least as large as 1.04 if H_0 is true. Again we do not reject H_0.

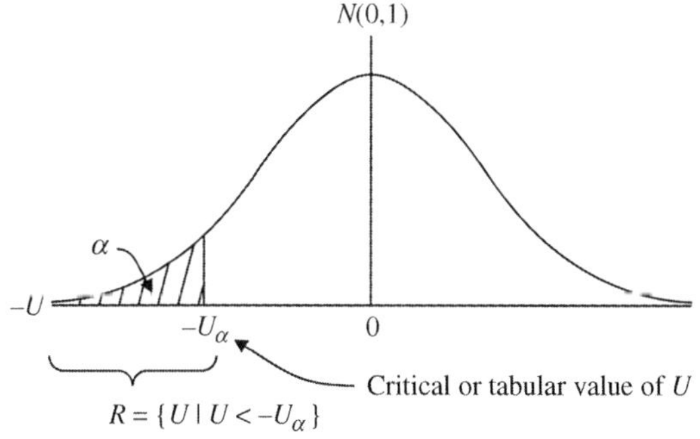

FIGURE 10.13 Lower-tail critical region.

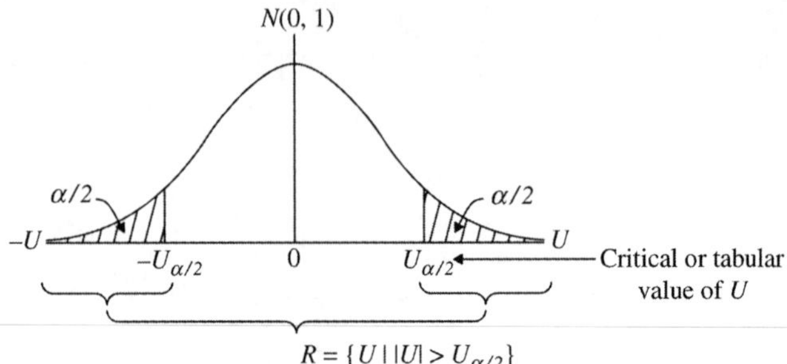

FIGURE 10.14 Two-tail critical region.

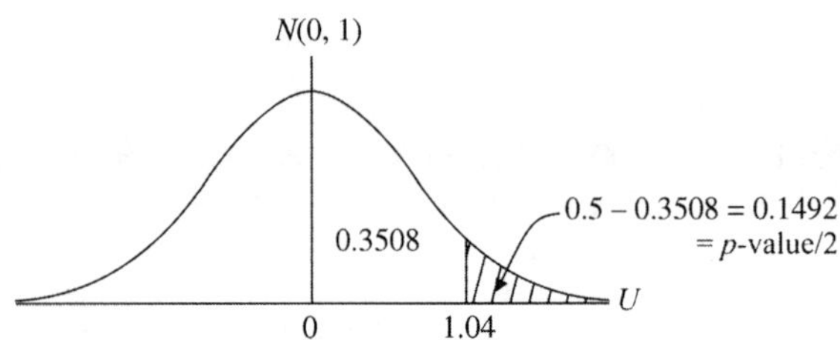

FIGURE 10.15 Two-tail p-value.

EXERCISES

1 Given the following sets of null and alternative hypotheses, determine the location of the critical region, that is, is the test left-tailed, right-tailed, or two-tailed? [Caution: Look before you leap.]

a. $H_o : \mu = 10, H_1 : \mu > 10$
b. $H_o : \mu = 70, H_1 : \mu \neq 70$
c. $H_o : p = 0.05, H_1 : p < 0.05$
d. $H_o : p > 0.15, H_1 : p = 0.15$

2 For each of the following sets of circumstances, state the null and alternative hypotheses; and state what it would mean to commit a Type I Error (respectively, a Type II Error).

a. In 2005, about 15% of all registered voters were of Hispanic origin. A sociologist believes that this percentage has increased since then. What would it mean if the null hypothesis were rejected?

b. Last year, the mean contribution to the United Way charities in a certain city was \$200. A reporter for *Philanthropy Today* magazine feels that it has changed since then. What does it mean if the null hypothesis is not rejected?

c. According to the Department of Transportation, last year the mean price of a gallon of regular gasoline was \$2.50. Because of increased demand, the department feels that the mean price has increased since then. What does it mean if the null hypothesis is not rejected?

d. Last year, reading comprehension scores in Mr. Smith's fourth grade class averaged 85 points. Due to lack of reading practice, video games, TV, and so on, Mr. Smith feels that this year's scores might have decreased a bit. What does it mean if the null hypothesis is rejected?

e. According to the Department of Health and Human Services, last year about 35% of adults were considered to be overweight. A case worker thinks that this percentage has increased since then. What does it mean if the null hypothesis is not rejected?

3 Suppose we extract a random sample of size $n=24$ from a normal population with $\sigma=12$. We find that $\bar{X}=47.5$. Test $H_o: \mu=50$, against $H_1: \mu<50$ for $\alpha=0.05$. Determine the critical region and locate it graphically. Also, determine the test's p-value.

4 A random sample of size $n=25$ is taken from a normal population with $\sigma=6$. The sample mean was found to be $\bar{X}=44$. For $\alpha=0.01$, test $H_o: \mu=40$ versus $H_1: \mu>40$. Determine the critical region and locate it graphically. What is the p-value for the test?

5 Given a random sample of size $n=30$ obtained from a normal population with $\sigma=8$, test $H_o: \mu=100$ versus $H_1: \mu\neq 100$ at the 1% level. Assume the sample mean to be $\bar{X}=105$. Determine the critical region and locate it graphically. What is the p-value?

6 The SAT scores at the Mitchell-Ross Regional High School have been found to be normally distributed with $\sigma=105$. The coaching staff feels that the student–athletes (both male and female combined) do just as well on the SAT exam as the non-athletes. The average score on the SAT for non-athletes is 565. The coaching staff selects a random sample of $n=35$ student–athletes from the school's records and finds that the mean SAT score for this group is $\bar{X}=545$. For $\alpha=0.05$, is there evidence to support the coaching staff's contention? Find the critical region. What is the test's p-value?

7 For each of the following sets of values, test the indicated hypothesis. Also, find the p-value for each test. (Assume that sampling has been undertaken from a normal population.)

a. $H_o: \mu-55,\ H_1: \mu<55;\ n=25, \bar{X}=50, s=10,\ \alpha=0.10$

b. $H_o: \mu=20,\ H_1: \mu\neq 20;\ n=25,\ \bar{X}=28,\ s=3,\ \alpha=0.05$

c. $H_o: \mu=320,\ H_1: \mu>320;\ n=9,\ \bar{X}=325.4,\ s=3,\ \alpha=0.10$

d. $H_o: \mu=1500,\ H_1: \mu\neq 1500;\ n=16,\ \bar{X}=1555,\ s=4,\ \alpha=0.05$

e. $H_o: \mu = 75,\ H_1: \mu \neq 75;\ n = 9,\ \bar{X} = 76,\ \Sigma(X_i - \bar{X})^2 = 32,\ \alpha = 0.10$
f. $H_o: \mu = 17,\ H_1: \mu < 17;\ n = 25,\ \Sigma X_i = 350,\ \Sigma X_i^2 = 6,600,\ \alpha = 0.05$

8 The average worker at XYZ Fabrication usually takes 180 sec to complete a specific task. John, a new employee, is given this particular task to perform. His times, in seconds, appear below. Can we conclude, at the 1% level of significance, that John's mean time is not equal to the 180 sec mean time of all other employees? What is the p-value for the test? (Assume completion time to be normally distributed.)

170, 179, 193, 204, 181, 208
192, 215, 165, 174, 183, 182

9 The following are the weights, in pounds, of a sample of large chocolate bunnies produced for display purposes only at Sweet Tooth Candy Shops. Should we accept the chocolatier's statement that the mean weight is not greater than 6 lbs? Use $\alpha = 0.05$. Also, find the p-value for the test. (Assume the weights to be normally distributed.)

6.56, 6.15, 6.99, 6.97, 6.83
5.95, 6.54, 6.30, 7.08, 6.11

10 The mean yield per acre of barley at the Farmer's Cooperative last year was reported to be 33.5 bushels. A member of the cooperative claims that the mean yield was a bit higher than that reported. He randomly samples 35 acres on his and nearby farms and obtains $\bar{X} = 37.1$ bushels with $s = 2.5$ bushels. Was the average that was reported too low? Use $\alpha = 0.05$. What is the test's p-value?

11 For each of the following sets of sample values, test the indicated hypothesis. Also, determine the p-value for the test:

a. $H_o: p = 0.35,\ H_1: p > 0.35;\ n = 200,\ X = 75,\ \alpha = 0.05$
b. $H_o: p = 0.50,\ H_1: p < 0.50;\ n = 150,\ X = 70,\ \alpha = 0.10$
c. $H_o: p = 0.75,\ H_1: p \neq 0.75;\ n = 500,\ X = 440,\ \alpha = 0.05$
d. $H_o: p = 0.63,\ H_1: p < 0.63;\ n = 250,\ X = 124,\ \alpha = 0.01$
e. $H_o: p = 0.44,\ H_1: p \neq 0.44;\ n = 1000,\ X = 420,\ \alpha = 0.05$

12 In a sample of $n = 400$ seeds, 347 of them germinated. At the 5% level of significance, can we reject the claim that at least 90% of the seeds will germinate? What is the p-value for the test?

13 In a sample of $n = 64$ ceramic pots we found eight with defects. Have we reason to believe that more than 10% of all such pots would show defects? Use $\alpha = 0.05$. Also, determine the p-value.

14 A sample of 144 voters contained 90 individuals who favored raising the drinking age from 18 to 21 years of age. Test the hypothesis that opinion is actually divided on this issue. Use $\alpha = 0.05$. Find the p-value.

15 *Zap* is an ointment that reduces itching due to dry skin. Suppose the manufacturer of *Zap* claims that at least 94% of the users of *Zap* get relief within 8 min of an application. In clinical trials, 213 of 224 patients suffering from itching due to dry skin obtained relief after 8 min. Should we reject the manufacturer's claim for $\alpha = 0.01$? Determine the test's p-value.

16 Can the following set of $n = 40$ data values be viewed as constituting a random sample taken from some population? Use $\alpha = 0.05$.

Ordered Data Values

130	147	75	100	92	102	120	122	176	131
77	89	141	91	73	97	138	140	80	100
89	99	107	81	102	91	98	79	106	120
87	91	101	92	107	131	140	99	98	105

17 A beginning college course in history has 453 enrollees. Over the course of the semester, the class absences were recorded as

12, 9, 7, 16, 8, 10, 17, 5, 9, 22, 18, 15, 7, 24
20, 13, 11, 8, 8, 14, 10, 9, 8, 7, 16, 15

Does this information provide evidence, at the 5% level, that the median number of class absences differs from 15? Use the Wilcoxon signed-rank test.

18 The useful lifetime of a certain type of leather cutting tool is thought to be normally distributed. A random sample of 10 such tools is selected and the tools are allowed to operate until the leather shows signs of fraying. The failure times, in hours, are

1400	1620	1100	1500	1450
1725	1560	1650	1700	1600

For $\alpha = 0.05$, use the Lilliefors test to determine if these sample values can be viewed as having been selected from a normal distribution. What is the p-value for this test?

APPENDIX 10.A ASSESSING THE RANDOMNESS OF A SAMPLE

One of the key assumptions made during our discussion of estimation and testing was that we were engaging in a process of simple random sampling from a normal population. Appendix 7.A considered the issue of assessing the normality of a simple random sample. Now, we look to whether or not we can view a given data set "as if" it was drawn in a random fashion.

The type of test that we shall employ to assess the randomness of a set of observations is the *runs test*. As the name implies, this test is based on the notion of a *run*—an unbroken sequence of identical outcomes/elements (denoted by, say, letters) that are preceded or followed by different outcomes or no outcomes at all. For instance, suppose a particular piece of machinery produces the following ordered sequence of defective (d) and non-defective (n) items:

nnnn/dd/n/d/nnn/d/nnn/d

According to the definition of a run, this sequence has eight runs, where a slash is used to separate individual runs. Thus each sequence of n's and d's, uninterrupted by the other letter, identifies a run. It is important to note that: (1) the order in which the observations occur must be preserved so that the various subsequences of runs can be identified; and (2) the length of each run is irrelevant.

As will be evidenced below, a runs test is used to determine if the elements within an ordered sequence occur in a random fashion, where each element within the sequence takes on one of two possible values (e.g., the n's and d's used above). The total number of runs exhibited by a given sequence of outcomes usually serves as a good indicator of its possible "lack of randomness." That is, either too few or too many runs provide us with evidence supporting the lack of randomness in a set of observations.

Suppose we observe an ordered sequence of n observations and we are able to identify "two types of outcomes or attributes," that we shall generally denote by the letters a and b. Suppose further that in the entire sequence of n items we have n_1 a's and n_2 b's, with $n_1 + n_2 = n$. Additionally, let the random variable R denote the number of runs occurring in the ordered sequence of n items.

In the test for sample randomness to follow, the null hypothesis will always be H_0: the order of the sample data is random, and the alternative hypothesis will appear as H_1: the order of the sample data is not random. Moreover, if either n_1 or n_2 exceeds 15, the sampling distribution of R can be shown to be approximately normal with

$$E(R) = \frac{2n_1n_2}{n} + 1, \tag{10.A.1}$$

$$V(R) = \frac{2n_1n_2(2n_1n_2 - n)}{n^2(n-1)}, \tag{10.A.2}$$

and $n = n_1 + n_2$. Then

$$Z_R = \frac{R - E(R)}{\sqrt{V(R)}} \tag{10.A.3}$$

is $N(0,1)$ and thus, for a test conducted at the $100\alpha\%$ level, we will reject H_0 if $|Z_R| > Z_{\alpha/2}$.

Given a set of "numerical values" or values measured on an interval or ratio scale, let us look to the specification of a runs test for randomness that employs runs above and below the sample median. To this end, let us first find the median of the data set. Once it is located, we must then determine if each observation in the ordered sequence is above or below the median. Using the same type of letter code employed earlier, let a represent the label assigned to a value falling above the median, and let b denote the label assigned to a value located below the median. (We shall discard any numerical values equal to the median.) Here too n_1 = total number of a's, n_2 = total number of b's, and R = total number of runs above and below the median. Once the values of n_1, n_2, and R have been determined, Equations (10.A.1)–(10.A.3) can be used to test the hypothesis of randomness.

Example 10.A.1

Suppose the following ordered sequence of $n = 40$ observations (Table 10.A.1) have been taken from a given population and we want to, say, determine a confidence interval for the population mean. Can we proceed "as if" this data set has been drawn in a random fashion?

Arranging these observations in an increasing sequence enables us to determine that the median is 41. Items above 41 will be coded with an a, and items below 41 will be labeled with a b. The ordered sequence of a's and b's appears as

aa/bb/aaaaa/bb/aaaa/bb/aaaaa/bbbbbbbbb/aa/bbbb

(Note that three 41's have been discarded since they coincide with the median.) Thus $n_1 = 18$, $n_2 = 19$, and $R = 10$. For $\alpha = 0.05$, the critical region appears as $\{Z_R \| Z_R| > Z_{0.025} = 1.96\}$, and, from Equations (10.A.1)–(10.A.3):

$$E(R) = \frac{2(18)(19)}{37} + 1 = 19.48,$$

$$V(R) = \frac{2(18)(19)[2(18)(19) - 37]}{37^2(36)} = 8.98,$$

and

$$|Z_R| = \left| \frac{10 - 19.48}{\sqrt{8.98}} \right| = 3.17.$$

TABLE 10.A.1 Ordered Data Values

68 69 32 35 59 57 57 58 61 32 35 41 61 67 68 60 32
36 59 60 41 42 50 60 35 30 29 30 22 27 22 25 37 50
61 2 31 31 41 30

Since $|Z_R| = 3.17 > 1.96$, we will reject H_0: the order of the sample data is random, in favor of H_1: the order of the sample observations is not random, at the $\alpha = 0.05$ level.

APPENDIX 10.B WILCOXON SIGNED RANK TEST (OF A MEDIAN)

Suppose we have evidence that points to a symmetric population distribution. Then a *Wilcoxon signed rank test for the median* is in order. Additionally, with the population distribution symmetric, this test for the median is "equivalent to a test for the mean." To execute the Wilcoxon signed rank test, let us extract a random sample of size n with values $X_1, X_2, \ldots, X_n$ from a continuous symmetric population and test the following hypotheses pertaining to the population median MED:

Case 1	Case 2	Case 3
H_0: $MED = MED_o$ H_1: $MED \neq MED_o$	H_0: $MED = MED_o$ H_1: $MED > MED_o$	H_0: $MED = MED_o$ H_1: $MED < MED_o$

where MED_o is the null value of MED.

To perform the Wilcoxon signed rank test, let us consider the following sequence of steps:

1. Subtract the null value MED_o from each X_i to obtain $Y_i = X_i - MED_o$, $i = 1, \ldots, n$. (If any $Y_i = 0$, then eliminate X_i and reduce n accordingly.)
2. Rank the Y_i's in order of increasing absolute value. (If any of the nonzero Y_i's are tied in value, then these tied Y_i's are given the average rank.)
3. Restore to the rank values $1, \ldots, n$ the algebraic sign of the associated difference Y_i. Then the ranks with the appropriate signs attached are called the *signed ranks* R_i, $i = 1, \ldots, n$, where R_i^+ denotes the rank carrying a positive sign.

Let us specify our test statistic as

$$W^+ = \sum_{i=1}^{n} R_i^+, \tag{10.B.1}$$

the sum of the positive signed ranks. Then for $n > 15$, a reasonably good approximate test can be performed using the standard normal distribution. In this regard, it can be shown that the standardized test statistic under H_0 is

$$Z_{w^+} = \frac{w^+ - \frac{m(m+1)}{4}}{\sqrt{\frac{m(m+1)(2m+1)}{24}}} \text{ is } N(0,\ 1), \tag{10.B.2}$$

where m is the "final" number of observations that are ranked. Given Equation (10.B.2), our decision rule for rejecting H_0 in favor of H_1 at the 100α % level of significance is:

Case 1: Reject H_0 if $|Z_{w^+}| > Z_{\frac{\alpha}{2}}$ (10.B.3a)

Case 2: Reject H_0 if $Z_{w^+} > Z_\alpha$ (10.B.3b)

Case 3: Reject H_0 if $Z_{w^+} < -Z_\alpha$ (10.B.3c)

Example 10.B.1

The daily ridership on an early morning commuter bus has been recorded over a 4-week period (Table 10.B.1). For $\alpha = 0.05$, does the median number of passengers per day differ significantly from 20?

Here H_0: $MED = MED_o = 20$ is to be tested against H_1: $MED \neq 20$. From Table 10.B.2, it is readily determined that $m = 21$ and

$$W^+ = \sum_{j=1}^{8} R_j^+ = 6.5 + 1 + 13.5 + 20 + 6.5 + 13.5 + 21 + 6.5$$
$$= 88.5.$$

(Note that 5 is given an average rank of 6.5[2], 10 is given an average rank of 13.5, and 11 has an average rank of 18.) Then from Equation (10.B.2)

$$|Z_{W^+}| = \left| \frac{88.5 - \frac{21(22)}{4}}{\sqrt{\frac{21(22)(43)}{24}}} \right| = 0.938.$$

Since $|Z_{W^+}|$ does not exceed $Z_{\frac{\alpha}{2}} = Z_{0.025} = 1.96$, we cannot reject H_0 for $\alpha = 0.05$—there is insufficient evidence for us to conclude that the median number of passengers per day differs significantly from 20 at the 5% level.

In Sections 10.1–10.9 of this chapter, we have been engaged in testing *parametric hypotheses*—we utilized data measured on a quantitative (interval or ratio) scale and we assumed some knowledge about the "form" of the population distribution (e.g.,

TABLE 10.B.1 Daily Ridership on an AM Commuter Bus (Number of Passengers Per Day)

Week 1	20	25	21	20	30	35
Week 2	15	16	10	25	30	40
Week 3	10	10	9	11	15	20
Week 4	10	9	9	15	17	25

[2] $|Y_i| = 1$ gets rank 1, $|Y_i| = 3$ gets rank 2, and $|Y_i| = 4$ has rank 3. There are six $|Y_i| = 5$ values that will occupy the rank positions 4, 5, 6, 7, 8, and 9. Hence the average rank of $|Y_i| = 5$ is $(4 + 5 + 6 + 7 + 8 + 9)/6 = 39/6 = 6.5$. The next largest $|Y_i|$ value is 9—and it gets a rank of 10, and so on.

TABLE 10.B.2 Ranks and Signed Ranks

$Y_i = X_i - 20$	$\lvert Y_i \rvert$	Rank of $\lvert Y_i \rvert$	Signed Ranks R_i
0	—	—	—
5	5	6.5	6.5
1	1	1.0	1.0
0	—	—	—
10	10	13.5	13.5
15	15	20.0	20.0
−5	5	6.5	−6.5
−4	4	3.0	−3.0
−10	10	13.5	−13.5
5	5	6.5	6.5
10	10	13.5	13.5
20	20	21.0	21.0
−10	10	13.5	−13.5
−10	10	13.5	−13.5
−11	11	18.0	−18.0
−9	9	10.0	−10.0
−5	5	6.5	−6.5
0	—	—	—
−10	10	13.5	−13.5
−11	11	18.0	−18.0
−11	11	18.0	−18.0
−5	5	6.5	−6.5
−3	3	2.0	−2.0
5	5	6.5	6.5

normality) and that our observations were drawn independently of each other (under simple random sampling). However, if we are unsure about the legitimacy of any of these assumptions, then a *non-parametric (distribution-free)* technique may be appropriate—a hypothesis testing technique that does not require that the probability distribution of some population characteristic assume any specific functional form. Two examples of this latter category of tests are Examples 10.A.1 and 10.B.1. In fact, for the Wilcoxon signed rank test, only the symmetry of the population was assumed—no assumption concerning the normality of the population was made.

The non-parametric Wilcoxon signed rank test has a parametric counterpart—it is equivalent to the t test of H_0: $\mu = \mu_o$ (which is made under the assumption of normality of the population distribution). It is important to note that:

1. The null hypothesis formulated in a non-parametric test is not as precise as the null hypothesis for a parametric test.
2. For small samples, the normality assumption is quite heroic; for such samples a non-parametric test may be appropriate.
3. When both parametric and non-parametric techniques apply, a non-parametric test may require a larger sample size than a parametric test.

APPENDIX 10.C LILLIEFORS GOODNESS-OF-FIT TEST FOR NORMALITY

In Appendix 7.B, we developed the concept of a normal probability plot as a graphical device for determining if a sample data set could be treated "as if" it was extracted from a normal population. Let us now consider an alternative and more formal "test for normality." Specifically, we shall test for the goodness-of-fit of a set of sample observations to a normal distribution with unknown mean and standard deviation—the *Lilliefors Test*. To set the stage for the development of this procedure, let us first review the concept of a *cumulative distribution* function (*CDF*) and then develop the notion of a sample or *empirical cumulative distribution function* (*ECDF*).

We previously defined the *CDF* of a discrete random variable X (Section 5.1) as

$$F\left(X_j\right) = P(X \leq X_J) = \sum\nolimits_{k \leq J} f(X_k),$$

where f is a probability mass function. The *ECDF*, denoted as $S(X)$, is formed by ordering the set of sample values from the smallest to largest (we form the set of order statistics $X_{(1)}, X_{(2)}, \ldots, X_{(n)}$) and then plotting the cumulative relative frequencies. Clearly, $S(X)$ is a discrete random variable that exhibits the proportion of the sample values that are less than or equal to X. It plots as a step function (Fig. 10.C.1)—it jumps or increases by at least $\frac{1}{n}$ at each of its points of discontinuity (where the latter occur at $X_{(1)}, X_{(2)}, \ldots, X_{(n)}$) as we move along the X-axis from left to right. (For a discrete random variable X, if two or more observations are equal in value, the *ECDF* jumps $\frac{1}{n}$ times the number of data points equal to that particular value.)

For instance, given a random sample of $n = 10$ observations on the random variable X: 3, 1, 3, 2, 4, 9, 3, 5, 7, 5, we first form the set of order statistics:

$$X_{(1)} = min\, X_i = 1,\ 2,\ 3,\ 3,\ 3,\ 4,\ 5,\ 5,\ 7,\ 9 = X_{(10)} = max\, X_i.$$

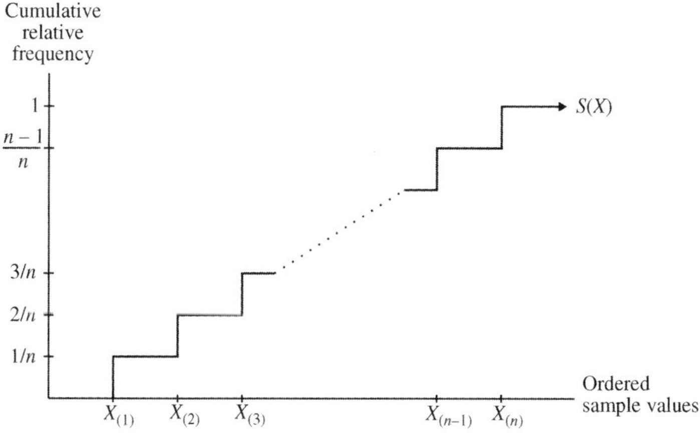

FIGURE 10.C.1 ECDF.

TABLE 10.C.1 ECDF

X	Relative Frequency $f(X)$	$S(X)$
1	0.10	0.10
2	0.10	0.20
3	0.30	0.50
4	0.10	0.60
5	0.20	0.80
7	0.10	0.90
9	0.10	1.00
	1.00	

Table 10.C.1 presents the relative frequencies and the values of the *ECDF*. A plot of the same appears in Fig. 10.C.2. In sum, $S(X)$ is the *ECDF* of the ordered $X_{(i)}$, $i = 1, \ldots, n$, taken from the population *CDF*, $F(X)$.

In the test that follows, we shall compare the values of a hypothetical *CDF*, $F_o(X)$ (called the *null CDF*), with the values of $S(X)$, the *ECDF*, for a set of ordered sample values, where $S(X)$ serves as an estimator of the population *CDF*, $F(X)$, for all $X_{(i)}$, $i = 1, \ldots, n$. For large n, the absolute differences between $S(X)$ and $F(X)$ or $|S(X) - F(X)|$ (the vertical distance between $S(X)$ and $F(X)$ at X) should be small save for sampling variation. In this regard, we test H_0:$F(X)=F_o(X)$ versus H_1: $F(X)$

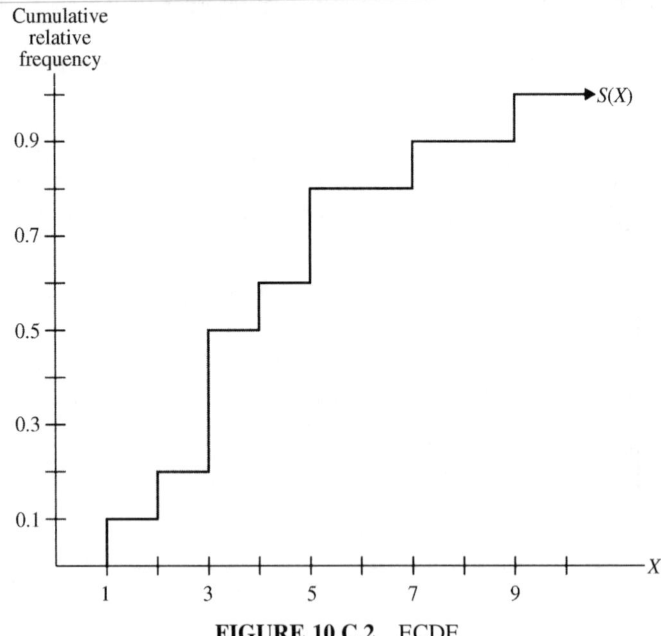

FIGURE 10.C.2 ECDF.

$\neq F_o(X)$ for at least one X. If H_0 is true, the absolute differences $|S(X) - F_o(X)|$ should be relatively small and attributable to only sampling fluctuations. Hence the random variable

$$D = \max_{X} |S(X) - F(X)| \tag{10.C.1}$$

will serve as a measure of goodness-of-fit between $S(X)$ and $F(X)$. Here D is the maximum of all vertical distances determined at each (ordered) sample value.

The Lilliefors Test procedure for normality now follows:

1. Assume that the values $X_1, X_2, \ldots, X_n$ constitute a random sample from some unknown population and the hypothesized *CDF*, F_o (X), is continuous.
2. We test H_0: the random sample is from a normal *CDF* with unknown mean and standard deviation, against H_1: the random sample is not from a normal *CDF*.
3. From the order statistics $X_{(i)}$, $i=1, \ldots, n$, form their standardized values

$$Z_{(i)} = \frac{X_{(i)} - \bar{X}}{s}, \quad i = 1, \ldots, n. \tag{10.C.2}$$

   Then the two-sided test statistic is

$$D = max\left|S(Z) - F\left(Z_{(i)};\ 0,\ 1\right)\right|, \tag{10.C.3}$$

   where $S(Z)$ is the *ECFD* of the $Z_{(i)}$'s and $F(Z_{(i)};0,1)$ is the standard normal *CDF*.
4. Reject H_0 if $D > d_{n,1-\alpha}$, where $d_{n,1-\alpha}$ is the $100(1-\alpha)\%$ quantile of the approximate sampling distribution of D (Table A.7). Otherwise, do not reject H_0.

TABLE 10.C.2 Calculation of Lilliefors D

$X_{(i)}$, $i=1, \ldots, 10$	$Z_{(i)}$, $i=1, \ldots, 10$	$S(Z)$	$F(Z_{(i)};0,1)$	$\lvert S(Z) - F\left(Z_{(I)};\ 0,\ 1\right)\rvert$
100	−2.55	0.10	0.0054	0.0946
145	−0.38	0.20	0.3520	0.1520
148	−0.23	0.30	0.4090	0.1090
150	−0.14	0.40	0.4443	0.0443
156	0.15	0.50	0.5596	0.0596
160	0.34	0.60	0.6331	0.0331
162	0.46	0.70	0.6772	0.0228
165	0.59	0.80	0.7224	0.0776
170	0.83	0.90	0.7967	0.1033
172	0.93	1.00	0.8238	0.1762

Example 10.C.1

For $1 - \alpha = 0.95$, determine if the following values of a random variable X can be viewed as having been chosen from a normal distribution: 145, 150, 170, 100, 172, 160, 165, 156, 148, 162. For this data set, $\bar{X} = 152.80$ and $s = 20.65$. The ordered sample values appear in Table 10.C.2 along with the $Z_{(i)}$'s (e.g. for $X_{(1)} = 100$, $Z_{(1)} = (100 - 152.80)/20.65 = -2{,}55$). From Equation (10.C.3), $D = 0.1762$. Since the critical region $R = \{D|D > 0.262\}$ (see Table A.7), we cannot reject H_0 at the 5% level of significance. Thus the normal CDF seems to be a reasonable representation of the unknown population CDF.

11

COMPARING TWO POPULATION MEANS AND TWO POPULATION PROPORTIONS

11.1 CONFIDENCE INTERVALS FOR THE DIFFERENCE OF MEANS WHEN SAMPLING FROM TWO INDEPENDENT NORMAL POPULATIONS

Let $\{X_1, X_2, \ldots, X_{n_X}\}$ and $\{Y_1, Y_2, \ldots, Y_{n_Y}\}$ be random samples taken from independent normal distributions with means μ_X and μ_Y and variances σ_X^2 and σ_Y^2, respectively. Our goal is to obtain a $100(1-\alpha)\%$ confidence interval for the difference of means $\mu_X - \mu_Y$. To accomplish this task, we must first examine the properties of the sampling distribution of the difference between two sample means.

Suppose X and Y are independent random variables, where X is $N(\mu_X, \sigma_X)$ and Y is $N(\mu_Y, \sigma_Y)$. For $\bar{X} - \bar{Y}$ to serve as an estimator for $\mu_X - \mu_Y$, we need to study the properties of the distribution of $\bar{X} - \bar{Y}$ given that we are sampling from two independent normal populations. The following special cases present themselves.

11.1.1 Sampling from Two Independent Normal Populations with Equal and Known Variances

We *first assume that the population variances are equal and known or* $\sigma_X^2 = \sigma_Y^2 = \sigma^2$ is known. Then, from Chapter 7, $\bar{X}$ is $N(\mu_X, \sigma/\sqrt{n_X})$ and $\bar{Y}$ is $N(\mu_Y, \sigma/\sqrt{n_Y})$.

Statistical Inference: A Short Course, First Edition. Michael J. Panik.

With $\bar{X}$ and $\bar{Y}$ independent normally distributed random variables, it can be shown that $\bar{X} - \bar{Y}$ is normally distributed with mean $\mu_X - \mu_Y$ and variance $(\sigma_X^2/n_X) + (\sigma_Y^2/n_Y) = \sigma^2\left(\frac{1}{n_X} + \frac{1}{n_Y}\right)$. Then with σ^2 known, the quantity

$$Z_{\Delta\mu} = \frac{(X - Y) - (\mu_X - \mu_Y)}{\sigma\sqrt{\frac{1}{n_X} + \frac{1}{n_Y}}} \text{ is } N\,(0, 1). \tag{11.1}$$

Hence a *100 (1 − α)% confidence interval for* $\mu_X - \mu_Y$ *when* $\sigma_X^2 = \sigma_Y^2 = \sigma^2$ *is known* is

$$(\bar{X} - \bar{Y}) \pm Z_{\alpha/2}\sigma\sqrt{\frac{1}{n_X} + \frac{1}{n_Y}}. \tag{11.2}$$

11.1.2 Sampling from Two Independent Normal Populations with Unequal but Known Variances

If *the population variances are unequal but still assumed to be known, then the* quantity

$$Z'_{\Delta\mu} = \frac{(\bar{X} - \bar{Y}) - \mu_X - \mu_Y}{\sqrt{\frac{\sigma_X^2}{n_X} + \frac{\sigma_Y^2}{n_Y}}} \text{ is } N(0, 1). \tag{11.3}$$

In this instance, a *100(1 − α)% confidence interval for* $\mu_X - \mu_Y$ *when* σ_X^2 and σ_X^2 *are known* is

$$(\bar{X} - \bar{Y}) \pm Z_{\alpha/2}\sqrt{\frac{\sigma_X^2}{n_X} + \frac{\sigma_Y^2}{n_Y}}. \tag{11.4}$$

11.1.3 Sampling from Two Independent Normal Populations with Equal but Unknown Variances

If *the common variance* $\sigma^2\left(=\sigma_X^2 = \sigma_Y^2\right)$ *is unknown*, then we need to examine the distribution of $\bar{X} - \bar{Y}$ under sampling from two independent normal populations with unknown but equal variances. Let the individual X and Y sample variances be written as

$$s_X^2 = \frac{\sum_{i=1}^{n_X}(X_i - \bar{X})^2}{n_X - 1} \text{ and } s_Y^2 = \frac{\sum_{i=1}^{n_Y}(Y_i - \bar{Y})^2}{n_Y - 1}$$

respectively, and let

$$s_p = \sqrt{\frac{(n_X - 1)s_X^2 + (n_Y - 1)s_Y^2}{k}}, \quad k = n_X + n_Y - 2. \tag{11.5}$$

denote the *pooled estimator* of the common standard deviation σ. Then for n_X and n_Y each less than or equal to 30, it can be demonstrated that the quantity

$$T_{\Delta\mu} = \frac{(\bar{X} - \bar{Y}) - (\mu_X - \mu_Y)}{s_p\sqrt{\frac{1}{n_X} + \frac{1}{n_Y}}} \tag{11.6}$$

follows a t distribution with *pooled degrees of freedom* $k = n_X + n_Y - 2$. Hence a *100(1 – α)% confidence interval for* $\mu_X - \mu_Y$ *when* σ_X^2 and σ_Y^2 *are unknown but equal* is

$$(\bar{X} - \bar{Y}) \pm t_{\alpha/2,k}\ s_p\sqrt{\frac{1}{n_X} + \frac{1}{n_Y}}. \tag{11.7}$$

Note that if both n_X and n_Y exceed 30, Equation (11.7) is replaced by

$$(\bar{X} - \bar{Y}) \pm Z_{\alpha/2} s_p\sqrt{\frac{1}{n_X} + \frac{1}{n_Y}}. \tag{11.7.1}$$

11.1.4 Sampling from Two Independent Normal Populations with Unequal and Unknown Variances

It can be shown that when σ_X^2 and σ_Y^2 in $Z'_{\Delta\mu}$ (Eq. 11.3) are replaced by their estimators s_X^2 and s_Y^2, respectively, and both n_X and n_Y are each less than or equal to 30, the resulting quantity

$$T'_{\Delta\mu} = \frac{(\bar{X} - \bar{Y}) - (\mu_X - \mu_Y)}{\sqrt{\frac{s_X^2}{n_X} + \frac{s_Y^2}{n_Y}}} \tag{11.8}$$

is approximately t distributed with degrees of freedom given approximately by

$$\phi = \frac{\left(\frac{s_X^2}{n_X} + \frac{s_Y^2}{n_Y}\right)^2}{\left(\frac{s_X^2}{n_X}\right)^2\left(\frac{1}{n_X+1}\right) + \left(\frac{s_Y^2}{n_Y}\right)^2\left(\frac{1}{n_Y+1}\right)} - 2. \tag{11.9}$$

(If ϕ is not an integer, it must be rounded to the nearest integer value.) Then a *100 (1 – α)% confidence interval for* $\mu_X - \mu_Y$ *when* σ_X^2 and σ_Y^2 *are unknown and unequal* is

$$(\bar{X} - \bar{Y}) \pm t_{\alpha/2,\phi}\sqrt{\frac{s_X^2}{n_X} + \frac{s_Y^2}{n_Y}}. \tag{11.10}$$

By way of some summary comments, it is important to remember that these two-sample procedures for the confidence interval estimation of $\mu_X - \mu_Y$ rely upon the following assumptions:

1. *Independent random samples*: This assumption is essential—it must hold if we are to make probability statements about $\mu_X - \mu_Y$.
2. *Normal populations*: This is a fairly weak assumption—if the samples are sufficiently large, then, via the Central Limit Theorem, substantial departures from normality will not greatly affect our confidence probabilities.
3. *Equal variances*: If the population variances are not the same (whether they are known or not), then our confidence interval for $\mu_X - \mu_Y$ is only approximate.

The confidence interval results for this section are summarized in Fig. 11.1.

Example 11.1 [based on Equation (11.2)]

Suppose that for independent *X* and *Y* populations we have, respectively,

$$X \text{ is } N(\mu_X, \sigma_X) = N(?, 3.7),$$
$$Y \text{ is } N(\mu_Y, \sigma_Y) = N(?, 3.7).$$

Clearly, each population has an unknown mean with $\sigma_X = \sigma_Y = \sigma = 3.7$. Our goal is to estimate $\mu_X - \mu_Y$. From a random sample of size $n_X = 40$ taken from the *X* population we obtain $\bar{X} = 7$; and from a random sample of size $n_Y = 45$ from the *Y* population we get $\bar{Y} = 5$. Suppose we choose $\alpha = 0.05$. Then from Equation (11.2),

$$(7 - 5) \pm 1.96(3.7)\sqrt{\frac{1}{40} + \frac{1}{45}}$$

or 2 ± 1.5722. Hence our 95% confidence interval for $\mu_X - \mu_Y$ is (0.4278, 3.5722), that is, we may be 95% confident that the difference in population means $\mu_X - \mu_Y$ lies between 0.4278 and 3.5722. How precisely have we estimated $\mu_X - \mu_Y$? We are within ± 1.5722 units of $\mu_X - \mu_Y$ with 95% reliability.

Example 11.2 [based on Equation (11.4)]

Suppose that for the independent *X* and *Y* populations we now have

$$X \text{ is } N(\mu_X, \sigma_X) = N(?, 3.7),$$
$$Y \text{ is } N(\mu_Y, \sigma_Y) = N(?, 4.2)$$

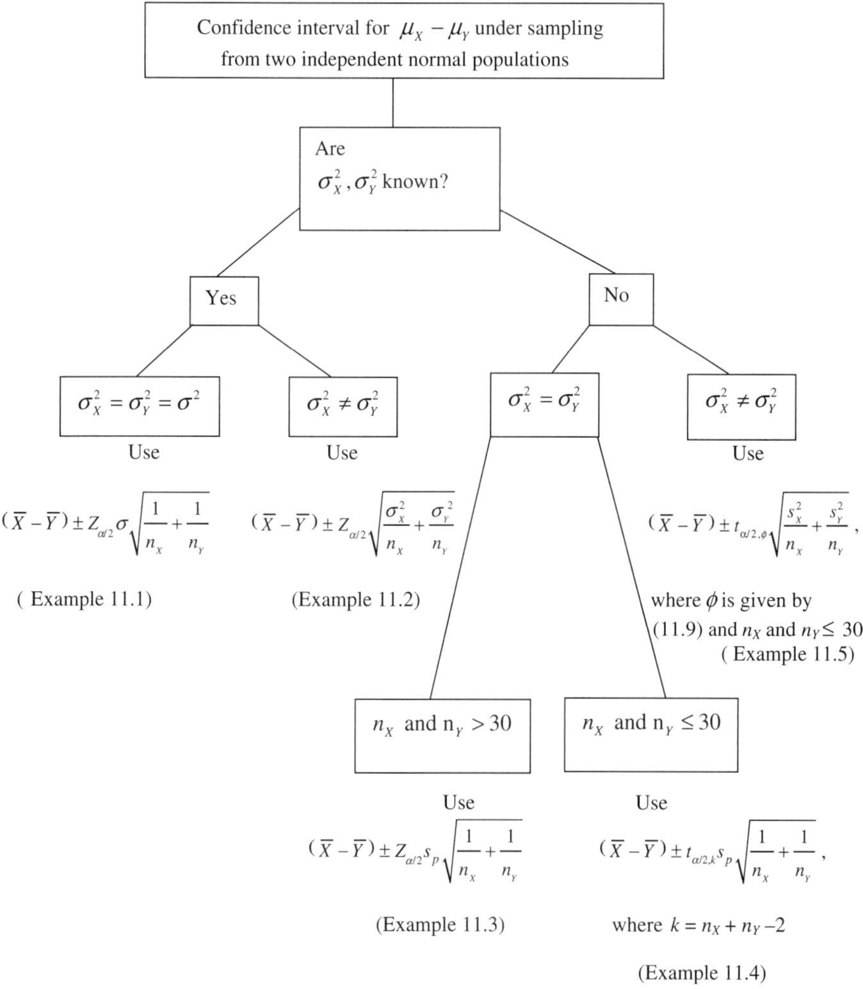

FIGURE 11.1 Confidence intervals for $\mu_X - \mu_Y$.

respectively. Again each population has an unknown mean but now, although σ_X and σ_Y are still known, $\sigma_X \neq \sigma_Y$. As in the preceding example problem, let

$$n_X = 40, \quad n_Y = 45, \quad \bar{X} = 7, \quad \bar{Y} = 5$$

and $\alpha = 0.05$. Then from Equation (11.4),

$$(7 - 5) \pm 1.96\sqrt{\frac{(3.7)^2}{40} + \frac{(4.2)^2}{45}}$$

or 2 ± 1.6796. Thus our 95% confidence interval for $\mu_X - \mu_Y$ is (0.3204, 3.6796). Hence we may be 95% confident that $\bar{X} - \bar{Y}$ will not differ from $\mu_X - \mu_Y$ by more than ± 1.6796 units (our "error bound" interpretation).

Example 11.3 [based on Equation (11.7.1)]

Suppose we extract random samples of size $n_X = 45$ and $n_Y = 50$, respectively, from the independent normally distributed X and Y populations and we find that

$$\bar{X} = 15, \quad \sum_{i=1}^{45} (X_i - \bar{X})^2 = 3{,}467,$$

$$\bar{Y} = 12.7, \quad \sum_{i=1}^{50} (Y_i - \bar{Y})^2 = 5{,}500.$$

Then

$$s_X^2 = \frac{3{,}467}{44} = 78.79, \quad s_Y^2 = \frac{5{,}500}{49} = 112.24$$

and, via Equation (11.5),

$$s_p = \sqrt{\frac{44(78.79) + 49(112.24)}{45 + 50 - 2}} = 9.8191.$$

Let us choose $\alpha = 0.05$. Given that we have "large samples" from the X and Y populations, our 95% confidence interval estimate for $\mu_X - \mu_Y$ is, from Equation (11.7.1),

$$2.3 \pm 1.96(9.8191)\sqrt{\frac{1}{45} + \frac{1}{50}}$$

or 2.3 ± 3.9530. Hence our 95% confidence interval for $\mu_X - \mu_Y$ is (−1.6530, 6.2530), that is, we may be 95% confident that $\mu_X - \mu_Y$ lies between −1.6530 and 6.2530.

Example 11.4 [based on Equation (11.7)]

Let us take random samples of size $n_X = 15$ and $n_Y = 20$ from independent normal X and Y populations, respectively. From these samples, we calculate

$$\bar{X} = 22, \quad \sum_{i=1}^{15} (X_i - \bar{X})^2 = 2{,}262,$$

$$\bar{Y} = 25, \quad \sum_{i=1}^{20} (Y_i - \bar{Y})^2 = 3{,}016.$$

Then

$$s_X^2 = \frac{2{,}262}{14} = 161.57, \quad s_Y^2 = \frac{3{,}016}{19} = 158.89$$

and, from Equation (11.5),

$$s_p = \sqrt{\frac{14(161.57) + 19(158.89)}{15 + 20 - 2}} = 12.846.$$

For $\alpha = 0.05$,

$$t_{\alpha/2,k} = t_{0.025,32} = 2.037$$

and thus, from Equation (11.7), our 95% "small-sample" confidence limits for $\mu_X - \mu_Y$ appear as

$$-3 \pm 2.037(12.846)\sqrt{\frac{1}{15} + \frac{1}{20}}$$

or -3 ± 8.9388. Hence a 95% confidence interval for $\mu_X - \mu_Y$ is $(-11.9388, 5.9388)$, that is, we may be 95% confident that the difference $\mu_X - \mu_Y$ lies between $-11.\ 94$ and 5.94.

Example 11.5 [based on Equation (11.10)]

Suppose we extract random samples from independent normal X and Y populations, respectively:

X Sample Values	Y Sample Values
10, 6, 13, 5, 11, 10, 14	13, 14, 9, 3, 1, 10, 4, 8, 8

Our objective is to determine a 95% confidence interval for $\mu_X - \mu_Y$. Here

$$n_X = 7, \quad \sum X_i = 69, \quad \sum X_i^2 = 747,$$

$$n_Y = 9, \quad \sum Y_i = 70, \quad \sum Y_i^2 = 700.$$

Then

$$\bar{X} = 9.86, \quad s_X^2 = \frac{\sum X_i^2}{n_X - 1} - \frac{\left(\sum X_i\right)^2}{n_X(n_X - 1)} = 11.14,$$

$$\bar{Y} = 7.78, \quad s_Y^2 = \frac{\sum Y_i^2}{n_Y - 1} - \frac{\left(\sum Y_i\right)^2}{n_Y(n_Y - 1)} = 19.44.$$

With $\alpha = 0.05$ and two obviously "small samples,"

$$t_{a/2,\phi} = t_{0.025,12} = 2.179,$$

where, from Equation (11.9),

$$\phi = \frac{\left(\frac{11.14}{7} + \frac{19.44}{9}\right)^2}{\left(\frac{11.14}{7}\right)^2\left(\frac{1}{8}\right) + \left(\frac{19.44}{9}\right)^2\left(\frac{1}{10}\right)} - 2 \approx 12.$$

Then from Equation (11.10), a 95% confidence interval for $\mu_X - \mu_Y$ is

$$2.08 \pm 2.179\sqrt{\frac{11.14}{7} + \frac{19.44}{9}}$$

or 2.08 ± 4.243. Hence we may be 95% confident that $\mu_X - \mu_Y$ lies within the interval (–2.163, 6.323). How precisely have we estimated $\mu_X - \mu_Y$? We are within ± 4.243 units of $\mu_X - \mu_Y$ with 95% reliability. Alternatively stated, we may be 95% confident that $\bar{X} - \bar{Y}$ will not differ from $\mu_X - \mu_Y$ by more than ± 4.243 units.

11.2 CONFIDENCE INTERVALS FOR THE DIFFERENCE OF MEANS WHEN SAMPLING FROM TWO DEPENDENT POPULATIONS: PAIRED COMPARISONS

The inferential techniques developed in the preceding section were based upon random samples taken from two "independent" normal populations. What approach should be taken when the samples are either intrinsically or purposefully designed to be "dependent?" A common source of dependence is when the samples from the two populations are *paired*—each observation in the first sample is related in some particular way to exactly one observation in the second sample, so that the two samples are obviously not independent.

Why is pairing advantageous? A glance back at, say, Equation (11.7) reveals that the precision of the interval estimate of $\mu_X - \mu_Y$, for fixed sample sizes n_X and n_Y, varies inversely with s_p^2, the pooled estimate of the common variance σ^2. Here s_p^2 serves as a measure of the unexplained variation among experimental units or subjects that receive similar treatments. One way to possibly reduce this variation, and thus increase the precision of our estimate of $\mu_X - \mu_Y$, is to pair the sample observations. In this regard, if the variation in the treatment outcomes between the members of any pair is less than the variation between corresponding members of different pairs, then the precision of our estimate of $\mu_X - \mu_Y$ can be enhanced. And this can be accomplished by randomizing the two treatments over the two members of each pair in a fashion such that each treatment is applied to one and only one member of each pair. The result of this *paired experiment* is that we can obtain, for each pair of outcomes, an estimate of the difference between treatment effects, and thus variation between pairs is not included in our estimate of the common variance.

For instance, measurements are taken from the same experimental units at two different points in time and the means of each set of measurements are subsequently compared. In this regard, suppose a researcher conducts a before and after weighing of a group of people to determine the effectiveness of a new diet pill. Here the weights will be *paired*—for a particular individual, his/her weight at the start of the experiment (this measurement is an observation in the first sample) will be compared to his/her weight at the end of the experiment (this latter data point is a member of the second sample). Or experimental units in the two samples may be paired to eliminate extraneous factors or effects that are deemed unimportant for the purpose at hand. For instance, the director of a physical therapy program in a nursing home may be interested in testing a new exercise method that uses a video-based routine. Two new groups, each comprising, say, 10 individuals, are being formed so that the new approach can be tested. The "old" exercise method will be used in the first group and the video-based method will be used by the second group. The director selects 10 pairs of individuals, where the persons comprising each pair are required to be as similar as possible so that the effects of any extraneous factors can be minimized. After making sure that the two members of any pair are as similar as possible (save for the exercise method), the director can flip a coin and determine which person gets to use the video-based exercise format. Again the two samples are dependent since the assignments of the individuals to the two exercise methods were not completely random but were in pairs.

In general, grouping experimental units into pairs (by design) and thus according to overall commonality of extraneous factors, enables us to eliminate much of the outcome variation due to these factors from our estimate of the common variance σ^2, thus increasing the precision of our estimate of $\mu_X - \mu_Y$. And as previously stated, the pairs must be chosen so that the outcome variation among the pairs exceeds that occurring between the experimental units within the pairs.

As we shall now see, the result of a paired observation experiment (in which the two samples are not chosen independently and at random) reduces to an application of a single–sample technique. In this regard, suppose our paired experiment yields the pairs of observations (X_i, Y_i), $i = 1, \ldots, n$, where the sample value X_i is drawn from the first population and the sample value Y_i comes from the second population. (Think of the X_i's as depicting a set of sample outcomes for the first treatment and the Y_i's as representing a set of sample observations for the second or follow-up treatment.)

For the ith pair (X_i, Y_i), let $D_i = X_i - Y_i$, $i = 1, \ldots, n$, represent the ith observation on the random variable D. Each D_i thus provides us with a measure of the difference between the effectiveness of the two treatments. Then the best estimators for the mean and variance of D are

$$\bar{D} = \frac{\sum_{i=1}^{n} D_i}{n}, \quad s_D^2 = \frac{\sum_{i=1}^{n} (D_i - \bar{D})^2}{n-1} \tag{11.11}$$

respectively. Here $\bar{D}$ estimates the mean difference between the effects of the first and second treatments while s_D^2 estimates the variance of the differences in treatment effects and excludes any variation due to extraneous factors (if pairing has been effective).

If the D_i's, $i = 1, \ldots, n$, constitute a single random sample from a normal population with mean $\mu_D = \mu_X - \mu_Y$ and variance σ_D^2, then the random variable $\bar{D}$ is distributed as $N(\mu_D, \sigma_D/\sqrt{n})$. Hence, for $n \leq 30$, the statistic

$$T = \frac{\bar{D} - \mu_D}{s_{\bar{D}}} = \frac{\bar{D} - \mu_D}{s_D/\sqrt{n}} \tag{11.12}$$

follows a t distribution with $n-1$ degrees of freedom. Then a *100(1 − α)% confidence interval for* $\mu_D = \mu_X - \mu_Y$ is

$$\bar{D} \pm t_{\alpha/2,n-1}\frac{s_D}{\sqrt{n}}. \tag{11.13}$$

Example 11.6

A group of 15 bankers was asked to rate their own financial institution's prospects for growth and profitability over the next year on a scale from 1 to 5 (with 5 being the highest rating). After the rating was conducted, the group watched a broadcast of the President's State of the Union message to Congress. At the conclusion of the President's speech, all 15 were again asked to rate their institution's economic prospects on the same 1 to 5 scale. The results of both ratings appear in Table 11.1.

How precisely have we estimated the true difference between the before- and after-mean ratings of the future economic climate by the group? Let $\alpha = 0.05$. From the differences D_i, $i = 1,\ldots,15$, appearing in Table 11.1, we have

$$\bar{D} = \frac{\sum_{i=1}^{15} D_i}{n} = -\frac{4}{15} = -0.266,$$

$$s_D^2 = \frac{\sum_{i=1}^{15} D_i^2}{n-1} - \frac{\left(\sum_{i=1}^{15} D_i\right)^2}{n(n-1)} = \frac{14}{14} - \frac{(-4)^2}{15(14)} = 0.924,$$

TABLE 11.1 Ratings of Economic Prospects by Bankers

Before (X)	After (Y)	Difference ($D = X - Y$)
4	4	0
4	5	−1
3	3	0
5	5	0
5	5	0
5	4	1
2	4	−2
3	5	−2
4	5	−1
3	3	0
3	4	−1
5	4	1
5	4	1
4	4	0
4	4	0

and thus, from Equation (11.13) (here $t_{\alpha/2,n-1} = t_{0.025,14} = 2.145$), our 95% confidence interval for μ_D is

$$-0.266 \pm 2.145\left(\frac{0.9613}{\sqrt{15}}\right)$$

or -0.266 ± 0.5324. Hence we may be 95% confident that μ_D lies between -0.7984 and 0.2664, that is, we are within ± 0.5324 units of $\mu_D = \mu_X - \mu_Y$ with 95% reliability.

11.3 CONFIDENCE INTERVALS FOR THE DIFFERENCE OF PROPORTIONS WHEN SAMPLING FROM TWO INDEPENDENT BINOMIAL POPULATIONS

Let $\{X_1, X_2, \ldots, X_{n_X}\}$ and $\{Y_1, Y_2, \ldots, Y_{n_Y}\}$ be random samples drawn from two independent binomial populations, where p_X and p_Y are the proportions of successes in the first and second binomial populations, respectively. Additionally, let X and Y be independent random variables representing the observed number of successes in the samples of size n_X and n_Y, respectively.

To determine a $100(1-\alpha)\%$ confidence interval for the difference in proportions $p_X - p_Y$, let us first discuss the characteristics of the sampling distribution of the difference between two sample proportions. We found in Section 9.1 that the best estimators for p_X and p_Y are *the sample proportions* or *observed relative frequencies of successes* $\hat{p}_X = X/n_X$ and $\hat{p}_Y = Y/n_Y$, respectively, with

$$E(\hat{p}_X) = p_X, \quad V(\hat{p}_X) = \frac{p_X(1-p_X)}{n_X},$$
$$E(\hat{p}_Y) = p_Y, \quad V(\hat{p}_Y) = \frac{p_Y(1-p_Y)}{n_Y}.$$

Hence the best estimator for $p_X - p_Y$ is $\hat{p}_X - \hat{p}_Y$; and the best estimators for $V(\hat{p}_X)$ and $V(\hat{p}_Y)$ are, respectively, $s^2(\hat{p}_X) = \frac{\hat{p}_X(1-\hat{p}_X)}{n_X}$ and $s^2(\hat{p}_Y) = \frac{\hat{p}_Y(1-\hat{p}_Y)}{n_Y}$.

Then it can be shown that, for "large samples," the quantity

$$Z_{\Delta p} = \frac{(\hat{p}_X - \hat{p}_Y) - (p_X - p_Y)}{\sqrt{s^2(\hat{p}_X) + s^2(\hat{p}_Y)}} \text{ is approximately } N(0,1). \tag{11.14}$$

On the basis of this result, a *large sample (approximate) 100(1 – α)% confidence interval for* $p_X - p_Y$ is

$$(\hat{p}_X - \hat{p}_Y) \pm Z_{\alpha/2}\sqrt{\frac{\hat{p}_X(1-\hat{p}_X)}{n_X} + \frac{\hat{p}_Y(1-\hat{p}_Y)}{n_Y}}. \tag{11.15}$$

It was mentioned above that, for large samples, Equation (11.14) holds approximately. In fact, a good approximation to normality obtains if n_X and n_Y are each ≥ 25. This method should not be employed if both n_X and n_Y are each <25.

Example 11.7

A random sample was obtained from each of two independent binomial populations, yielding the following results: $\hat{p}_X = X/n_X = 18/30 = 0.600$ and $\hat{p}_Y = Y/n_Y = 10/35 = 0.286$. Hence our best estimate of $p_X - p_Y$, the difference between the two population proportions of successes, is $\hat{p}_X - \hat{p}_Y = 0.314$. How precisely has $p_X - p_Y$ been estimated? Using Equation (11.15) and, for $\alpha = 0.05$ (thus $Z_{\alpha/2} = Z_{0.025} = 1.96$), the 95% confidence limits for $p_X - p_Y$ are

$$0.314 \pm 1.96\sqrt{\frac{0.600(0.400)}{30} + \frac{0.286(0.714)}{35}}$$

or 0.314 ± 0.2303. Hence we may be 95% confident that $p_X - p_Y$ lies within the interval (0.0837, 0.5443), that is, we are within ± 0.2303 units of the true difference $p_X - p_Y$ with 95% reliability.

11.4 STATISTICAL HYPOTHESIS TESTS FOR THE DIFFERENCE OF MEANS WHEN SAMPLING FROM TWO INDEPENDENT NORMAL POPULATIONS

Let $\{X_1, X_2, \ldots, X_{n_X}\}$ and $\{Y_1, Y_2, \ldots, Y_{n_Y}\}$ be random samples taken from independent normal distributions with means μ_X and μ_Y and variances σ_X^2 and σ_Y^2, respectively. Our objective is to test the difference between the two population means μ_X and μ_Y, where this difference will be denoted as $\mu_X - \mu_Y = \delta_o$. So if $\mu_X = \mu_Y$, then $\delta_o = 0$; if $\mu_X > \mu_Y$, then $\delta_o > 0$; and if $\mu_X < \mu_Y$, then $\delta_o < 0$. In this regard, we may test the null hypothesis $H_0 : \mu_X - \mu_Y = \delta_o$ against any one of the following three alternative hypotheses:

Case 1	Case 2	Case 3
$H_0: \mu_X - \mu_Y = \delta_o$	$H_0: \mu_X - \mu_Y = \delta_o$	$H_0: \mu_X - \mu_Y = \delta_o$
$H_1: \mu_X - \mu_Y \neq \delta_o$	$H_1: \mu_X - \mu_Y > \delta_o$	$H_1: \mu_X - \mu_Y < \delta_o$

To construct a test of $H_0: \mu_X - \mu_Y = \delta_o$ against any of the preceding alternatives, we must exploit the characteristics of the sampling distribution of the difference between the two sample means $\bar{X}$ and $\bar{Y}$ or the sampling distribution of $\bar{X} - \bar{Y}$. In this regard, suppose that X and Y are independent random variables, where X is $N(\mu_X, \sigma_X)$ and Y is $N(\mu_Y, \sigma_Y)$ and that $\bar{X} - \bar{Y}$ is our estimator for $\mu_X - \mu_Y$. As evidenced in Section 11.1 above, the form of the distribution of $\bar{X} - \bar{Y}$ must be specified under a variety of assumptions concerning the variances of X and Y.

11.4.1 Population Variances Equal and Known

Given that $\sigma_X^2 = \sigma_Y^2 = \sigma^2$ is known, the quantity

$$Z_{\delta_o} = \frac{(\bar{X} - \bar{Y}) - \delta_o}{\sigma\sqrt{\frac{1}{n_X} + \frac{1}{n_Y}}} \text{ is } N(0, 1). \tag{11.16}$$

Then for a test conducted at the $100\alpha\%$ level of significance, the appropriate decision rules for rejecting $H_0 : \mu_X - \mu_Y = \delta_o$ relative to H_1 are as follows:

$$\text{Case 1: Reject } H_0 \text{ if } |Z_{\delta_o}| > Z_{\alpha/2} \tag{11.17a}$$
$$\text{Case 2: Reject } H_0 \text{ if } Z_{\delta_o} > Z_{\alpha} \tag{11.17b}$$
$$\text{Case 3: Reject } H_0 \text{ if } Z_{\delta_o} < -Z_{\alpha} \tag{11.17c}$$

11.4.2 Population Variances Unequal but Known

If σ_X^2 and σ_Y^2 are known and $\sigma_X^2 \neq \sigma_Y^2$, then the quantity

$$Z'_{\delta_o} = \frac{(\bar{X} - \bar{Y}) - \delta_o}{\sqrt{\frac{\sigma_X^2}{n_X} + \frac{\sigma_Y^2}{n_Y}}} \quad \text{is} \quad N(0, 1). \tag{11.18}$$

Then for a given $100\alpha\%$ level of significance, the set of decision rules for rejecting $H_0 : \mu_X - \mu_Y = \delta_o$ relative to H_1 are as follows:

$$\text{Case 1: Reject } H_0 \text{ if } |Z'_{\delta_o}| > Z_{\alpha/2} \tag{11.19a}$$
$$\text{Case 2: Reject } H_0 \text{ if } Z'_{\delta_o} > Z_{\alpha} \tag{11.19b}$$
$$\text{Case 3: Reject } H_0 \text{ if } Z'_{\delta_o} < -Z_{\alpha} \tag{11.19c}$$

11.4.3 Population Variances Equal and Unknown

If $\sigma^2 \left(= \sigma_X^2 = \sigma_Y^2\right)$ is unknown and n_X and n_Y are each less than or equal to 30, the quantity

$$T_{\delta_o} = \frac{(\bar{X} - \bar{Y}) - \delta_o}{s_p\sqrt{\frac{1}{n_X} + \frac{1}{n_Y}}} \tag{11.20}$$

is t distributed with $k = n_X + n_Y - 2$ degrees of freedom and

$$s_p = \sqrt{\frac{(n_X - 1)s_X^2 + (n_Y - 1)s_Y^2}{k}} \tag{11.21}$$

is the pooled estimator of the common standard deviation σ, with

$$s_X^2 = \frac{\sum_{i=1}^{n_X}(X_i - \bar{X})^2}{n_X - 1}, \quad s_Y^2 = \frac{\sum_{i=1}^{n_Y}(Y_i - \bar{Y})^2}{n_Y - 1}.$$

Then at the 100α% level of significance, our decision rules for rejecting $H_0 : \mu_X - \mu_Y = \delta_o$ relative to H_1 are as follows:

$$\text{Case 1: Reject } H_0 \text{ if } |T_{\delta_o}| > t_{\alpha/2,k} \tag{11.22a}$$

$$\text{Case 2: Reject } H_0 \text{ if } T_{\delta_o} > t_{\alpha,k} \tag{11.22b}$$

$$\text{Case 3: Reject } H_0 \text{ if } T_{\delta_o} < -t_{\alpha,k} \tag{11.22c}$$

If both n_X and n_Y exceed 30, then Equation (11.20) is replaced by

$$Z''_{\delta_o} = \frac{(\bar{X} - \bar{Y}) - \delta_o}{s_p\sqrt{\frac{1}{n_X} + \frac{1}{n_Y}}} \text{ is } N(0,1) \tag{11.20.1}$$

so that, at the 100α% level of significance:

$$\text{Case 1: Reject } H_0 \text{ if } |Z''_{\delta_o}| > Z_{\alpha/2} \tag{11.22.1a}$$

$$\text{Case 2: Reject } H_0 \text{ if } Z''_{\delta_o} > Z_{\alpha} \tag{11.22.1b}$$

$$\text{Case 3: Reject } H_0 \text{ if } Z''_{\delta_o} < -Z_{\alpha} \tag{11.22.1c}$$

It is important to note that the preceding t test is based upon the following set of assumptions:

1. Both population distributions are normal.
2. The two random samples (along with the observations within each sample) are independent.
3. The population variances are equal.

That assumption 2 holds is essential and, for large samples, the violation of assumptions 1 and 3 is not all that serious. However, if the sample sizes are small and assumption 3 seems untenable, then, as will be indicated in section 11.4.4, the pooled estimator of σ (Eq. 11.21) can be replaced by separate (unbiased) estimators for σ_X^2 and σ_Y^2 and an approximation to degrees of freedom can be employed.

11.4.4 Population Variances Unequal and Unknown (an Approximate Test)

If the sample sizes n_X and n_Y are small and the population variances σ_X^2 and σ_Y^2 are unknown and unequal, the quantity

$$T'_{\delta_o} = \frac{(\bar{X} - \bar{Y}) - \delta_o}{\sqrt{\frac{s_X^2}{n_X} + \frac{s_Y^2}{n_Y}}} \tag{11.23}$$

is approximately t distributed with degrees of freedom given by

$$\phi = \frac{\left(\frac{s_X^2}{n_X} + \frac{s_Y^2}{n_Y}\right)^2}{\left(\frac{s_X^2}{n_X}\right)^2 \left(\frac{1}{n_X+1}\right) + \left(\frac{s_Y^2}{n_Y}\right)^2 \left(\frac{1}{n_Y+1}\right)} - 2, \tag{11.24}$$

where s_X^2 and s_Y^2 are given in Equation (11.21) and ϕ must be rounded to the nearest integer value. When using Equations (11.23) and (11.24) to test $H_o : \mu_X - \mu_Y = \delta_o$ versus H_1 at the $100(1-\alpha)\%$ level of significance, the decision rules for rejecting H_0 relative to H_1 are:

$$\text{Case 1: Reject } H_0 \text{ if } |T'_{\delta_o}| > t_{\alpha/2,\phi} \tag{11.25a}$$
$$\text{Case 2: Reject } H_0 \text{ if } T'_{\delta_o} > t_{\alpha,\phi} \tag{11.25b}$$
$$\text{Case 3: Reject } H_0 \text{ if } T'_{\delta_o} < -t_{\alpha,\phi} \tag{11.25c}$$

The various hypothesis tests presented in this section are summarized in Fig. 11.2.

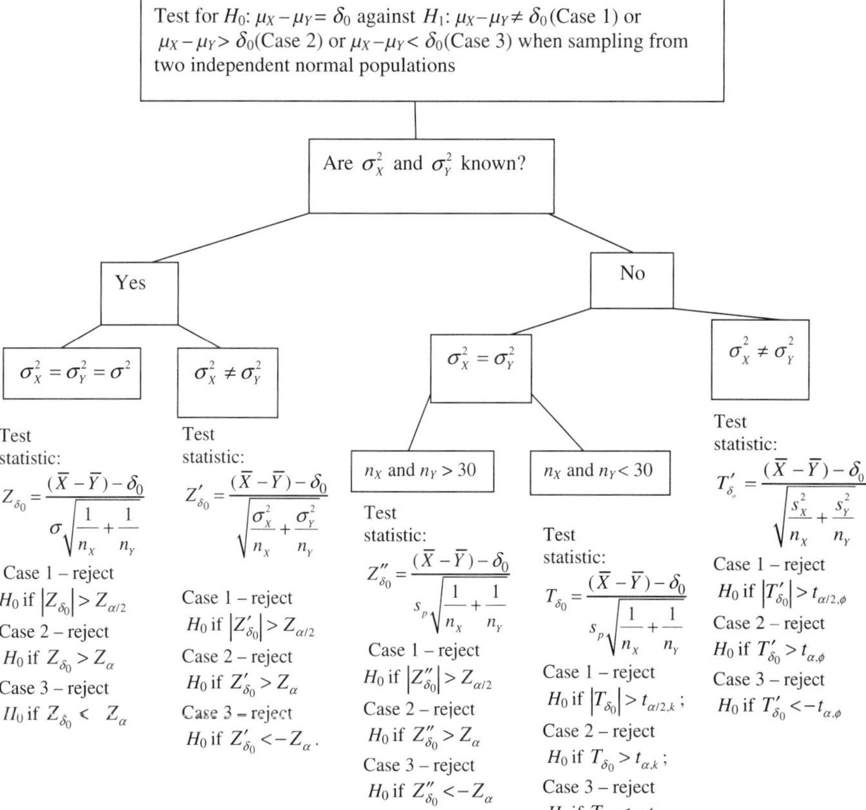

FIGURE 11.2 Hypothesis tests for $\mu_x - \mu_y$.

Example 11.8 [based on Equation (11.16)]

Suppose that random samples of size $n_X = 35$ and $n_Y = 48$ are taken from independent X and Y populations, respectively, with the result that $\bar{X} = 17$ and $\bar{Y} = 14$, where

$$X \text{ is } N(\mu_X, \sigma_X) = N(?, 1.6),$$
$$Y \text{ is } N(\mu_Y, \sigma_Y) = N(?, 1.6).$$

For $\alpha = 0.05$, let us test $H_0 : \mu_X - \mu_Y = \delta_o = 0$ (there is no difference between the X and Y population means), against $H_1 : \mu_X - \mu_Y > 0$ (the mean of the X population is larger than that of the Y population). Since the population variances are known and equal, Equation (11.16) gives

$$Z_{\delta_o} = \frac{17 - 14}{1.6\sqrt{\frac{1}{35} + \frac{1}{48}}} = 8.438.$$

And since the critical region is $R = \left\{ Z_{\delta_o} | Z_{\delta_o} > Z_\alpha = Z_{0.05} = 1.645 \right\}$ [see Equation (11.17.b)], this result has us reject H_0 in favor of H_1 at the 5% level, that is, μ_X lies significantly above μ_Y for $\alpha = 0.05$.

Example 11.9 [based on Equation (11.18)]

Given the independent X and Y populations

$$X \text{ is } N(\mu_X, \sigma_X) = N(?, 4.1),$$
$$Y \text{ is } N(\mu_Y, \sigma_Y) = N(?, 3.7),$$

suppose from random samples of size $n_X = 50$ and $n_Y = 45$, we find that $\bar{X} = 24$ and $\bar{Y} = 26$, respectively. For $\alpha = 0.01$, is there any reason to believe that the population means are different? Here we are to test $H_0 : \mu_X - \mu_Y = \delta_o = 0$ (no difference between μ_X and μ_Y) versus $H_1 : \mu_X - \mu_Y \neq 0$ (there is a difference between μ_X and μ_Y). From Equation (11.18) (the population variances are unequal but known),

$$Z'_{\delta_o} = \frac{24 - 26}{\sqrt{\frac{4.1}{50} + \frac{3.7}{45}}} = -4.94.$$

With $R = \left\{ Z'_{\delta_o} || Z'_{\delta_o} | > Z_{\alpha/2} = Z_{0.005} = 2.58 \right\}$ [see Equation (11.19a)], we can safely reject the null hypothesis of equality of means at the 1% level of significance.

Example 11.10 [based on Equation (11.20)]

Suppose we extract random samples of size $n_X = 10$ and $n_Y = 15$, respectively, from independent and normally distributed X and Y populations that are assumed to have unknown but equal variances and we find that:

$$\bar{X} = 20, \quad \sum_{i=1}^{10} (X_i - \bar{X})^2 = 237,$$

$$\bar{Y} = 17.5, \quad \sum_{i=1}^{15} (Y_i - \bar{Y})^2 = 251.$$

Let us test $H_0: \mu_X - \mu_Y = \delta_o = 0$, against $H_1: \mu_X - \mu_Y \neq 0$ for $\alpha = 0.05$. To accomplish this we first find:

$$s_X^2 = \frac{237}{9} = 26.33, \quad s_Y^2 = \frac{251}{14} = 17.92,$$

and, from Equation (11.21),

$$s_p = \sqrt{\frac{9(26.33) + 14(17.92)}{10 + 15 - 2}} = 4.61.$$

Then from Equation (11.20),

$$T_{\delta_o} = \frac{20 - 17.5}{4.61\sqrt{\frac{1}{10} + \frac{1}{15}}} = 1.33.$$

With $t_{\alpha/2,k} = t_{0.025,23} = 2.069$, the critical region appears as $R = \left\{ T_{\delta_o} \| T_{\delta_o} | > 2.069 \right\}$ [see Equation (11.22a)]. Clearly, we cannot reject H_0 at the 5% level of significance, that is, the X and Y population means are not significantly different for $\alpha = 0.05$.

Example 11.11 [based on Equation (11.23)]

In the preceding example problem, we made the assumption that the X and Y populations had "unknown but equal" variances. If the assumption of equality of variances is untenable (they are still taken to be unknown), then we must use Equations (11.23) and (11.24) to conduct our hypothesis test. In this regard, let us employ the sample information given in Example 11.10 to find

$$T'_{\delta_o} = \frac{20 - 17.5}{\sqrt{\frac{26.33}{10} + \frac{17.92}{15}}} = 1.28$$

and

$$\phi = \frac{\left(\frac{26.33}{10} + \frac{17.92}{15}\right)^2}{\left(\frac{26.33}{10}\right)^2\left(\frac{1}{11}\right) + \left(\frac{17.92}{15}\right)^2\left(\frac{1}{16}\right)} - 2 \approx 20.$$

Let us again test $H_0 : \mu_X - \mu_Y = \delta_o = 0$, against $H_1 : \mu_X - \mu_Y \neq 0$ with $\alpha = 0.05$. Hence $t_{\alpha/2,\phi} = t_{0.025,20} = 2.085$ so that $R = \left\{T'_{\delta_o} || T'_{\delta_o}| > 2.086\right\}$. Again, we cannot reject H_0 at the 5% level of significance.

If the normality assumption made in the preceding collection of tests is deemed "too heroic," the reader can consider a couple of nonparametric tests designed to assess differences among two population distributions offered in Appendices 11.A and 11.B.

11.5 HYPOTHESIS TESTS FOR THE DIFFERENCE OF MEANS WHEN SAMPLING FROM TWO DEPENDENT POPULATIONS: PAIRED COMPARISONS

A key assumption made in the preceding section was that random samples were taken from two "independent" (normal) populations. However, if the samples from the two populations are paired (each observation in the first sample is related in some particular way to exactly one observation in the second sample), then, as noted in Section 11.2 above, the samples are not independent. As we shall now see, the analysis of a paired observation experiment reduces to the application of a single-sample approach to hypothesis testing.

Following the approach of Section 11.2, let us assume that our paired experiment yields n pairs of observations denoted by $(X_1, Y_1), (X_2, Y_2), \ldots, (X_n, Y_n)$, where X_i is drawn from the first population and Y_i is drawn from the second population, $i = 1, \ldots, n$. For the ith pair (X_i, Y_i), let $D_i = X_i - Y_i, i = 1, \ldots, n$, where D_i can be viewed as the ith observation on the random variable D. Here each $D_i = X_i - Y_i, i = 1, \ldots, n$, provides us with a measure of the difference between, say, the effectiveness of two separate treatments, where the X_i's depict a set of sample outcomes for the first treatment and the Y_i's represent a set of sample outcomes for a second or follow-up treatment.

If the D_i's, $i = 1, \ldots, n$, constitute a single random sample from a normal population with mean $\mu_D = \mu_X - \mu_Y$ and variance σ_D^2, then the random variable $\bar{D}$ has a sampling distribution that is $N(\mu_D, \sigma_D/\sqrt{n})$ and thus the statistic

$$T = \frac{\bar{D} - \mu_D}{s_{\bar{D}}} = \frac{\bar{D} - \mu_D}{s_D/\sqrt{n}} \tag{11.26}$$

follows a t distribution with $n - 1$ degrees of freedom, where

$$\bar{D} = \frac{\sum_{i=1}^{n} D_i}{n} \text{ and } s_D^2 = \frac{\sum_{i=1}^{n} (D_i - \bar{D})^2}{n - 1}.$$

Here $\bar{D}$ serves as an estimator of the mean difference between, say, the effects of the first and second treatments, and s_D^2 is an estimator of the variance of the differences in treatment effects.

For a paired comparison experiment, we may test the null hypothesis $H_0 : \mu_D = \delta_o$ against any of the three alternative hypotheses:

Case 1	Case 2	Case 3
H_0: $\mu_D = \delta_o$ H_1: $\mu_D \neq \delta_o$	H_0: $\mu_D = \delta_o$ H_1: $\mu_D > \delta_o$	H_0: $\mu_D = \delta_o$ H_1: $\mu_D < \delta_o$

Given $H_0 : \mu_D = \delta_o$, the test statistic Equation (11.26) becomes

$$T''_{\delta_o} = \frac{\bar{D} - \delta_o}{s_D/\sqrt{n}} \tag{11.27}$$

and the appropriate $100\alpha\%$-level critical regions are:

$$\text{Case 1: Reject } H_0 \text{ if } \left|T''_{\delta_o}\right| > t_{\alpha/2,n-1} \tag{11.28a}$$

$$\text{Case 2: Reject } H_0 \text{ if } T''_{\delta_o} > t_{\alpha,n-1} \tag{11.28b}$$

$$\text{Case 3: Reject } H_0 \text{ if } T''_{\delta_o} < -t_{\alpha,n-1} \tag{11.28c}$$

Example 11.12

The ABC Bakery Supply Company claims that, because of a certain additive it puts in its flour, it takes at least 5 min less, on average, for dough made with its flour to rise relative to dough made with a certain competitor's flour. To support their claim, they offer to allow $n = 15$ bakers to bake two loaves of bread—one with their flour and one with the competitor's flour. The results appear in Table 11.2, where the X and Y values are expressed in minutes. Is there sufficient evidence to support ABC's claim at the 5% level? Let us test $H_0 : \mu_X - \mu_Y = \delta_o \geq 5$ against $H_1 : \mu_X - \mu_Y < 5$.

From the differences $D_i = X_i - Y_i$, $i = 1,\ldots,15$, appearing in Table 11.2, we can readily find

$$\bar{D} = \frac{82}{15} = 5.46, \quad s_D = \sqrt{\frac{\sum_{i=1}^{15} D_i^2}{n-1} - \frac{\left(\sum_{i=1}^{15} D_i\right)^2}{n(n-1)}} = \sqrt{\frac{938}{14} - \frac{(82)^2}{15 \times 14}} = 5.91.$$

Then, from Equation (11.26),

$$T''_{\delta_o} = \frac{\bar{D} - \delta_o}{s_D/\sqrt{n}} = \frac{5.46 - 5}{5.91/\sqrt{15}} = 0.3014.$$

Since $t_{\alpha,n-1} = t_{0.05,14} = 1.761$ and 0.3014 is not a member of the critical region $R = \left\{T''_{\delta_o} | T''_{\delta_o} > t_{\alpha,n-1} = 1.761\right\}$, we cannot reject the null hypothesis at the 5% level—ABC's claim cannot be rejected at this level of significance.

TABLE 11.2 Effectiveness of ABC's Additive to Flour (Minutes)

Time Taken for the Competitor's Dough to Rise (X)	Time Taken for the ABC's Dough to Rise (Y)	Difference ($D_i = X_i - Y_i$)
72	70	2
72	72	0
71	61	10
69	67	2
67	65	2
85	80	5
85	82	3
80	71	9
70	60	10
72	60	12
75	70	5
75	80	−5
73	60	13
67	50	17
85	88	−3
		82

If the D_i's, $i = 1, \ldots, n$, in the preceding test cannot be treated as having been drawn from a normal population, then an alternative nonparametric test for comparing dependent populations is offered in Appendix 11.C.

11.6 HYPOTHESIS TESTS FOR THE DIFFERENCE OF PROPORTIONS WHEN SAMPLING FROM TWO INDEPENDENT BINOMIAL POPULATIONS

Let $\{X_1, X_2, \ldots, X_{n_X}\}$ and $\{Y_1, Y_2, \ldots, Y_{n_Y}\}$ be random samples taken from two independent binomial populations with p_X and p_Y representing the proportions of successes in the first and second binomial populations. In addition, let X and Y be independent random variables depicting the observed number of successes in the samples of sizes n_X and n_Y, respectively. To test hypotheses regarding the differences in population proportions $p_X - p_Y$, we must review the characteristics of the sampling distribution of the difference between two sample proportions. (Remember that these tests are only approximate and hold only for large samples.)

We know that the best estimators for p_X and p_Y are the sample proportions of successes $\hat{p}_X = X/n_X$ and $\hat{p}_Y = Y/n_Y$, respectively, with

$$E(\hat{p}_X) = p_X, \quad V(\hat{p}_X) = \frac{p_X(1 - p_X)}{n_X},$$

$$E(\hat{p}_Y) = p_Y, \quad V(\hat{p}_Y) = \frac{p_Y(1 - p_Y)}{n_Y}.$$

Hence the best estimator for $p_X - p_Y$ is $\hat{p}_X - \hat{p}_Y$; and the best estimators for $V(\hat{p}_X)$ and $V(\hat{p}_Y)$ are, respectively,

$$s^2(\hat{p}_X) = \frac{\hat{p}_X(1-\hat{p}_X)}{n_X} \text{ and } s^2(\hat{p}_Y) = \frac{\hat{p}_Y(1-\hat{p}_Y)}{n_Y}.$$

Then it can be demonstrated that, for "large samples," the quantity

$$\begin{aligned} Z_{\Delta p} &= \frac{(\hat{p}_X - \hat{p}_Y) - (p_X - p_Y)}{\sqrt{s^2(\hat{p}_X) + s^2(\hat{p}_Y)}} \\ &= \frac{(\hat{p}_X - \hat{p}_Y) - (p_X - p_Y)}{\sqrt{\dfrac{\hat{p}_X(1-\hat{p}_X)}{n_X} + \dfrac{\hat{p}_Y(1-\hat{p}_Y)}{n_Y}}} \text{ is approximately } N(0,1). \end{aligned} \tag{11.29}$$

Let us test the null hypothesis $H_0 : p_X - p_Y = \delta_o$ against any of the three alternative hypotheses given by the following:

Case 1	Case 2	Case 3
H_0: $p_X - p_Y = \delta_o$ H_1: $p_X - p_Y \neq \delta_o$	H_0: $p_X - p_Y = \delta_o$ H_1: $p_X - p_Y > \delta_o$	H_0: $p_X - p_Y = \delta_o$ H_1: $p_X - p_Y < \delta_o$

Under $H_0 : p_X - p_Y = \delta_o \neq 0$, the quantity

$$Z_{\delta_o} = \frac{(\hat{p}_X - \hat{p}_Y) - \delta_o}{\sqrt{\frac{\hat{p}_X(1-\hat{p}_X)}{n_X} + \frac{\hat{p}_Y(1-\hat{p}_Y)}{n_Y}}} \text{ is approximately } N(0,1). \tag{11.29.1}$$

Then for a test conducted at the $100\alpha\%$ level, the decision rules for rejecting H_0 relative to H_1 are as follows:

Case 1: Reject H_0 if $|Z_{\delta_o}| > Z_{\alpha/2}$ (11.30a)
Case 2: Reject H_0 if $Z_{\delta_o} > Z_\alpha$ (11.30b)
Case 3: Reject H_0 if $Z_{\delta_o} < -Z_\alpha$ (11.30c)

If $p_X - p_Y = \delta_o = 0$ is true (the two population proportions p_X and p_Y are assumed equal), then we may take p as their common value. In this instance, the best estimator of the common proportion p is the *pooled estimator*:

$$p = \frac{X+Y}{n_X + n_Y}.$$

where, as stated earlier, X and Y are the observed number of successes in the two independent random samples. Then

$$s^2(\hat{p}_X - \hat{p}_Y) = \frac{\hat{p}(1-\hat{p})}{n_X} + \frac{\hat{p}(1-\hat{p})}{n_Y} = \hat{p}(1-\hat{p})\left(\frac{1}{n_X} + \frac{1}{n_Y}\right),$$

so that Equation (11.29.1) becomes

$$Z'_{\delta_o} = \frac{\hat{p}_X - \hat{p}_Y}{\sqrt{\hat{p}(1-\hat{p})\left(\frac{1}{n_X} + \frac{1}{n_Y}\right)}}. \tag{11.29.2}$$

Then we will reject the null hypothesis of equal population proportions at the $100\alpha\%$ level of significance if:

$$\text{Case 1: Reject } H_0 \text{ if } |Z'_{\delta_o}| > Z_{\alpha/2} \tag{11.31a}$$

$$\text{Case 2: Reject } H_0 \text{ if } Z'_{\delta_o} > Z_{\alpha} \tag{11.31b}$$

$$\text{Case 3: Reject } H_0 \text{ if } Z'_{\delta_o} < -Z_{\alpha} \tag{11.31c}$$

Example 11.13

A random sample of $n_X = 200$ high school seniors was taken from an urban high school and it was found that $X = 75$ of them were cigarette smokers. A second random sample of $n_Y = 180$ high school seniors was taken from a suburban high school and it was found that $Y = 42$ of them were cigarette smokers. Is there any compelling sample evidence to indicate that one group's proportion of smokers is any different than the other's? To answer this question, let us test $H_0 : p_X - p_Y = \delta_o = 0$ against $H_1 : p_X - p_Y \neq 0$ for $\alpha = 0.05$. Then based upon our sample results:

$$\hat{p}_X = \frac{X}{n_X} = \frac{75}{200} = 0.375, \quad \hat{p}_Y = \frac{Y}{n_Y} = \frac{42}{180} = 0.233,$$

$$\hat{p} = \frac{X+Y}{n_X + n_Y} = \frac{117}{380} = 0.308$$

and, from Equation (11.29.2),

$$Z'_{\delta_o} = \frac{0.375 - 0.233}{\sqrt{0.308(0.692)\left(\frac{1}{200} + \frac{1}{180}\right)}} = \frac{0.142}{0.0484} = 2.933.$$

For $\alpha = 0.05$, the critical region is $R = \{Z'_{\delta_o} | Z'_{\delta_o} > Z_{\alpha/2} = 1.96\}$. Since $2.933 > 1.96$, we will reject the null hypothesis at the 5% level—the proportion of cigarette smokers at the urban high school lies significantly above that of its suburban counterpart for $\alpha = 0.05$.

EXERCISES

1 The independent and normally distributed random variables X and Y have known standard deviations of $\sigma_x = 9$ and $\sigma_y = 10$, respectively. A sample of 36 observations from the X population yielded $\bar{X} = 40$, and a sample of 49 observations from the Y population resulted in $\bar{Y} = 35$. Using Equation (11.4), find a 99% confidence interval for the mean difference $\mu_X - \mu_Y$.

2 The American Automobile Assn. (AAA) compared the mean prices per gallon of gasoline in two different metropolitan areas. The mean price per gallon in area X was $\bar{X} = \$3.04(n_X = 40)$ and the mean price per gallon in area Y was $\bar{Y} = \$2.72(n_Y = 35)$. If prior studies reveal that $\sigma_X = 0.10$ and $\sigma_Y = 0.08$, find a 95% confidence interval for the mean difference $\mu_X - \mu_Y$. What assumptions are being made?

3 An experiment was designed to compare the mean service lives of two types of tire chains for winter driving under icy conditions. It was found that the difference between the means was 2,250 miles, and the pooled estimate of variance was 625,000. There were 20 chain sets of each type. Determine a 95% confidence interval for the difference between the mean service lives of these types of tire chains. What assumptions are being made?

4 To compare the durabilities of two patching materials for highway use, $n_X = n_Y = 12$ patches, each 4 ft by 10 ft, were laid down on an interstate highway. The order was decided at random. After a 1-month trial, wear indicators were examined (the higher the reading, the more wear the patch exhibits). Assuming that the observations have been drawn from independent normal populations, find a 95% confidence interval for the difference between the mean wear of the two patch materials. Use a pooled estimate of the variance.

Material A	9.4	12.5	11.3	11.7	8.7	9.9	9.6	11.5	10.3	10.6	9.6	9.7
Material B	11.6	7.2	9.4	8.4	9.7	7.0	10.4	8.2	6.9	12.7	7.3	9.2

5 Given the following information pertaining to two random samples (A and B) drawn from two separate independent normal populations, find a 95% confidence interval for the difference of means $\mu_X - \mu_Y$. [Use Equation (11.10).]

Sample A	$n_X = 14,$	$\bar{X} = 7.84,$	$s_X^2 = 1.03$
Sample B	$n_Y = 14,$	$\bar{Y} = 8.48,$	$s_Y^2 = 1.01$

6 Resolve Problem 11.4 using Equation (11.10).

7 In a recent medical study, researchers performed a randomized double-blind study to measure the effects of a new drug to combat anxiety attacks. A total of 115 patients suffering from anxiety attacks were randomly divided into two groups:

group 1 (the experimental group that generated the X values) received 10 mg per day of the drug, while group 2 (the control group that produced the Y values) received the inert placebo. (Neither the subjects nor the researchers knew who was in the experimental or control group—the double-blind property.) The effectiveness of the drug was measured using a generally accepted scoring technique with the net improvement in the score recorded. The results are presented below. Find a 95% confidence interval about $\mu_X - \mu_Y$. Use Equation (11.10).

Group 1	$n_X = 55,$	$\bar{X} = 14.89,$	$s_X^2 = 156.25$
Group 2	$n_Y = 60,$	$\bar{Y} = 8.17,$	$s_Y^2 = 161.29$

8 Determine whether the following sampling plans are independent or dependent:

a. A research team wishes to determine the effects of alcohol on a subject's reaction time to a flash of light. (When the light flashes, the subject presses a button and his/her reaction time is recorded.) One hundred individuals of legal drinking age are divided into two groups: group 1 subjects drink 3 ounces of alcohol while group 2 participants drink a placebo. Both drinks look and taste the same and individuals participating in the study do not know who gets the alcohol and who gets the placebo. Each person is given 1 min to consume the drink and 30 min later the reaction times of the subjects in each group are recorded.

b. The local Grange wants to determine if there is any significant improvement in crop yield due to the application of a new pesticide. They divide a large plot of land with uniform soil quality into 40 subplots. They then randomly selected 20 plots to be sprayed with the new pesticide while the remaining plots were not sprayed. Crop yield at the end of the growing season was recorded on each subplot in order to determine if there was a significant difference in average yield.

c. A marriage counselor decides to compare the drinking habits of married couples. He obtains a random sample of 45 married couples in which both the husband and wife drink, and he has each member of the couple fill out a questionnaire on their drinking customs.

d. A sociologist contends that daughters seem to be taller than their mothers. To test this hypothesis, she randomly selects 25 mothers who have adult female children and records the height (in inches) of both the mother and her daughter.

e. An automotive engineer believes that his new fuel additive will increase gas mileage significantly. To test his claim, 15 individuals have been randomly selected to drive their own cars on a closed course. Each vehicle received 10 gal of gasoline and was driven until it ran out of gas, with the number of miles driven recorded. This exercise was repeated with an additional 10 gal of gasoline. A random device was used to determine whether the additive was in the first 10 gal or in the second 10 gal of gas, and the driver did not know when the additive was put into the tank.

9 A metallurgist wants to test the effectiveness of a rust inhibitor on a certain grade of steel. He has collected before-treatment (X) and after-treatment (Y) data on 12 uniform sheets of steel. The mean difference D had an average of $\bar{D} = 0.258$ and a variance of $s_D^2 = 0.1259$. Find a 95% confidence interval for the mean difference $\mu_D = \mu_X - \mu_Y$.

10 Two different labor-intensive manufacturing methods (call then X and Y) have been proposed at the ABC Corp. To minimize production cost, ABC Corp. wants to adopt the method with the smaller average completion time. A random sample of $n = 6$ workers is selected, with each worker first using one method and then using the other. The order in which each worker uses the two methods is determined by a flip of a coin. In Table E.11.10, each worker generates a pair of data values—completion time for method X (in minutes) and completion time for method Y (in minutes). Find a 95% confidence interval for the mean difference $\mu_D = \mu_X - \mu_Y$. What assumption must be made?

TABLE E.11.10 Completion Times

Worker	*X* Completion Time	*Y* Completion Time
1	7.0	6.5
2	6.2	5.9
3	6.0	6.0
4	6.4	5.8
5	6.0	5.4
6	5.0	5.2

11 A marketing manager used a sample of $n = 8$ individuals to rate the attractiveness of its current packaging of a product and a new proposed packaging. The attractiveness rating was based on a scale from 1 to 10 (10 being the highest rating). The data set produced by the study immediately follows. Find a 99% confidence interval for the mean difference $\mu_D = \mu_X - \mu_Y$. What assumption is being made?

	Attractiveness Rating	
Individual	New Packaging (X)	Current Packaging (Y)
1	3	5
2	9	8
3	7	5
4	6	6
5	4	3
6	7	7
7	6	4
8	6	5

12 Samples of size 200 ($=n_X=n_Y$) are taken from each of two independent binomial populations. They contained $X=104$ and $Y=96$ successes, respectively. Find a 95% confidence interval for p_X-p_Y.

13 A sample of 500 people was classified as being health conscious or not. Among $n_X=300$ individuals classified as health conscious, it was found that $X=48$ regularly eat yogurt for breakfast. Among the $n_Y=200$ persons who were classified as not health conscious, $Y=52$ regularly eat yogurt for breakfast. Find a 99% confidence interval for the difference in the proportions p_X-p_Y of persons who regularly eat yogurt for breakfast.

14 A random sample containing $n_X=467$ males and $n_Y=433$ females under the age of 50 yrs old exhibited $X=8$ males that were hard of hearing while only $Y=1$ female possessed this problem. Find a 90% confidence interval for the difference p_X-p_Y.

15 For Exercise 11.1, test $H_0: \mu_X-\mu_Y=0$, against $H_1: \mu_X-\mu_Y>0$ for $\alpha=0.05$. What is the p-value?

16 For Exercise 11.2, does μ_X lie significantly above μ_Y at the $\alpha=0.10$ level? What is the test's p-value?

17 For Exercise 11.3, test $H_0: \mu_X-\mu_Y=2000$ versus $H_1: \mu_X-\mu_Y\neq 0$ for $\alpha=0.05$. What is the p-value?

18 For Exercise 11.4, test $H_0: \mu_X-\mu_Y=0$ versus $H_1: \mu_X-\mu_Y\neq 0$ for $\alpha=0.05$. What is the p-value?

19 For Exercise 11.5, test $H_0: \mu_X-\mu_Y=0$, against $H_1: \mu_X-\mu_Y<0$ for $\alpha=0.05$. Determine the p-value.

20 For Exercise 11.7, test $H_0: \mu_X-\mu_Y=7$, against $H_1: \mu_X-\mu_Y<7$ at the 5% level. Determine the test's p-value.

21 For Exercise 11.9, test $\mu_D=0$ versus $H_1: \mu_D>0$ for $\alpha=0.05$. Also, find the test's p-value.

22 For Exercise 11.10, test $H_0: \mu_D=0$, against $H_1: \mu_D\neq 0$ for $\alpha=0.01$. Find the test's p-value.

23 For Exercise 11.11, test the hypothesis that the mean attractiveness rating for the new packaging is at least as good as the mean attractiveness rating for the old packaging. Use $\alpha=0.05$. What is the p-value?

24 For Exercise 11.12, test $H_0: p_1-p_2=0$ versus $H_1: p_X-p_Y>0$ at the 5% level of significance. Determine the p-value.

25 For Exercise 11.13, test $H_0: p_1-p_2=0$ versus $H_1: p_X-p_Y\neq 0$ for $\alpha=0.01$. Find the p-value for the test.

26 For Exercise 11.14, test $H_0: p_X-p_Y=0$ versus $H_1: p_X-p_Y\neq 0$ for $\alpha=0.05$. What is the p-value?

27 Random samples of size $n_1 = n_2 = 20$ are taken from two grinding processes, with the weight (in pounds) of each item sampled recorded (see Table E.11.27). Use the two-sample runs test to determine if there is any significant difference between the two processes at the 5% level.

TABLE E.11.27 Grinding Processes

Process 1: 125, 125, 130, 129, 128, 124, 127, 128, 123, 130 124, 126, 125, 131, 125, 132, 125, 124, 127, 129
Process 2: 125, 129, 128, 128, 130, 129, 134, 126, 123, 127, 131, 130, 128, 126, 127, 125, 128, 130, 131, 126

28 Apply the Mann–Whitney rank sum test to the problem 11.27 data set. Does the same conclusion emerge?

29 A random sample of $n = 26$ newly married couples were asked (independently) to select the ideal number of children they would like to have (either by birth or by adoption). The responses are provided in Table E.11.29. Use the Wilcoxon signed rank test to determine if there is a statistically significant difference between the husbands' and wives' numbers quoted. Use $\alpha = 0.05$.

TABLE E.11.29 Ideal Number of Children

Husband: 2, 0, 3, 5, 7, 1, 0, 2, 10, 5, 2, 0, 1 3, 5, 2, 3, 0, 1, 2, 0, 2, 1, 2, 2, 1
Wife: 4, 1, 4, 2, 5, 2, 3, 4, 3, 3, 4, 2, 3, 2, 2, 1, 2, 1, 0, 2, 3, 3, 2, 3, 3, 3

APPENDIX 11.A RUNS TEST FOR TWO INDEPENDENT SAMPLES

The single-sample runs test, presented in Appendix 10. A, can also be used to compare the identity of two population distributions given that we have two independent random samples of sizes n_1 and n_2, respectively. Let us assume that the underlying population characteristic that these samples represented can be described by a variable that follows a continuous distribution. Hence the following test applies to interval- or ratio- scale data.

Let us assign a letter code to each of the $n = n_1 + n_2$ observations in these samples. For instance, any observation in sample 1 can be marked with the letter a, and any observation in sample 2 can be marked with the letter b. Next, let us rank all n observations according to the magnitude of their scores, with an a placed below each observation belonging to sample 1 and a b placed beneath each observation belonging to sample 2. We thus have an ordered sequence of a's and b's so that we can now conduct a test for the randomness of this arrangement.

We now consider the runs or clusterings of the a's and b's. If the two samples have been drawn from identical populations, then we should expect to see many runs since the n observations from the two samples should be completely intermingled when placed in numerical order. But if the two populations are not identical (e.g., they differ with respect to location or central tendency), then we shall expect fewer runs in the ordered arrangement.

On the basis of this discussion, let us state the null and alternative hypotheses as follows:

H_0: The population distributions are identical.

H_1: Too few runs (the two samples come from populations having, say, unequal means).

Clearly, this alternative hypothesis implies a one-tailed critical region.

Let R denote the total number of runs appearing in the joint sequence of a's and b's. Then it can be shown that for large values of n_1 and n_2, the sampling distribution of R is approximately normal with

$$E(R) = \frac{2n_1n_2}{n} + 1, \quad V(R) = \frac{2n_1n_2(2n_1n_2 - n)}{n^2(n-1)}. \tag{11.A.1}$$

(Actually, the approximation to normality is quite good when both n_1 and n_2 exceed 10.) Hence, under H_0, the distribution of the standardized R statistic or

$$Z_R = \frac{R - E(R)}{\sqrt{V(R)}} \text{ is } N(0, 1). \tag{11.A.2}$$

For a test conducted at the 100α% level of significance, we will reject H_0 when $Z_R < -Z_\alpha$.

Example 11.A.1

Suppose the following two samples (Table 11.A.1) have been drawn independently and at random from two distinct continuous populations. Can we conclude via a runs test with $\alpha = 0.05$ that these samples come from identical populations?

TABLE 11.A.1 Sample 1 and Sample 2 Values

Sample 1 ($n_1 = 12$)	15.2	9.0	17.4	18.0	12.0	13.1
	20.6	9.1	10.8	15.3	19.0	7.5
Sample 2 ($n_2 = 15$)	19.5	22.0	17.1	19.1	15.6	20.2
	21.7	21.9	30.0	15.1	17.3	17.5
	31.0	35.2	33.4			

TABLE 11.A.2 Joint Ranking of Samples 1 and 2 ($n = n_1 + n_2 = 27$)

7.5	9.0	9.1	10.8	12.0	13.1	15.1	15.2	15.3	15.6	17.1	17.3
a	*a*	*a*	*a*	*a*	*a/*	*b/*	*a*	*a/*	*b*	*b*	*b/*
17.4	17.5	18.0	19.0	19.1	19.5	20.2	20.6	21.7	21.9	22.0	30.0
a/	*b/*	*a*	*a/*	*b*	*b*	*b/*	*a/*	*b*	*b*	*b*	*b*
31.0	33.4	35.2									
b	*b*	*b*									

The joint ranking of these samples appears in Table 11.A.2.

Upon counting the number of runs exhibited in Table 11.A.2 we see that $R = 10$. Then from Equation (11.A.1), we have

$$E(R) = \frac{2(12)(15)}{27} + 1 = 14.33,$$

$$V(R) = \frac{2(12)(15)[2(12)(15) - 27]}{(27)^2(26)} = 6.33,$$

and thus, from Equation (11.A.2),

$$Z_R = \frac{10 - 14.33}{\sqrt{6.33}} = -1.722.$$

Since $Z_R = -1.722 < Z_{0.05} = -1.645$, we can reject the null hypothesis of identical population distributions at the 5% level; it appears that the mean of population 1 lies significantly below the mean of population 2.

This version of the runs test is actually consistent with a whole assortment of differences among two continuous population distributions (and not just differences in central tendency). If we reject the null hypothesis that the populations are identical, there may be many reasons why the populations differ—reasons that are not explicitly incorporated in the alternative hypothesis. So while rejection of the null hypothesis leads us to believe that the populations are not identical, we may not have a clue as to how they actually differ. In this regard, this runs test should be applied when all other types of two-sample tests have either been conducted or deemed inappropriate.

APPENDIX 11.B MANN–WHITNEY (RANK SUM) TEST FOR TWO INDEPENDENT POPULATIONS

The Mann–Whitney (M–W) test, like the runs test of Appendix 11.A, is designed to compare the identity of two population distributions by examining the characteristics of two independent random samples of sizes n_1 and n_2, respectively, where n_1 and n_2

are taken to be "large." Here, too, we need only assume that the population distributions are continuous and the observations are measured on an interval or ratio scale. However, unlike the runs test, the M–W procedure exploits the numerical ranks of the observations once they have been jointly arranged in an increasing sequence.

In this regard, suppose we arrange the $n = n_1 + n_2$ sample values in an increasing order of magnitude and assign them ranks 1, . . ., n while keeping track of the source sample from which each observation was selected for ranking, for example, an observation taken from sample 1 can be tagged with, say, letter a, and an observation selected from sample 2 gets tagged with letter b. (If ties in the rankings occur) simply assign each of the tied values the average of the ranks that would have been assigned to these observations in the absence of a tie.)

What can the rankings tell us about the population distributions? Let R_1 and R_2 denote the rank sums for the first and second samples, respectively. If the observations were selected from identical populations, then R_1 and R_2 should be approximately equal in value. However, if the data points in, say, population 1 tend to be larger than those in population 2 (given equal sample sizes), then obviously R_1 should be appreciable larger than R_2, thus providing evidence that the populations differ in some fashion (typically in location or central tendency). Hence very large or very small values of R_1 imply a separation of rankings of the sample 1 versus sample 2 observations, thus providing evidence of a shift in the location of one population relative to the other. (Note that it is immaterial which sample is designated as sample 1.)

The M–W test specifies the null hypothesis as

$$H_0\text{: The population distributions are identical.}$$

This hypothesis may be tested against the following alternatives:

Case 1	Case 2	Case 3
H_0: Population distributions identical H_1: The populations differ in location	H_0: Population distributions identical H_1: Population 1 located to the right of population 2	H_0: Population distributions identical H_1: Population 1 located to the left of population 2

To conduct a large-sample M–W rank-sum test, let us employ the test statistic U = number of observations in sample 1 that precede each observations in sample 2, given that all $n = n_1 + n_2$ observations have been jointly arranged in an increasing sequence. For example, given the following two samples containing $n_1 = n_2 = 4$ observations each,

Sample 1:	20	25	13	30
Sample 2:	12	17	23	28

the combined ordered arrangement is

12	13	17	20	23	25	28	30
b	*a*	*b*	*a*	*b*	*a*	*b*	*a*

Since the smallest sample 2 observation is 12, $U_1 = 0$ observations for sample 1 precede it. The sample 2 value of 17 is preceded only by $U_2 =$ one sample 1 value; the sample 2 value of 23 is preceded by $U_3 =$ two sample 1 values; and the sample 2 value of 28 is preceded by $U_4 =$ three sample 1 values. Hence the total number of observations in sample 1 that precede each observation in sample 2 is $U = U_1 + U_2 + U_3 + U_4 = 0 + 1 + 2 + 3 = 6$. Clearly, the value of U depends upon how the a's and b's are distributed under the ranking process.

Given continuous population distributions, it can be shown that

$$U = n_1 n_2 + \frac{n_1(n_1+1)}{2} - R_1, \tag{11.B.1}$$

where, as defined above, R_1 is the sum of the ranks assigned to observations in the first sample. Since under the null hypothesis of identical population distributions we have

$$E(R_1) = \frac{n_1(n+1)}{2} \text{ and } V(R_1) = \frac{n_1 n_2(n+1)}{12},$$

it follows that

$$E(U) = \frac{n_1 n_2}{2} \text{ and } V(U) = \frac{n_1 n_2(n+1)}{12}. \tag{11.B.2}$$

Then as n_1 and n_2 increase without bound, asymptotically,

$$Z_U = \frac{U - E(U)}{\sqrt{V(U)}} \text{ is } N(0, 1). \tag{11.B.3}$$

(The approximation to normality is quite good for $n_1 \geq 10$ and $n_2 \geq 10$.)

For a test conducted at the $100\alpha\%$ level, the decision rules for rejecting H_0 relative to H_1 are as follows:

Case 1: Reject H_0 if $|Z_U| > Z_{\alpha/2}$ (11.B.4a)

Case 2: Reject H_0 if $Z_U < -Z_\alpha$ (11.B.4b)

Case 3: Reject H_0 if $Z_U > Z_\alpha$ (11.B.4c)

Note that for the Case 2 alternative, U will be small when R_1 is large. Hence the critical region will be under the left-hand tail of the standard normal distribution.

And if R_1 is small, then U will be large so that, for the Case 3 alternative, the critical region is actually under the right-hand tail of the standard normal curve.

Example 11.B.1

A manufacturing company wishes to compare the "downtimes" (the time, in minutes, during a working day in which a piece of machinery is inoperative due to mechanical failure) of two different labeling machines that it is interested in purchasing and currently has on a trial basis. Machine 1 is run for $n_1 = 15$ days and machine 2 can only be run for $n_2 = 10$ days. The recorded downtimes are as follows:

Machine 1: 57, 65, 64, 70, 60, 61, 72, 73, 63, 62, 60, 68, 64, 71, 59
Machine 2: 66, 71, 70, 62, 67, 73, 71, 69, 70, 68

Given that these machines operate independently, let us test H_0: the machine population distributions are identical, against H_1: the machine populations differ in location, at the $\alpha = 0.05$ level. Looking at Table 11.B.1, column 1 specifies the observation numbers assigned to the pooled sample values listed in increasing order of magnitude in columns 2 and 3. The machine ranks are then reported in columns 4 and 5; that is, although ranks from 1 to 25 are assigned to the pooled sample values, they are separately identified for each sample. (Note that observations 3 and 4 are tied for machine 1 so that the average rank assigned to this "within-sequence" tie is 3.5. A "between-sequence" tie occurs for observations 6 and 7. Hence each of these observations is given an average rank of 6.5. Observations 17, 18, and 19 involve both a within-sequence and a between-sequence tie. The average rank assigned to each of these tied values is 18. Other ties are handled in like fashion.)

The sum of the ranks for machine 1 is $R_1 = 160.5$. Then from Equations (11.B.1)–(11.B.3),

$$U = 150 + \frac{15(16)}{2} - 160.5 = 109.5,$$

$$E(U) = \frac{15(10)}{2} = 75,$$

$$V(U) = \frac{15(10)(26)}{12} = 325,$$

and thus

$$|Z_U| = \left| \frac{109.5 - 75}{18.03} \right| = 1.91.$$

Since 1.91 is not a member of the critical region $R = \{Z_U \| Z_U| > Z_{\alpha/2} = Z_{0.025} = 1.96\}$, we cannot reject the null hypothesis at the 5% level. It thus appears that the two machines exhibit identical population distributions at the 5% level of significance.

TABLE 11.B.1 Downtimes in Minutes

(1) Pooled Observation Number	(2) Machine 1	(3) Machine 2	(4) Machine 1 Ranks	(5) Machine 2 Ranks
1	57		1.0	
2	59		2.0	
3	60		3.5	
4	60		3.5	
5	61		5.0	
6	62		6.5	
7		62		6.5
8	63		8.0	
9	64		9.5	
10	64		9.5	
11	65		11.0	
12		66		12.0
13		67		13.0
14		68		14.5
15	68		14.5	
16		69		16.0
17		70		18.0
18		70		18.0
19	70		18.0	
20		71		21.0
21		71		21.0
22	71		21.0	
23	72		23.0	
24		73		24.5
25	73		24.5	
			160.5	

The preceding M–W test is one of the best nonparametric tests for differences in location (and consequently for differences in means and medians). The parametric equivalent to the M–W test is the t test of $H_0: \mu_1 - \mu_2 = 0$, provided that the two populations are normal. In fact, for continuous and independent distributions, the M–W test might perform better than the t test in the absence of normality. If in the M–W test we reject the null hypothesis of "identical population distributions," we are not told exactly how the populations actually differ. But if we reject the null hypothesis of equal population means for the two-sample t test, then we know that the populations differ specifically with respect to their means.

APPENDIX 11.C WILCOXON SIGNED RANK TEST WHEN SAMPLING FROM TWO DEPENDENT POPULATIONS: PAIRED COMPARISONS

As in Sections 11.2 and 11.5, let us assume that a paired experiment yields the n pairs of observations $(X_1, Y_1), \ldots, (X_n, Y_n)$, where X_i and Y_i, $i = 1, \ldots, n$, are members of the same pair (X_i, Y_i), with X_i drawn from population 1 and Y_i drawn from population 2. For the ith pair (X_i, Y_i), let $D_i = X_i - Y_i$, $i = 1, \ldots, n$, where D_i is the ith observation on the random variable D. It is assumed that the population probability distributions are continuous, with the measurements taken on an interval or ratio scale since both the signs of the D_i's as well as their ranks will be utilized. Moreover, the D_i's are taken to be independent random variables that follow a distribution that is symmetrical about a common median.

A comparison of the members of each pair (X_i, Y_i) will render a "+" sign, a "−" sign, or a zero value. In this regard, when the value of an element from population 1 exceeds the value of its paired element from population 2, we will assign a "+" sign to the pair (X_i, Y_i) and $D_i > 0$. If the value of an element from population 1 falls short of the value of its paired counterpart from population 2, then the pair is assigned a "−" sign and $D_i < 0$. Ties obviously occur if $D_i = 0$, in which case the pair (X_i, Y_i) producing the tie is eliminated from the sample.

The Wilcoxon signed rank test involving matched pairs is constructed so as to test the null hypothesis H_0: the population distributions are identical. (The Wilcoxon signed rank procedure can also be considered as a test for symmetry only if we make the assumption that the D_i's are randomly drawn from a continuous distribution.) This hypothesis is then tested against any of the following alternative hypotheses:

Case 1	Case 2	Case 3
H_0: Population distributions identical	H_0: Population distributions identical	H_0: Population distributions identical
H_1: The populations differ in location	H_1: Population 1 located to the right of population 2	H_1: Population 1 located to the left of population 2

To perform the Wilcoxon signed rank test, let us execute the following sequence of steps:

1. Determine $D_i = X_i - Y_i$ for all sample points (X_i, Y_i), $i = 1, \ldots, n$. (Discard any zero-valued D_is.)
2. Rank the D_is in order of increasing absolute value. (If any of the D_is are tied in value, then the tied D_is are given the average rank.)
3. Restore to the rank values $1, \ldots, n$ the algebraic signs of the associated difference D_i. The ranks with the appropriate signs are the *signed ranks* R_i, $i = 1, \ldots, n$. There are only two types: R_i^+ is a rank carrying a "+" sign and R_i^- denotes a rank carrying a "−" sign. Furthermore, the quantity $W^+ = \Sigma_{i=1}^{n} R_i^+$ is the sum of the positive signed ranks and will serve as our test statistic.

For $n > 15$, an approximate test using the standard normal distribution can be performed. Our standardized test statistic under H_0 is

$$Z_{W^+} = \frac{W^+ - m(m+1)/4}{\sqrt{m(m+1)(2m+1)/24}} \text{ is } N(0,1), \tag{11.C.1}$$

where m is the final number of observations that are ranked. For a given level of significance α, our decision rule for rejecting H_0 in favor of H_1 is provided by the following:

$$\text{Case 1: Reject } H_0 \text{ if } |Z_{W^+}| > Z_{\alpha/2} \tag{11.C.2a}$$

$$\text{Case 2: Reject } H_0 \text{ if } Z_{W^+} > Z_{\alpha} \tag{11.C.2b}$$

$$\text{Case 3: Reject } H_0 \text{ if } Z_{W^+} < -Z_{\alpha} \tag{11.C.2c}$$

It is important to note that if we want to explicitly make inferences about the median MD of the D_is, then we can specify the null hypothesis as H_0: $MD = MD_o$, where MD_o is the null value of the median. Then the absolute differences $|D_i| = |X_i - Y_i - MD_o|$ may be ranked and signed as above. This process assumes, of course, that each D_i is drawn independently from a population of differences that is continuous and symmetric about its median.

Example 11.C.1

Suppose two different interior wall paints are to be compared to determine which one dries faster. Both paints are touted by their manufacturers as being "fast drying." Twenty pairs of painters are chosen, having been matched by years of experience and

TABLE 11.C.1 Paint Drying Times (in Minutes)

Pair	Paint 1	Paint 2	D_i	$\|D_i\|$	Signed Rank of D_i
1	55	50	5	5	16.0
2	53	50	3	3	14.5
3	47	47	0	0	tie
4	49	48	1	1	5.0
5	50	51	−1	1	−5.0
6	49	50	−1	1	−5.0
7	48	49	−1	1	−5.0
8	47	49	−2	2	−11.5
9	52	50	2	2	11.5
10	51	52	−1	1	−5.0
11	50	51	−1	1	−5.0
12	55	52	3	3	14.5
13	58	50	8	8	17.0
14	50	49	1	1	5.0
15	47	49	−2	2	−11.5
16	49	47	2	2	11.5
17	48	47	1	1	5.0
18	50	50	0	0	tie
19	50	51	−1	1	−5.0
20	50	50	0	0	tie

skill level. The times to drying (in minutes) of walls of similar sizes and textures are presented in Table 11.C.1. For $\alpha = 0.05$, determine if the underlying distributions of drying times differ significantly in terms of time to drying. Clearly, Case 1 applies.

Given three ties, n is reduced to $m = 17$, with $W^+ = 95$, and thus, via Equation (11.C.1),

$$Z_{W^+} = \frac{95 - 17(18)/4}{\sqrt{17(18)(35)/24}} = 0.88.$$

Since 0.88 does not fall within the critical region $R = \{Z_{W^+} | |Z_{W^+}| > 1.96\}$, we cannot reject the null hypothesis; at the 5% level of significance, the two drying time population distributions are essentially identical.

The preceding Wilcoxon signed rank test is the test of choice among nonparametric tests that hypothesize identical population distributions under paired comparisons. Given a sample of matched pairs, the test is useful in determining whether the medians of the population distributions are identical, or whether one such distribution is located to the right or left of the other. In the instance where the assumption of normality cannot be justified, the nonparametric Wilcoxon signed rank test should be utilized instead of the parametric t test for matched pairs since the latter test explicitly assumes normality.

12

BIVARIATE REGRESSION AND CORRELATION

12.1 INTRODUCING AN ADDITIONAL DIMENSION TO OUR STATISTICAL ANALYSIS

Throughout most of this book, we dealt exclusively with a single quantitative variable X with observations $X_1, \ldots, X_n$, that is, we worked with so-called *univariate data*. However, as everyone knows, the world is not one-dimensional. Observations on X can also be studied relative to, or in contrast with, observations on another quantitative variable Y with values $Y_1, \ldots, Y_n$. In this regard, the X versus Y data values can be represented as a set of ordered pairs $(X_i, Y_i), i = 1, \ldots, n$, in the X-Y-plane (Fig. 12.1). This type of two-dimensional graph is termed a *scatter diagram* involving *bivariate data*. Such data will now constitute a sample of size n.

A scatter diagram can be used to "detect" a relationship between the X and Y variables. For instance, Fig. 12.2a shows a *direct relationship* between X and Y (both variables increase or decrease together); Fig. 12.2b exhibits an *inverse relationship* between X and Y (these variables move in opposite directions); Fig. 12.2c displays a *nonlinear relationship* between X and Y; and, in Fig. 12.2d, there is *no determinate relationship* between these variables.

Statistical Inference: A Short Course, First Edition. Michael J. Panik.

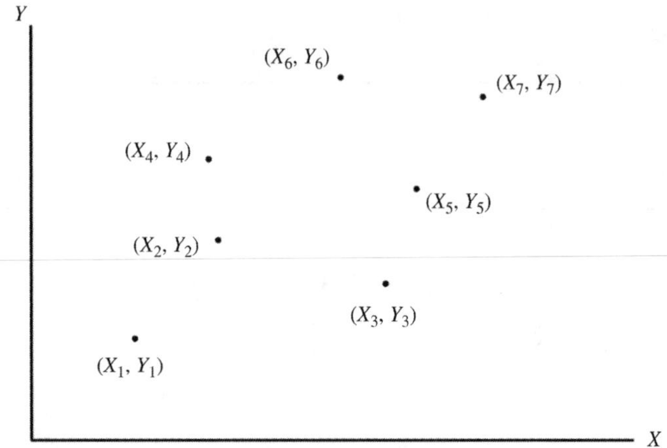

FIGURE 12.1 Scatter diagram for (X_i, Y_i), $i = 1, \ldots, 7$.

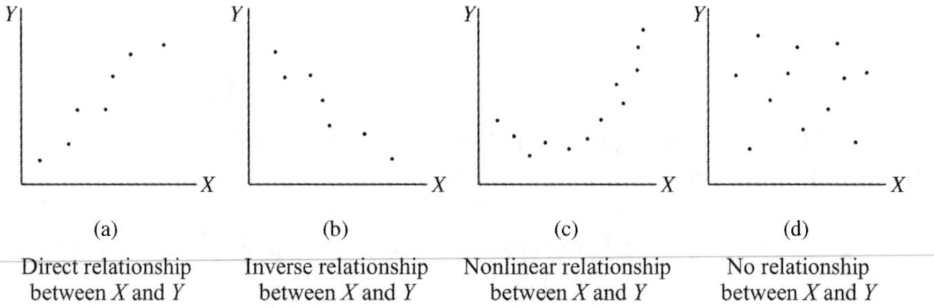

FIGURE 12.2 Relationships between X and Y.

12.2 LINEAR RELATIONSHIPS

One of the most common types of relationships detected between the X and Y variables is a *linear relationship* (e.g., the type of relationship exhibited in Figs. 12.2a and 12.2b).

12.2.1 Exact Linear Relationships

Suppose we have the equation $Y = 2 + 3X$, where Y is the *dependent variable* and X is the *independent variable* (Fig. 12.3). Here "2" is the *vertical intercept*—the value of Y obtained when $X = 0$. The coefficient "3" attached to X is the slope (= rise/run) of the line—when X increases by one unit, Y increases by three units. More generally, the *slope* of this linear equation is the rate of change in Y per unit change in X. The expression $Y = 2 + 3\ X$ is "not a statistical equation"—it is an *exact mathematical equation*. There is very little (if anything) that a course in statistics can contribute to the understanding of this equation.

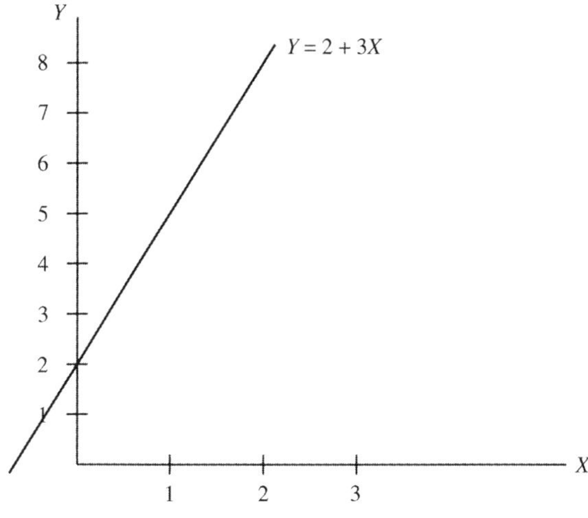

FIGURE 12.3 A linear equation.

Let us examine Fig. 12.2a again. We noted previously that this scatter diagram involving, say, n sample points exhibits a direct linear relationship between X and Y. In fact, because not all of the points (X_i, Y_i), $i = 1, \ldots, n$, lie on a linear equation passing through them, these points could "not" have been generated by an "exact" linear equation. This observation opens the door for the specification of a statistical model (called a *regression model*) that can describe the "goings-on" in Fig. 12.2a. These "goings-on" highlight the regression problem. In particular:

1. What is the equation of the line that best fits this scatter of points?
2. How strong is the linear relationship between X and Y?
3. Is the resulting linear equation statistically significant or did it emerge primarily because of chance factors?

Let us term variable Y the *response* or *dependent* or *explained* variable. The variable X is then taken to be the *explanatory* or *independent* or *regressor* variable. (Here Y can also be thought of as the *predicted* variable and X is thus the *predictor* variable.) The process of finding the equation that best fits the scatter of sample points (e.g., Fig. 12.2a) and assessing its meaningfulness is called *regression modeling*.

To solve the regression problem, let us first express our sample of n observations on the variable Y as $Y_1, \ldots, Y_n$, where

$$Y_i = \text{systematic component} + \text{random component}, \; i = 1, \ldots, n. \tag{12.1}$$

Here the *systematic* or *deterministic part* of Y_i reflects a particular (observable) behavioral hypothesis, whereas the *random part* is unobservable and arises because of some combination of sampling, measurement, and specification error. Hence this random component depicts the influence on Y_i of many omitted variables, each presumably exerting an individually small effect.

Let us specify the systematic component of Y_i in Equation (12.1) as linear or $\beta_0 + \beta_1 X_i$, where X_i is taken to be non-random or predetermined. The random component of Y_i in Equation (12.1) will be denoted as ε_i. Hence Equation (12.1) can be rewritten as follows:

$$Y_i = \beta_0 + \beta_1 X_i + \varepsilon_i \, , i = 1 \, , \, \ldots \, , n \tag{12.2}$$

or

$$Y = \beta_0 + \beta_1 X + \varepsilon, \tag{12.3}$$

where ε is a random variable. How do we know that ε is operating on Y? After all, ε is not observable. We know that ε impacts Y simply because, in, say, Fig. 12.2a, not all of the sample points lie on a straight line—ε_i operates to pull the sample point (X_i, Y_i) away from the line $Y = \beta_0 + \beta_1 X$. Hence Equation (12.3) is a *statistical (probability) model* or *equation* and not an exact mathematical equation. If all of the sample points were on a straight line, there would be nothing to estimate and thus no need for statistical analysis.

Let us assume that $E(\varepsilon) = 0$ (positive deviations from $\beta_0 + \beta_1 X$ are just as likely to occur as negative ones so that the random variable ε has a mean of zero). With the X_i's held fixed, Y is a random variable because ε is random. Then from Equation (12.3),

$$E(Y|X) = E(\beta_0 + \beta_1 X) + E(\varepsilon) \; = \beta_0 + \beta_1 X \text{ (since } E(\varepsilon) = 0\text{)}. \tag{12.4}$$

Hence Equation (12.4) serves as the *population regression equation*. How does the interpretation of this equation differ from that of an exact mathematical equation? Remembering that an expectation is an "average:"

$$E(Y|0) = \beta_0 \text{ (thus } \beta_0 \text{ is the "average" value of } Y \text{ when } X = 0\text{);}$$

β_1 is the "average" rate of change in Y per unit increase in X (i.e., when X increases by one unit, Y changes by β_1 units "on the average").

How might an equation such as Equation (12.3) arise? Let us look to the determinants of, say, the quantity demand (Y) of some consumer product. Specifically, we can express quantity demanded as a function (f) of a variety of arguments, for example,

Quantity demanded = f (the product's own price, the prices of related goods, income, tastes, wealth, expectations, seasonal factors, and so on).

We obviously cannot be expected to account for the effects of all possible determinants of quantity demanded in this equation. However, let us focus on the most important explanatory variable, namely the product's own price (X). To account for the effects of the other explanatory variables, let us consolidate all of them into the term ε. The role of ε is thus to account for the "net" effect of all excluded factors on quantity demanded (Y). Since the impact of ε on Y cannot be predicted, ε is treated as a random variable or *random error term*. Hence we are left with the expression

$$Y = f(X, \varepsilon).$$

If the systematic portion of Y varies linearly with X, then the preceding expression becomes

$$Y = \beta_0 + \beta_1 X + \varepsilon.$$

One of our objectives is to obtain "good" estimates of β_0 and β_1. Since the impact of ε on Y can mask a multitude of sins, "good" estimates of β_0 and β_1 obtain only if ε is "well behaved." That is, ε must satisfy certain distributional assumptions such as: a zero mean, a constant variance, successive ε_i's are not associated with each other, and the ε_i's are not associated with the X_i's, $i = 1, \ldots, n$. While extensive diagnostic testing of these assumptions concerning ε is beyond the scope of this text, suffice it to say that "there are no free lunches in regression analysis;" estimates of β_0 and β_1 are useful or reliable only if ε does not display any aberrant behavior.

12.3 ESTIMATING THE SLOPE AND INTERCEPT OF THE POPULATION REGRESSION LINE

Let us develop the following notation. The *population regression line* will be written as

$$E(Y|X) = \beta_0 + \beta_1 X \text{ (unobserved)} \tag{12.4}$$

and the *sample regression line* will appear as

$$\hat{Y} = \hat{\beta}_0 + \hat{\beta}_1 X \text{ (estimated from sample data)}, \tag{12.5}$$

where $\hat{\beta}_0$ is the sample estimate of β_0 and $\hat{\beta}_1$ is the sample estimate of β_1. The sample regression line is illustrated in Fig. 12.4. Note that there are two values of Y that occur at each X_i value—the observed value of Y or Y_i, and the estimated value of Y or $\hat{Y}_i \left(= \hat{\beta}_0 + \hat{\beta}_1 X_i\right)$. Their difference, $e_i = Y_i - \hat{Y}_i$, will be termed the ith *residual* or ith *deviation from the sample regression line*. This residual is "observed" and serves as an estimate of ε_i, which is purely random and thus "unobserved."

As we shall now see, these deviation e_i from the sample regression line are of paramount importance. For instance, if a sample point (X_i, Y_i) lies above the sample regression line, then $e_i > 0$; and if that point lies below the sample regression line, then $e_i < 0$. (If a sample point lies directly on the sample regression line, then clearly,

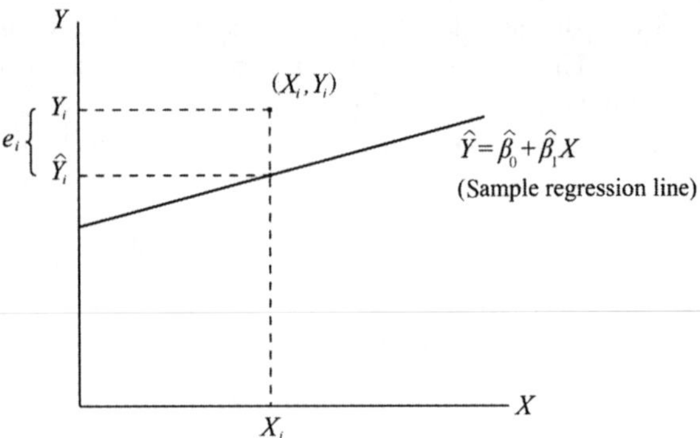

FIGURE 12.4 The sample regression equation.

$e_i = 0$.) In this regard, it should be intuitively clear that if $\hat{Y} = \hat{\beta}_0 + \hat{\beta}_1 X$ is the line that best fits the scatter of sample points, then $\Sigma e_i = 0$. Moreover, the magnitude of Σe_i^2 varies directly with the spread of the sample points about the sample regression line, for example, if Σe_i^2 is large, then the sample points are scattered widely about the sample regression line. Should we want Σe_i^2 to be large or small? The answer is, "the smaller the better." In this regard, how should $\hat{\beta}_0$ and $\hat{\beta}_1$ be chosen? Our choice of $\hat{\beta}_0$ and $\hat{\beta}_1$ will be guided by the *Principle*[1] *of Least Squares*: To obtain the "line of best fit," choose $\hat{\beta}_0$ and $\hat{\beta}_1$ so as to minimize

$$\begin{aligned}\sum e_i^2 = \sum (Y_i - \hat{Y}_i)^2 &= \sum \left(Y_i - \hat{\beta}_0 - \hat{\beta}_1 X_i\right)^2 \\ &= f\left(\hat{\beta}_0, \hat{\beta}_1\right),\end{aligned} \tag{12.6}$$

that is, we want to minimize the sum of the squared deviations about the sample regression line.

Once the implied minimization is carried out (through calculus techniques), we get the so-called set of *least squares normal equations*:

$$n\hat{\beta}_0 + \hat{\beta}_1 \sum X_i = \sum Y_i, \tag{12.7a}$$

$$\hat{\beta}_0 \sum X_i + \hat{\beta}_1 \sum X_i^2 = \sum X_i Y_i. \tag{12.7b}$$

While the unknowns $\hat{\beta}_0$ and $\hat{\beta}_1$ can be obtained by solving system Equation (12.7) simultaneously, let us take an alternative path to enlightenment. If we divide both sides of Equation (12.7a) by n, then we obtain

$$\hat{\beta}_0 + \hat{\beta}_1 \bar{X} = \bar{Y}.$$

This equation expresses the fact that the least squares line of best fit passes through the point of means ($\bar{X}, \bar{Y}$) of the sample data set (Fig. 12.5).

[1] A *principle* is a rule of conduct or action.

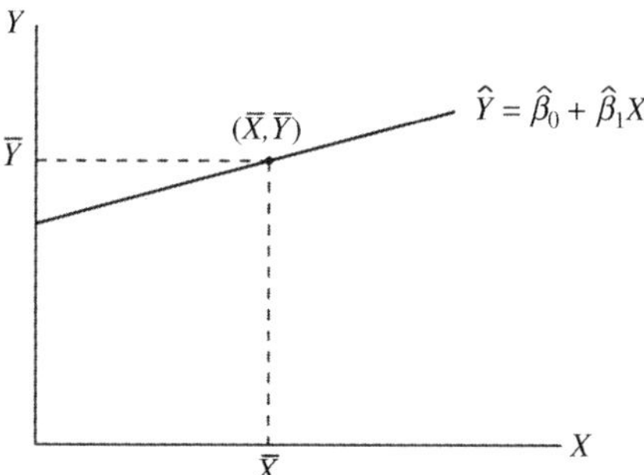

FIGURE 12.5 The least squares line of best fit passes through the point of means $(\bar{X}, \bar{Y})$.

This notion then allows us to express the observations on the variables X and Y in terms of *deviations from their respective means*:

$$x_i = X_i - \bar{X},$$

$$y_i = Y_i - \bar{Y}.$$

Armed with these considerations, we can write

$$\hat{\beta}_1 = \frac{\sum (X_i - \bar{X})(Y_i - \bar{Y})}{\sum (X_i - \bar{X})^2} = \frac{\sum x_i y_i}{\sum x_i^2}, \tag{12.9a}$$

$$\hat{\beta}_0 = \bar{Y} - \hat{\beta}_1 \bar{X} \text{ [from Equation (12.8)]}. \tag{12.9b}$$

Example 12.1

Given the following set of observations on the variables X and Y (Table 12.1), determine the equation of the (regression) line that best fits the scatter of points (X_i, Y_i), $i = 1, \ldots, 12$ (Fig. 12.6). That is, we need to find $\hat{\beta}_1$ and $\hat{\beta}_0$ using Equation (12.9).

Our work array appears as Table 12.2. With $n = 12$, $\Sigma X_i = 66$, and $\Sigma Y_i = 60$, it follows that $\bar{X} = 5.5$ and $\bar{Y} = 5$. On the basis of these values, we may determine the entries in columns 3–7. Then from Equation (12.9), $\hat{\beta}_1 = 68/99 = 0.6869$ and $\hat{\beta}_0 = 5 - 0.6869\ (5.5) = 1.2221$. Hence our estimated (sample) equation for the line of best fit is

$$\hat{Y} = \hat{\beta}_0 + \hat{\beta}_1 X = 1.2221 + 0.6869X.$$

How are we to interpret these coefficients?

$\hat{\beta}_0 = 1.2221$ —this is the average value of Y when $X = 0$

$\hat{\beta}_1 = 0.6869$ —when X increases by one unit, Y increases by 0.6869 units on the average

TABLE 12.1 Observations on X and Y

X	Y
1	2
2	1
2	2
3	4
5	4
5	5
6	7
7	5
8	6
8	7
9	8
10	9

One of the purposes of regression analysis is prediction. In this regard, given a new value of X, say $X_o = 12$, what is the predicted value of Y? To answer this question, all we have to do is substitute $X_o = 12$ into the regression equation to get the predicted Y at X_o, denoted $\hat{Y}_o$. That is,

$$\hat{Y}_o = 1.2221 + 0.6968(12) = 9.4649.$$

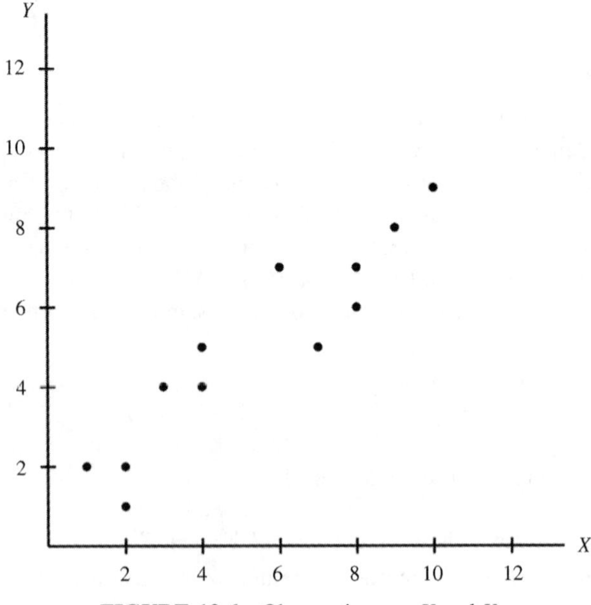

FIGURE 12.6 Observations on X and Y.

TABLE 12.2 Calculations for Least Squares Regression Line

(1)	(2)	(3)	(4)	(5)	(6)	(7)
X	Y	$x_i = X_i - \bar{X}$ $= X_i - 5.5$	$y_i = Y_i - \bar{Y}$ $= Y_i - 5$	x_i^2	y_i^2	$x_i y_i$
1	2	−4.5	−3	20.25	9	13.5
2	1	−3.5	−4	12.25	16	14.0
2	2	−3.5	−3	12.25	9	10.5
3	4	−2.5	−1	6.25	1	2.5
5	4	−0.5	−1	0.25	1	0.5
5	5	−0.5	0	0.25	0	0.0
6	7	0.5	2	0.25	4	1.0
7	5	1.5	0	2.25	0	0.0
8	6	2.5	1	6.25	1	2.5
9	8	3.5	3	12.25	9	10.5
10	9	4.5	4	20.25	16	18.0
$66(\Sigma X_i)$	$60(\Sigma Y_i)$	$0(\Sigma x_i)$	$0(\Sigma y_i)$	$99(\Sigma x_i^2)$	$70(\Sigma y_i^2)$	$68(\Sigma x_i y_i)$

A couple of points to ponder:

1. Equation (12.3) is called a "linear regression equation" simply because the variable X appears to the first power? Not so. Technically speaking, Equation (12.3) is termed "linear" because both $\hat{\beta}_0$ and $\hat{\beta}_1$ enter the equation in a linear fashion—each has, as indicated, an exponent of one. (For instance, $Y = \beta_0 + \beta_1 X + \beta_2 X^2 + \varepsilon$ is a linear regression equation while $Y = \log \beta_0 + \beta_1 X$ is nonlinear (since $\log \beta_0$ is nonlinear).
2. Why are the "least squares estimators" $\hat{\beta}_0$ and $\hat{\beta}_1$ in Equation (12.9) used to determine the values of the population parameters β_0 and β_1, respectively? What's the "big deal" about the least squares fit to the data points? According to the famous *Gauss–Markov Theorem*, if ε is "well behaved," then the least squares estimators are *BLUE*, that is, they are the

   *B*est *L*inear *U*nbiased *E*stimators of β_0 and β_1.

   "Unbiased" means that, on the average, $\hat{\beta}_0$ and $\hat{\beta}_1$ are "on target" ($E(\hat{\beta}_0) = \beta_0$ and $E(\hat{\beta}_1) = \beta_1$ so that the sampling distributions of $\hat{\beta}_0$ and $\hat{\beta}_1$ are centered on β_0 and β_1, respectively). "Best" means that, out of the class of all unbiased linear estimators for β_0 and β_1, the least squares estimators have "minimum variance," that is, their sampling distributions tend to be more closely concentrated about β_0 and β_1 than those of any alternative unbiased linear estimators. Clearly, least squares estimators are not used on a lark—there are compelling statistical reasons for their use.

12.4 DECOMPOSITION OF THE SAMPLE VARIATION IN *Y*

Let us express the variance of the random variable Y as

$$V(Y) = \frac{\sum (Y_i - \bar{Y})^2}{n-1} = \frac{\sum y_i^2}{n-1}. \tag{12.10}$$

We noted above that there were two components to Y_i—a systematic component reflecting the linear influence of X_i and a random component ε_i due to chance factors. A moment's reflection reveals that if most of the variation in Y is due to random factors, then the estimated least squares line is probably worthless as far as predictions are concerned. However, if most of the variation in Y is due to the linear influence of X, then we can obtain a meaningful regression equation. So the question arises—"How much of the variation in Y is due to the linear influence of X and how much can be attributed to random factors? We can answer this question with the aid of Fig. 12.7.

Since our goal is to explain the variation in Y, let's start with the numerator of Equation (12.10). That is, from Fig. 12.7,

$$\begin{aligned} Y_i - \bar{Y} &= Y_i - \hat{Y}_i + \hat{Y}_i - \bar{Y} \\ &= e_i + (\hat{Y}_i - \bar{Y}), \end{aligned} \tag{12.11}$$

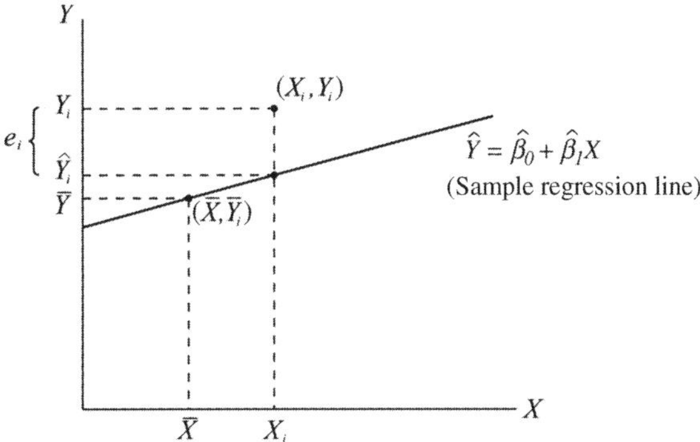

FIGURE 12.7 Decomposition of $Y_i - \bar{Y}$ into $\hat{Y}_i - \bar{Y}$ and e_i.

where $\hat{Y}_i - \bar{Y}$ is attributed to the linear influence of X (the regression of Y on X) and e_i is due to random factors. Then a substitution of Equation (12.11) into the numerator of Equation (12.10) gives

$$\sum (Y_i - \bar{Y})^2 = \sum e_i^2 + \sum (\hat{Y}_i - \bar{Y})^2 + \underbrace{2\sum e_i(\hat{Y}_i - \bar{Y})}_{=0} \tag{12.12}$$

or

$$\underset{\text{(SST)}}{\text{Total sum of squares}} = \underset{\text{(SSE)}}{\text{Error sum of squares}} + \underset{\text{(SSR)}}{\text{regression sum of squares.}} \tag{12.12.1}$$

Thus the total sum of squares can be neatly partitioned into two parts: (1) the regression sum of squares SSR (which reflects the variation in Y attributed to the linear influence of X); and (2) the error sum of squares SSE (which depicts the variation in Y ascribed to random factors). Alternatively stated, SSR is oftentimes called the *explained sum of squares* and SSE is termed the *unexplained sum of squares*.

How are these various sums of squares calculated? It can be shown that:

$$\begin{aligned} \text{SST} &= \sum y_i^2, \\ \text{SSR} &= \hat{\beta}_1 \sum x_i y_i, \\ \text{SSE} &= \text{SST} - \text{SSR} = \sum y_i^2 - \hat{\beta}_1 \sum x_i y_i. \end{aligned} \tag{12.13}$$

These various sums of squares, along with their associated degrees of freedom, constitute the *partitioned sums of squares table* (Table 12.3). (Degrees of freedom for SSE is $n-2$ since we needed the "two" estimates $\hat{\beta}_0$ and $\hat{\beta}_1$ to find SSE.)

TABLE 12.3 Partitioned Sums of Squares

Source of Variation in Y	Sums of Squares (SS)	Degrees of Freedom (d.f.)	Mean Square (MS) = SS/d.f.
Regression (explained variation)	$\text{SSR} = \hat{\beta}_1 \Sigma x_i y_i$	1	Regression MS: MSR = SSR/1 = $\hat{\beta}_1 \Sigma x_i y_i$
Error (unexplained variation)	$\text{SSE} = \Sigma y_i^2 - \hat{\beta}_1 \Sigma x_i y_i$	$n - 2$	Error MS: MSE = SSE/($n-2$) = $\left(\Sigma y_i^2 - \hat{\beta}_1 \Sigma x_i y_i\right)/(n-2)$
Total variation in Y	$\text{SST} = \Sigma y_i^2$	$n - 1$	$r^2 = \text{SSR}/\text{SST}$

An important application of these sums of squares is the specification of the *sample coefficient of determination*

$$r^2 = \frac{\text{SSR}}{\text{SST}} = \frac{\text{Explained SS}}{\text{Total SS}} = \frac{\hat{\beta}_1 \sum x_i y_i}{\sum y_i^2} \tag{12.14}$$

or, since SSR = SST − SSE,

$$r^2 = 1 - \frac{\text{SSE}}{\text{SST}} = 1 - \frac{\text{Unexplained SS}}{\text{Total SS}} = 1 - \frac{\sum y_i^2 - \hat{\beta}_1 \sum x_i y_i}{\sum y_i^2}. \tag{12.14.1}$$

Here r^2 serves as a measure of "goodness of fit;" it represents the proportion of variation in Y that can be explained by the linear influence of X. Clearly, $0 \leq r^2 \leq 1$.

Example 12.2

Given the results of Example 12.1, determine the partitioned sums of squares (Table 12.4) and the coefficient of determination.

Hence $r^2 = 0.6673$ informs us that about 67% of the variation in Y is explained by the linear influence of X; about 33% is unexplained or attributed to random factors.

12.5 MEAN, VARIANCE, AND SAMPLING DISTRIBUTION OF THE LEAST SQUARES ESTIMATORS $\hat{\beta}_0$ AND $\hat{\beta}_1$

We noted above (via the Gauss–Markov Theorem) that $E\left(\hat{\beta}_0\right) = \beta_0$ and $E\left(\hat{\beta}_1\right) = \beta_1$ (hence the sampling distributions of $\hat{\beta}_0$ and $\hat{\beta}_1$ are centered right on β_0 and β_1, respectively). In addition, it can be demonstrated that

$$V\left(\hat{\beta}_0\right) = \sigma_\varepsilon^2 \left(\frac{1}{n} + \frac{\bar{X}^2}{\sum x_i^2}\right), V\left(\hat{\beta}_1\right) = \frac{\sigma_\varepsilon^2}{\sum x_i^2}, \tag{12.15}$$

TABLE 12.4 Partitioned Sums of Squares

Source of Variation in Y	Sums of Squares (SS)	Degrees of Freedom (d.f.)	Mean Square (MS) = SS/d.f.
Regression	$SSR = \hat{\beta}_1 \Sigma x_i y_i$	1	$MSR = SSR/1$
			$= 46.7092$
	$= 0.6869\ (68)$		
	$= 46.7092$		
Error	$SSE = \Sigma y_i^2 - \hat{\beta}_1 \Sigma x_i y_i$	$n - 2 = 10$	$MSE = SSE/(n-2)$
	$= 70\text{-}46.7092$		$= 23.2908/10$
	$= 23.2908$		$= 2.3290$
Total	$SST = \Sigma y_i^2 = 70$	$n - 1 = 11$	$r^2 = \dfrac{SSR}{SST} = 0.6673$

where σ_ε^2 is the unknown variance of the random error term ε. Since σ_ε^2 is unknown and must be estimated from the sample data, let us use

$$S_\varepsilon^2 = \frac{\sum e_i^2}{n-2} = \frac{\sum (Y_i - \hat{Y}_i)^2}{n-2} = \frac{\sum y_i^2 - \hat{\beta}_1 \sum x_i y_i}{n-2} = \frac{SSE}{n-2} = MSE \tag{12.16}$$

as an unbiased estimator for σ_ε^2. If Equation (12.16) is substituted into Equation (12.15), then the "estimated" variances of $\hat{\beta}_0$ and $\hat{\beta}_1$ are, respectively,

$$S_{\hat{\beta}_0}^2 = S_\varepsilon^2 \left(\frac{1}{n} + \frac{\bar{X}^2}{\sum x_i^2} \right), S_{\hat{\beta}_1}^2 = \frac{S_\varepsilon^2}{\sum x_i^2}. \tag{12.17}$$

Given these values, the *estimated standard errors of the regression coefficients* are

$$S_{\hat{\beta}_0} = S_\varepsilon \sqrt{\frac{1}{n} + \frac{\bar{X}^2}{\sum x_i^2}}, S_{\hat{\beta}_1} = \frac{S_\varepsilon}{\sqrt{\sum x_i^2}}. \tag{12.18}$$

Here $S_{\hat{\beta}_0}$ is the estimated standard deviation of the sampling distribution of $\hat{\beta}_0$ while $S_{\hat{\beta}_1}$ is the estimated standard deviation of the sampling distribution of $\hat{\beta}_1$.
Additionally,

$$S_\varepsilon = \sqrt{\frac{\sum e_i^2}{n-2}} = \sqrt{MSE} \tag{12.16.1}$$

is the standard deviation of the e_i's and is termed the *standard error of estimate*—it serves as an estimate of the "degree of scatter" of the sample points about the estimated regression line. (Technically speaking, S_ε is the estimated standard deviation of all Y_i values around their mean $E(Y|X) = \beta_0 + \beta_1 X$ for given X_i's.) Note that both $S_{\hat{\beta}_o}$ and $S_{\hat{\beta}_1}$ vary directly with S_ε (the greater the dispersion of the sample points about the estimated regression line, the less precise our estimates of β_0 and β_1) and inversely with Σx_i^2 (the greater the dispersion of the X_i values about $\bar{X}$, the more precise are our estimates of β_0 and β_1).

Example 12.3

Given the Example 12.1 data set, let us find $S_{\hat{\beta}_0}$ and $S_{\hat{\beta}_1}$. We first need, from Equation (12.16.1) and Table 12.4,

$$S_\varepsilon = \sqrt{\frac{\text{SSE}}{n-2}} = \sqrt{\text{MSE}} = \sqrt{2.3290} = 1.5261.$$

Then from Equation (12.18) and Table 12.2,

$$S_{\hat{\beta}_0} = 1.5261\sqrt{\frac{1}{12} + \frac{(5.5)^2}{99}} = 0.9515, S_{\hat{\beta}_1} = \frac{1.5261}{\sqrt{99}} = 0.1533.$$

We noted earlier that the least squares estimators $\hat{\beta}_0$ and $\hat{\beta}_1$ are BLUE. If we couple this result with the assumption that the random error term ε is normally distributed, then

$$\frac{\hat{\beta}_0 - \beta_0}{\sqrt{V\left(\hat{\beta}_0\right)}} \text{ is } N(0,1) \text{ and } \frac{\hat{\beta}_1 - \beta_1}{\sqrt{V\left(\hat{\beta}_1\right)}} \text{ is } N(0,1). \tag{12.19}$$

If in Equation (12.19), we replace $\sqrt{V\left(\hat{\beta}_0\right)}$ and $\sqrt{V\left(\hat{\beta}_1\right)}$ by $S_{\hat{\beta}_0}$ and $S_{\hat{\beta}_1}$, respectively, then the resulting quantities follow a t distribution with $n-2$ degrees of freedom or

$$\frac{\hat{\beta}_0 - \beta_0}{S_{\hat{\beta}_0}} \text{ is } t_{n-2} \text{ and } \frac{\hat{\beta}_1 - \beta_1}{S_{\hat{\beta}_1}} \text{ is } t_{n-2}. \tag{12.20}$$

12.6 CONFIDENCE INTERVALS FOR β_0 AND β_1

To determine how precisely β_0 and β_1 have been estimated from the sample data, let us find $100(1-\alpha)\%$ confidence intervals for these parameters. To this end:

(a) a *100(1 – α)% confidence interval for* β_0 is

$$\hat{\beta}_0 \pm t_{\alpha/2,n-2} S_{\hat{\beta}_0}, \tag{12.21}$$

(b) a *100(1 – α)% confidence interval for* β_1 *is*

$$\hat{\beta}_1 \pm t_{\alpha/2,n-2} S_{\hat{\beta}_1}, \qquad (12.22)$$

where $-t_{\alpha/2,n-2}$ and $t_{\alpha/2,n-2}$ are, respectively, the lower and upper percentage points of the t distribution and $1-\alpha$ is the confidence coefficient. Hence Equation (12.21) informs us that we may be $100(1-\alpha)\%$ confident that the true (population) regression intercept β_0 lies between $\hat{\beta}_0 \pm t_{\alpha/2,n-2} S_{\hat{\beta}_0}$; and Equation (12.22) tells us that we may be $100(1-\alpha)\%$ confident that the true regression slope β_1 lies between $\hat{\beta}_1 \pm t_{\alpha/2,n-2} S_{\hat{\beta}}$.

Example 12.4

Find 95% confidence intervals for β_0 and β_1 using the results of Examples 12.1 and 12.2. Given $1-\alpha=0.95$, $\alpha/2=0.025$, and thus $t_{0.025,10}=2.228$. Then, via Equation (12.21),

$$1.2221 \pm 2.228\,(0.9515) \rightarrow 1.2221 \pm 2.1199 \rightarrow (-0.8978,\ 3.3420),$$

that is, we may be 95% confident that β_0 lies between -0.8978 and 3.3420. And from Equation (12.22),

$$0.6869 \pm 2.228\,(0.1533) \rightarrow 0.6869 \pm 0.3416 \rightarrow (0.3453,\ 1.0285),$$

so that we may be 95% confident that β_1 lies between 0.3453 and 1.0285. So how precisely have we estimated β_0 and β_1 ? We are within ± 2.1199 units of β_0 with 95% reliability; and we are within ± 0.3416 units of β_1 with 95% reliability. These results may be stated in yet another fashion by using the "error bound" concept: we may be 95% confident that $\hat{\beta}_0$ will not differ from β_0 by more than ± 2.1199 units; and we may be 95% confident that $\hat{\beta}_1$ will not differ from β_1 by more than ± 0.3416 units.

12.7 TESTING HYPOTHESES ABOUT β_0 AND β_1

How do we know that X and Y are truly linearly related? Looked at in another fashion, how do we know that X actually contributes significantly to the prediction of Y? Answers to these questions can be garnered by way of testing a particular null hypothesis against an appropriate alternative hypothesis.

The most common type of hypothesis test is that of "no linear relationship between X and Y" or $E(Y|X) = \beta_0$, that is, the conditional mean of Y given X does not depend linearly on X. Hence the implied null hypothesis of "no linear relationship between X and Y" is $H_0 : \beta_1 = 0$. Possible alternative hypotheses are:

(a) If we are uncertain about the relationship between X and Y (we have no clue as to whether or not these variables vary directly or inversely), then the appropriate alternative hypothesis is $H_1 : \beta_1 \neq 0$. Clearly, this alternative implies a two-tail test with $R = \left\{ t \middle| |t| = \left| \hat{\beta}_1 / S_{\hat{\beta}_1} \right| > t_{\alpha/2,n-2} \right\}$. If we reject H_0 in favor of H_1, then we can conclude that, at the $100\alpha\%$ level, there exists a statistically significant linear relationship between X and Y.

(b) If, say, *a priori* theory indicates that X and Y are directly related, then we choose $H_1 : \beta_1 > 0$. This implies a one-tail test involving the right-hand tail of the t distribution so that $R = \{t | t = \hat{\beta}_1 / S_{\hat{\beta}_1} > t_{\alpha,n-2}\}$. If we reject H_0 in favor of H_1, then we conclude that, at the $100\alpha\%$ level, there exists a statistically significant positive linear relationship between X and Y.

(c) If, *a priori* theory indicates that X and Y are inversely related, then choose $H_1 : \beta_1 < 0$. We now have a one-tail test involving the left-hand tail of the t distribution so that $R = \{t | t = \hat{\beta}_1 / S_{\hat{\beta}_1} < -t_{\alpha,n-2}\}$. If we reject H_0 in favor of H_1, then we conclude that, at the $100\alpha\%$ level, there exists a statistically significant negative linear relationship between X and Y.

(Note: if we do not reject H_0, we cannot legitimately conclude that X and Y are unrelated but only that there is no statistically significant "linear relationship" exhibited by the data. The true underlying relationship between X and Y may be highly nonlinear.)

As far as a test for the regression intercept β_0 is concerned, we typically test $H_0{:}\beta_0 = 0$ (the population regression line passes through the origin) against $H_1{:}\beta_0 \neq 0$ or $H_1{:}\beta_0 > 0$ or $H_1{:}\beta_0 < 0$. The appropriate t test statistics and critical regions are developed in a fashion similar to those specified above when testing hypotheses about β_1.

Example 12.5

Given the results obtained in Examples 12.1 and 12.2, determine if there exists a statistically significant linear relationship between X and Y for $\alpha = 0.05$. Here $H_0{:}\beta_1 = 0$ is tested against $H_1{:}\beta_1 \neq 0$. In addition, $t_{\alpha/2,n-2} = t_{0.025,10} = 2.228$ and

$$|t| = \left| \frac{\hat{\beta}_1}{S_{\hat{\beta}_1}} \right| = \left| \frac{0.6869}{0.1533} \right| = 4.481.$$

Since $R = \{t | |t| > 2.228\}$, we can readily reject H_0 in favor of H_1—there exists a statistically significant linear relationship between X and Y at the 5% level. (Here p-value < 0.001.)

Could we have reached this conclusion by examining the confidence interval for β_1 provided in Example 12.4? The answer is, yes. The 95% confidence interval for β_1 contains all null hypotheses "not rejected" by a two-tail test at the 5% level. Since zero is not a member of this confidence interval, the null hypothesis of $\beta_1 = 0$ is rejected.

Can we conclude that $\beta_1 = \beta_1^0 = 0.475$ at the $\alpha = 0.05$ level? Here we will test $H_0: \beta_1 = \beta_1^0 = 0.475$ against $H_1: \beta_1 \neq \beta_1^0 = 0.475$. Now our test statistic is

$$t = \frac{\hat{\beta}_1 - \beta_1^0}{S_{\hat{\beta}_1}} \tag{12.23}$$

with $R = \{t \| t| > t_{\alpha/2,n-2}\}$. For $\alpha = 0.05$, $t_{0.025,10} = 2.228$. Then, from Equation (12.23),

$$|t| = \left| \frac{0.6869 - 0.475}{0.1533} \right| = 1.3822.$$

Clearly, we are not within the critical region so that we cannot reject the null hypothesis at the 5% level of significance. (Here $0.05 < p$-value < 0.10.) Could we have reached this conclusion by examining the confidence interval for β_1?

12.8 PREDICTING THE AVERAGE VALUE OF *Y* GIVEN *X*

Suppose we are interested in predicting the average value of Y given $X = X_i$, $E(Y|X_i)$. With $\hat{\beta}_0$ and $\hat{\beta}_1$ unbiased estimators of β_0 and β_1, respectively, it follows that $E(\hat{Y}) = E(\hat{\beta}_0 + \hat{\beta}_1 X) = \beta_0 + \beta_1 X$, that is, an unbiased estimator for the population regression line is the sample regression line. So with $\hat{Y}_i = \hat{\beta}_0 + \hat{\beta}_1 X_i$ an estimator for $E(Y_i|X_i) = \beta_0 + \beta_1 X_i$, it can be shown that the *estimated standard deviation of* $\hat{\mathbf{Y}}_i$ may be computed as

$$S_{\hat{Y}_i} = S_\varepsilon \sqrt{\frac{1}{n} + \frac{(X_i - \bar{X})^2}{\sum x_i^2}}. \tag{12.24}$$

Under the assumption that ε is normally distributed and $\hat{\beta}_0$ and $\hat{\beta}_1$ are BLUE, it follows that

$$\frac{\hat{Y}_i - (\beta_0 + \beta_1 X_i)}{S_{\hat{Y}_i}} \text{ is } t_{n-2}$$

and thus a *100(1 − α)% confidence interval for* $E(Y_i|X_i) = \beta_0 + \beta_1 X_i$ is

$$\hat{Y}_i \pm t_{\alpha/2,n-2} S_{\hat{Y}_i}, \tag{12.25}$$

that is, we may be $100(1 - \alpha)\%$ confident that the true or population average value of Y_i given X_i is bounded by the interval specified by Equation (12.25).

The interval determined by Equation (12.25) can be calculated for any X_i value. And as we vary X_i, we generate a whole set of confidence intervals, with each one centered around a point on the sample regression line. The collection of all these

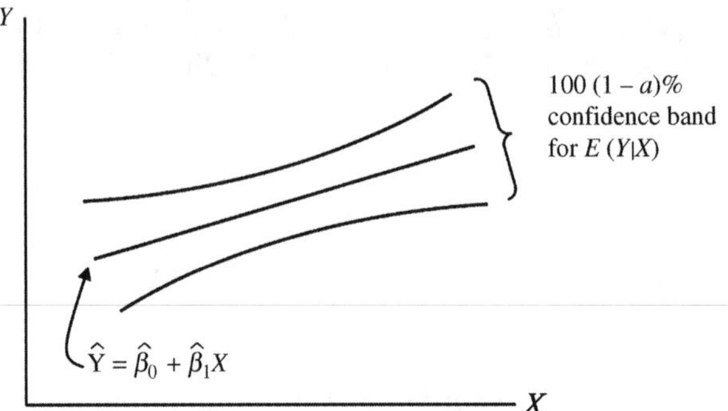

FIGURE 12.8 A 100(1 – α)% confidence band for $E(Y|X)$.

confidence intervals forms a *100(1 – α)% confidence band* about the estimated sample regression equation (Fig. 12.8). That is, if we use Equation (12.5) to calculate a confidence interval for each X_i, then the confidence band is actually the locus of the end points of these intervals. Note the bowed-out shape of the confidence band—we have greater precision in predicting $E(Y_i|X_i)$ near the center of the X_i's than at the extremes of the range of X values.

Example 12.6

Using the Table 12.2 data set, let us construct a 95% confidence interval for $E(Y_i|X_i)$ when $X_i = 6$. Here $\hat{Y}_i = \hat{\beta}_0 + \hat{\beta}_1 X_i = 1.2221 + 0.6869(6) = 5.3435$ and, from Equation (12.24),

$$S_{\hat{Y}_i} = 1.5261\sqrt{\frac{1}{12} + \frac{(6 - 5.5)^2}{99}} = 0.4470.$$

With $t_{\alpha/2, n-2} = t_{0.025, 10} = 2.228$, Equation (12.25) renders 5.3435 ± 0.9959. Hence we may be 95% confident that the average value of Y given $X_i = 6$ lies between 4.3476 and 6.3394.

12.9 THE PREDICTION OF A PARTICULAR VALUE OF *Y* GIVEN *X*

We now turn to the task of predicting a specific Y value from a given level of X. In particular, for $X = X_o$, let us predict or forecast the value of the random variable $Y_o = \beta_0 + \beta_1 X_o + \varepsilon_o$. If the true population regression line were known, the predictor of Y_o would be $E(Y_o|X_o) = \beta_0 + \beta_1 X_o$, a point on the population regression line corresponding to X_o. Since $E(Y_o|X_o)$ is unknown, and must be estimated from sample data, let us use the estimator $\hat{Y}_o = \hat{\beta}_0 + \hat{\beta}_1 X_o$, a point on the sample regression line. If $\hat{Y}_o$ is used to estimate Y_o, then the *forecast error* is the random variable $Y_o - \hat{Y}_o$; its mean is $E(Y_o - \hat{Y}_o) = 0$, that is, $\hat{Y}_o$ is an unbiased estimator for Y_o since $E(Y_o) = \hat{Y}_o$.

Moreover, it can be demonstrated that the *estimated standard deviation of the forecast error* is

$$S_{(Y_o-\hat{Y}_o)} = S_\varepsilon\sqrt{1+\frac{1}{n}+\frac{(X_o-\bar{X})^2}{\sum x_i^2}}. \tag{12.26}$$

If we next assume (as above) that ε is normally distributed (given that the least squares estimators $\hat{\beta}_0$ and $\hat{\beta}_1$ are BLUE), then ultimately

$$\frac{Y_o-\hat{Y}_o}{S_{(Y_o-\hat{Y}_o)}} \text{ is } t_{n-2}$$

and thus a *100(1 − α)% confidence interval for Y_o given X_o* is

$$\hat{Y}_o \pm t_{\alpha/2,n-2}S_{(Y_o-\hat{Y}_o)}, \tag{12.27}$$

that is, we may be 100(1 − α)% confident that the true value of Y_o given X_o lies within the interval determined by Equation (12.27).

To test the hypothesis that Y_o equals some specific or anticipated value Y_o^o, let us consider $H_0: Y_o = Y_o^o$. If H_0 is true, the statistic

$$t = \frac{\hat{Y}_o - Y_o^o}{S_{(Y_o-\hat{Y}_o)}} \text{ is } t_{n-2}. \tag{12.28}$$

Possible alternative hypotheses are:

(a) $H_1: Y_o \neq Y_o^o$. Here $R = \{t||t| > t_{\alpha/2,n-1}\}$
(b) $H_1: Y_o > Y_o^o$. Now $R = \{t|t > t_{\alpha,n-1}\}$
(c) $H_1: Y_o < Y_o^o$. Then $R = \{t|t < -t_{\alpha,n-1}\}$

Example 12.7

Let us again look to the sample data set provided in Table 12.2. We determined in Example 12.1 that when $X_o = 12$, $\hat{Y}_o = 9.4649$. Then from Equations (12.26) and (12.27), 95% *lower and upper prediction limits for Y_o given $X_o = 12$* (the true value of Y at X_o) are, respectively,

$$\hat{Y}_o - t_{\alpha/2,n-2}S_{(Y_o-\hat{Y}_o)} = 9.4649 - 2.228(2.0035) = 5.0011,$$

$$\hat{Y}_o + t_{\alpha/2,n-2}S_{(Y_o-\hat{Y}_o)} = 9.4649 + 2.228(2.0035) = 13.9287$$

(Here

$$S_{(Y_o-\hat{Y}_o)} = 1.5261\sqrt{1+\frac{1}{12}+\frac{(12-5.5)^2}{99}} = 2.0035).$$

Hence we may be 95% confident that the population value of Y given $X_o = 12$ lies between 5.0011 and 13.9287.

At the $\alpha = 0.05$ level, can we conclude that, for $X_o = 12$, the predicted value of Y exceeds 9? Here H_0: $Y_o = Y_o^o = 9$ is tested against H_1:$Y_o > Y_o^o = 9$. With $\hat{Y}_o = 9.4649$ and $t_{\alpha,n-2} = t_{0.05,10} = 1.812$, we see, via Equation (12.28), that

$$t = \frac{9.4649 - 9}{2.0035} = 0.2320$$

is not an element of $R = \left\{t \middle| t > 1.812\right\}$. Hence we cannot reject H_0 at the 5% level (p-value > 0.10); Y_o does not differ significantly from 9 for $\alpha = 0.05$.

One final point is in order. Looking to the structure of Equation (12.24) versus Equation (12.26), it is evident that $S_{(Y_o-\hat{Y}_o)} > S_{\hat{Y}_i}$. Hence, for a given $1-\alpha$ value, the prediction interval obtained from Equation (12.26) is wider than the prediction interval calculated from Equation (12.24). For any $X = X_o$, it should be intuitively clear that the prediction of an individual value of Y, namely Y_o, should have a larger error associated with it than the error arising from an estimate of the average value of Y given X_o, $E(Y|X_o)$.

12.10 CORRELATION ANALYSIS

In the area of correlation analysis, two separate cases or approaches present themselves:

Case A: Suppose X and Y are both random variables. Then the purpose of correlation analysis is to determine the degree of "covariability" between X and Y.

Case B: If only Y is taken to be a random variable and Y is regressed on X, with the values of non-random X taken to be fixed (as in the regression model), then the purpose of correlation analysis is to measure the "goodness of fit" of the sample linear regression equation to the scatter of observations on X and Y.

Let us consider these two situations in turn.

12.10.1 Case A: *X* and *Y* Random Variables

In this instance, we need to determine the direction as well as the strength (i.e., the degree of closeness) of the relationship between the random variables X and Y, where

X and Y follow a "joint bivariate distribution." This will be accomplished by first extracting a sample of points (X_i, Y_i), $i = 1,\ldots,n$, from the said distribution. Then once we compute the sample correlation coefficient, we can determine whether or not it serves as a "good" estimate of the underlying degree of covariation within the population.

To this end, let X and Y be random variables that follow a joint bivariate distribution. Let: $E(X)$ and $E(Y)$ depict the means of X and Y, respectively; $S(X)$ and $S(Y)$ represent the standard deviations of X and Y, respectively; and $COV(X,Y)$ denotes the covariance between X and Y.[2] Then the population correlation coefficient, which serves as a measure of linear association between X and Y, is defined as

$$\rho = \frac{\mathrm{COV}(X,Y)}{S(X)S(Y)} = \frac{E([X - E(X)][Y - E(Y)])}{S(X)S(Y)}. \tag{12.30}$$

Note that if we form the standardized variables $\tilde{X} = (X - E(X))/S(X)$ and $\tilde{Y} = (Y - E(Y))/S(Y)$, then Equation (12.30) can be rewritten as

$$\rho = \mathrm{COV}\left(\tilde{X}, \tilde{Y}\right), \tag{12.30.1}$$

that is, the *population correlation coefficient* is simply the covariance of the two standardized variables $\tilde{X}$ and $\tilde{Y}$.

As far as the properties of ρ are concerned:

(a) It is symmetrical with respect to X and Y (the correlation between X and Y is the same as the correlation between Y and X).
(b) It is dimensionless (a pure number).
(c) It is independent of units or of the measurement scale used.
(d) Since calculating ρ incorporates deviations from the means of X and Y, the origin of the population values is shifted to the means of X and Y.

The value of ρ exhibits both the direction and strength of the linear relationship between the random variables X and Y. That is, if $\rho > 0$, we have a *direct relationship* between X and Y—both variables tend to increase or decrease together; if $\rho < 0$, there exists an *inverse relationship* between X and Y—an increase in one variable is

[2] For observations (X_i, Y_i), $i = 1,\ldots,N$, a measure that depicts the joint variation between X and Y is the *covariance* of X and Y,

$$\mathrm{COV}(X,Y) = \frac{\Sigma_{i=1}^{N}(X_i - \mu_X)(Y_i - \mu_Y)}{N} = \frac{\Sigma_{i=1}^{N} X_i Y_i}{N} - \mu_X \mu_Y, \tag{12.29}$$

where μ_X and μ_Y are the means of the variables X and Y, respectively. For this measure:

(a) If higher values of X are associated with higher values of Y, then $COV(X,Y) > 0$.
(b) If higher values of X are associated with lower values of Y, then $COV(X, Y) < 0$.
(c) If X is neither higher nor lower for higher values of Y, then $COV(X,Y) = 0$.

accompanied by a decrease in the other. Clearly, the sign of ρ must be determined by the sign of COV(X,Y).

Looking to the range of values that ρ can assume, it can be shown that $-1 \leq \rho \leq 1$. In this regard:

(a) For $\rho = 1$ or -1, we have *perfect positive association or perfect negative association*, respectively.

(b) If $\rho = 0$, the variables are said to be *uncorrelated*, thus indicating the absence of any linear relationship between X and Y. (It is important to remember that ρ depicts the strength of the "linear relationship" between X and Y. If X and Y are independent random variables, then $\rho = 0$ since COV(X,Y) $= 0$. However, the converse of this statement is not true—we cannot infer that X and Y are independent if $\rho = 0$ since the true underlying relationship between X and Y may be highly nonlinear.)

(c) As ρ increases in value from 0 to 1, the strength of the linear relationship between X and Y concomitantly increases; and as ρ decreases in value from 0 to -1, the strength of the linear relationship between X and Y also increases.

(d) If two random variables are highly correlated (ρ near either 1 or -1), the association between them does not allow us to infer anything about "cause and effect"—there may be some unobserved variable W that causes movements in both X and Y (readers interested in exploring the problems connected with making causal inferences are directed to Appendix B of this chapter).

(e) Correlation does not allow us to predict values of Y from values of X or vice versa.

(f) No functional relationship between X and Y is involved in the specification of ρ.

12.10.1.1 Estimating the Population Correlation Coefficient ρ Suppose we have a sample of size n consisting of the pairs of observations (X_i, Y_i), $i = 1, \ldots, n$, taken from a bivariate population of X and Y values. Then we may estimate ρ by using the *sample correlation coefficient* (called the *Pearson product–moment correlation coefficient*)

$$r = \frac{\sum (X_i - \bar{X})(Y_i - \bar{Y})}{\sqrt{\sum (X_i - \bar{X})^2} \sqrt{\sum (Y_i - \bar{Y})^2}} = \frac{\sum x_i y_i}{\sqrt{\sum x_i^2} \sqrt{\sum y_i^2}}. \tag{12.31}$$

Our interpretation of r mirrors that offered earlier for ρ. That is, if $r = 1$ (Fig. 12.9a), then X and Y have perfect positive linear association. When $r = -1$ (Fig. 12.9b), X and Y display perfect negative linear association. In these two situations, the sample points lie on some imaginary positively or negatively sloped line, respectively. If $|r| < 1$, the X and Y random variables are linearly related, but to a lesser degree (Figs 12.9c and 12.9d). If $r = 0$ (Fig. 12.9e), then we conclude that X and Y are not linearly related. (Note that, if $r = 0$, we cannot conclude that X and Y are "not related." Their true underlying relationship may be highly nonlinear (Fig. 12.9f).)

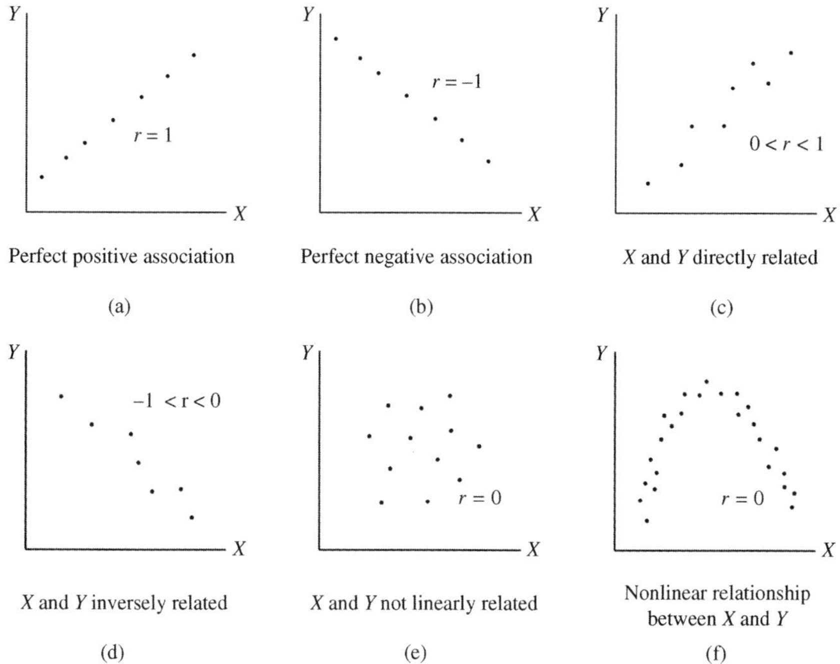

FIGURE 12.9 Sample correlation coefficient.

Example 12.8

The data set provided in Table 12.5 involves a random sample of $n = 10$ observations drawn from a joint bivariate distribution for the random variables X and Y. Are these variables linearly related? If so, in what direction? If the totals of columns 5, 6, and 7 are inserted into Equation (12.31), we obtain

$$r = \frac{\sum x_i y_i}{\sqrt{\sum x_i^2}\sqrt{\sum y_i^2}} = \frac{-20.80}{\sqrt{10.40}\sqrt{168.10}} = -0.4975.$$

With $r < 0$, we may conclude that X and Y vary inversely.

12.10.1.2 Inferences about the Population Correlation Coefficient ρ The population correlation coefficient ρ may be estimated by r [see Equation (12.31)] from a set of observations taken from any joint bivariate distribution relating the random variables X and Y. If we want to go a step further and test hypotheses about ρ, then we must strengthen our assumptions concerning the bivariate population at hand. Specifically, we must assume that X and Y follow a "joint bivariate normal distribution." We noted earlier that if X and Y are independent random variables, then $\rho = 0$. It was also mentioned that if X and Y follow a joint bivariate distribution and

TABLE 12.5 Observations on the Random Variables X and Y

(1)	(2)	(3)	(4)	(5)	(6)	(7)
		$x_i = X_i - \bar{X}$	$y_i = Y_i - \bar{Y}$			
X	Y	$X_i - 4.6$	$= Y_i - 85.3$	x_i^2	y_i^2	$x_i y_i$
5	90	0.4	4.7	0.16	22.09	1.88
4	91	−0.6	5.7	0.36	32.49	−3.42
4	87	−.06	1.7	0.36	2.89	−1.02
3	87	1.6	1.7	2.56	2.89	−2.72
3	86	−1.6	0.7	2.56	0.49	−1.12
5	79	0.4	−6.3	0.16	39.69	−2.52
6	78	1.4	−7.3	1.96	53.29	−10.22
5	83	0.4	−2.3	0.16	5.29	−0.92
5	88	0.4	2.7	0.16	7.29	1.08
6	84	1.4	−1.3	1.96	1.69	−1.82
46	853	0	0	10.40	168.10	−20.80

$\rho = 0$, then we cannot generally conclude that X and Y are independent. However, for a joint bivariate "normal" distribution, independence implies $\rho = 0$ and conversely, that is, zero covariance is equivalent to independence.

Given that X and Y follow a joint bivariate normal distribution, it follows that, under H_0: $\rho = 0$,

$$r \text{ is } N\left(E(r), \sqrt{V(r)}\right) = N\left(0, \sqrt{\frac{1-\rho^2}{n-2}}\right).$$

Then

$$\frac{r - E(r)}{\sqrt{V(r)}} = \frac{r\sqrt{n-2}}{\sqrt{1-\rho^2}} \text{ is } N(0, 1). \tag{12.32}$$

If r is used to estimate ρ in Equation (12.32), then the resulting quantity

$$T = \frac{r\sqrt{n-2}}{\sqrt{1-r^2}} \text{ is } t_{n-2}. \tag{12.33}$$

We noted above that "lack of linear association" is equivalent to independence if X and Y follow a joint bivariate normal distribution. Hence testing for lack of association is equivalent to testing for the independence of X and Y. In this regard, we may test the null hypothesis H_0: $\rho = 0$ against any of the following alternative hypotheses:

Case 1	Case 2	Case 3
H_0: $\rho = 0$	H_0: $\rho = 0$	H_0: $\rho = 0$
H_1: $\rho \neq 0$	H_1: $\rho > 0$	H_1: $\rho < 0$

The corresponding critical regions are determined as follows. At the 100α% level of significance, the decision rules for rejecting H_0 relative to H_1 are:

$$\text{Case 1: Reject } H_0 \text{ if } |T| > t_{\alpha/2,n-2} \qquad (12.34a)$$
$$\text{Case 2: Reject } H_0 \text{ if } T > t_{\alpha,n-2} \qquad (12.34b)$$
$$\text{Case 3: Reject } H_0 \text{ if } T < -t_{\alpha,n-2} \qquad (12.34c)$$

Example 12.9

Given the data set appearing in Table 12.5, does our estimate for ρ, $r = -0.4975$, lie significantly below zero at the $\alpha = 0.05$ level? From Equation (12.33),

$$T = \frac{-0.4975\sqrt{8}}{\sqrt{1 - 0.2475}} = -1.6221$$

is not a member of $R = \{T | T < -t_{0.05,8} = -1.86\}$. Hence we cannot reject H_0 in favor of H_1—at the 5% level, the population correlation coefficient does not lie significantly below zero. (Here 0.05 $<$ p-value $<$ 0.10.)

12.10.2 Case B: *X* Values Fixed, *Y* a Random Variable

We assumed in the preceding section that X and Y follow a joint bivariate normal distribution. Let us, for the moment, retain that assumption here. In fact, under this assumption, it can be shown that the population regression line has the form

$$E(Y|X) = \beta_0 + \beta_1 X = \beta_0 + \rho \frac{S(X)}{S(Y)} X. \qquad (12.35)$$

Then if X and Y follow a joint bivariate normal distribution, ρ^2 can be used as a measure of the goodness of fit of the population regression line to the data points (X_i, Y_i), $i = 1, \ldots, n$; it indicates the proportion of the variation in Y explained by the linear influence of X as well as the "degree of covariability" between the random variables X and Y. This same conclusion is valid if the assumption of normality is dropped and the random variables X and Y simply follow a joint bivariate distribution.

Now, if only Y is a random variable and the X values are held fixed (as was the case when we developed our basic linear regression model earlier), then, since COV(X,Y) does not exist, ρ^2 (or ρ) cannot serve as a measure of covariability—it is only indicative of the "goodness of fit" of the population regression line to the scatter of sample points. Here's the rub. If, under our previous set of regression assumptions (ε is "well behaved"), we have obtained r^2 according to

$$\frac{\text{SSR}}{\text{SST}} = \frac{\hat{\beta}_1 \sum x_i y_i}{\sum y_i^2}$$

(see Table 12.3), then we have also obtained r^2 as an estimate of ρ^2. However, r does not serve as an estimate of ρ since the latter is, strictly speaking, undefined.

EXERCISES

1 Given the following set of X and Y values:

a. Use Equations (12.9a) and (12.9b) to find the regression line that best fits the scatter of points (X_i, Y_i), $i = 1, \ldots, 5$.

b. What is the predicted value of Y given $X_o = 5$?

c. Verify that

$$\Sigma x_i^2 = \Sigma X_i^2 - (\Sigma X_i)^2/n$$
$$\Sigma y_i^2 = \Sigma Y_i^2 - (\Sigma Y_i)^2/n$$
$$\Sigma x_i y_i = \Sigma X_i Y_i - (\Sigma X_i)(\Sigma Y_i)/n.$$

X	2	4	6	8	10
Y	2	1	6	5	9

2 Given the following pairs of X and Y values:

a. Use Equations (12.9a) and (12.9b) to find the least squares regression line of best fit.

b. Determine the predicted value of Y given $X_o = 13$.

X	2	3	4	5	6	7	8
Y	1	3	5	8	9	11	12

3 For the following set of X and Y values:

a. Find the regression line of best fit via Equations (12.9a) and (12.9b).

b. What is the predicted value of Y given $X_o = 20$?

X	12	13	14	15	16	17	18
Y	−8	−6	−4	0	2	4	3

4 For the regression equation obtained in Exercise 12.1:

a. Determine the partitioned sums of squares table (Table 12.3).

b. Find and interpret the coefficient of determination r^2.

5 For the regression equation obtained in Exercise 12.2:

a. Calculate the partitioned sums of squares table (Table 12.3).
b. Find and interpret the coefficient of determination r^2.

6 For the regression equation obtained in Exercise 12.3:

a. Determine the partitioned sums of squares table (Table 12.3).
b. Calculate and interpret the coefficient of determination r^2.

7 Suppose $n = 30$, $\bar{X} = 0.5$, $\bar{Y} = 1$, $\Sigma x_i y_i = -30$, $\Sigma x_i^2 = 10$, and $\Sigma y_i^2 = 160$:

a. Find the equation for the regression of Y on X.
b. Determine the partitioned sums of squares table.
c. Interpret the value of r^2.

8 Given $n = 25$, $\Sigma X_i = 85$, $\Sigma Y_i = 50$, $\Sigma X_i^2 = 625$, $\Sigma Y_i^2 = 228$ and $\Sigma x_i y_i = 30$:

a. Find the equation for the regression of Y on X. [Hint: see part (c) of Problem 12.1]
b. Determine the partitioned sums of squares table.
c. Find and interpret r^2.

9 For the regression equation obtained in Exercise 12.7:

a. Find 95% confidence limits for β_0 and β_1.
b. Test $H_0 : \beta_0 = 0$ versus $H_1 : \beta_0 \neq 0$ for $\alpha = 0.05$. What is your conclusion?
c. Test $H_0 : \beta_1 = 0$ versus $H_1 : \beta_1 \neq 0$ for $\alpha = 0.05$. What is your conclusion?
d. Can we conclude that $\beta_1 = \beta_1^o = -3.7$ at the 5% level? Set $H_1 : \beta_1 \neq \beta_1^o = -3.7$.

10 For the regression equation obtained in Exercise 12.8:

a. Find 99% confidence limits for β_0 and β_1.
b. Test $H_0 : \beta_0 = 0$ versus $H_1 : \beta_0 > 0$ for $\alpha = 0.01$. What is your conclusion?
c. Test $H_0 : \beta_1 = 0$ versus $H_1 : \beta_1 < 0$ for $\alpha = 0.01$. What is your conclusion?
d. Can we conclude that $\beta_1 = \beta_1^o = -0.25$ at the 1% level? Set $H_1 : \beta_1 \neq \beta_1^o = -0.25$.

11 For the regression equation obtained in Exercise 12.1:

a. Find a 95% confidence interval for the average value of Y given $X_o = 8$.
b. Find a 95% confidence interval for predicting a single value of Y given $X_o = 8$.

12 For the regression equation obtained in Exercise 12.2:

a. Find a 95% confidence interval for the mean value of Y given $X_o = 6$.
b. Find a 95% prediction interval for a single value of Y given $X_o = 6$.

13 Given $n = 11$, $\bar{X} = 4$, $\bar{Y} = 20$, $\Sigma x_i^2 = 64$,$\Sigma y_i^2 = 1600$, and $\Sigma x_i y_i = 256$:

a. Determine the least squares regression equation.
b. What is the predicted value of Y for $X_o = 6$?
c. Find a 95% confidence interval for the average value of Y given $X_o = 6$.
d. Find a 95% prediction interval for a single value of Y given $X_o = 6$.

14 For the data set in Exercise 12.1:

a. Calculate the coefficient of correlation r [Equation (12.31)]. What assumption is being made?
b. Test $H_0 : \rho = 0$ versus $H_1 : \rho \neq 0$ for $\alpha = 0.05$. What is your conclusion? What assumption is being made?

15 For the data set in Exercise 12.2:

a. Calculate the correlation coefficient r via Equation (12.31). What assumption is being made?
b. Test $H_0 : \rho = 0$ versus $H_1 : \rho > 0$ for $\alpha = 0.01$. What is your conclusion? What assumption is being made?

16 For the data set in Exercise 12.3:

a. Find the correlation coefficient r. Under what assumption are your operating?
b. Test $H_0 : \rho = 0$ against $H_1 : \rho > 0$ for $\alpha = 0.01$. What is your conclusion? What assumption are you making?

APPENDIX 12.A ASSESSING NORMALITY (APPENDIX 7.B CONTINUED)

Appendix 7.B addressed the issue of determining if a simple random sample taken from an unknown population is normally distributed. The approach offered therein for assessing the normality of a data set was the normal probability plot of the ordered observed sample data values against their normal scores in the sample or their "expected" Z-scores (see Table 7.B.2 and Fig. 7.B.2). As Fig. 7.B.2 reveals, we do obtain a fairly good approximation to a straight line (the case of perfect normality). Is getting a "fairly good" approximation to a straight line obtained by simply "eyeballing" the data points (X_j, Z_{p_j}) (columns 3 and 4, respectively, in Table 7.B.2) a satisfactory way to proceed? Can we devise a more sophisticated approach to determining if a linear equation actually obtains? And if one does obtain, can we test its statistical significance? To answer these questions, one need only look to the developments in this chapter. The task of assessing normality can be addressed via the application of regression analysis.

Let us take the X_j and Z_{p_j} data values and regress Z_p on X so as to obtain the equation of the line that best fits the data points in Fig. 7.B.2. Then, once we estimate the least squares regression line, we can test its slope for statistical significance. If the slope is significantly positive, then we may conclude that our sample data set can be viewed "as if" it was extracted from a normal population.

The regression of Z_p on X results in the estimated linear equation

$$\hat{Z}_p = \hat{\beta}_0 + \hat{\beta}_1 X = \underset{(-10.52)}{-5.0475} + \underset{(10.67)}{0.2448}X,$$

where the numbers in parentheses under the estimated regression coefficients are their associated t values determined under the null hypotheses $\beta_0 = 0$ and $\beta_1 = 0$, respectively. Here the p-value for each of these estimated coefficients is <0.0001. Additionally, $r^2 = 0.9344$, that is, in excess of 93% of the variation in Z_p is explained by the linear influence of X. We thus have a highly significant fit to this data set. Clearly, we have gone beyond simply saying that, in graphical terms, the data points X_j, Z_{p_j} provide a "fairly good" approximation to a straight line.

APPENDIX 12.B ON MAKING CAUSAL INFERENCES[3]

12.B.1 Introduction

This appendix addresses the issue of causality. While my motivation for discussing this slippery topic comes from the well-known utterance "correlation does not imply causation," the scope and coverage of this discussion goes far beyond simply debunking the existence of any unequivocal connection between association and causation.

To gain a full appreciation of the essentials of causality, one has to make a fairly small investment in learning some important terminology—an investment that should yield a sizable payoff in terms of one's ability to think critically. That is, one has to become familiar with the rudiments of experimental design, logical relations, conditional statements, and the concepts of necessary and sufficient conditions. Given this grounding, the search for a working definition of causality and the link between logical implications and causality becomes a straightforward matter. It is then a short step to examining correlation as it relates to causality. Finally, we look to developing a counterfactual framework for assessing causality, and all this subsequently leads to the topic of testing for causal relations via experimental and observational techniques.

[3] The material in this section draws heavily from the presentations in Rubin (1974, 1977, 1978, 1980), Pearl (1988, 2009), Holland (1986), Croucher (2008), Morgan and Winship (2010), Collier, Sekhan, and Stark (2010), and Dechter, Geffner, and Halpern (2010).

While the principal focus of this book is statistics, it is the case that there exists a profound difference between statistical and causal concepts. Specifically:

(a) Statistical analysis: Employs estimation and testing procedures to determine the parameters of a static probability distribution via sampling from that distribution. Here one looks for associations, calculates the likelihood of events, makes predictions, and so on. It is assumed that the experimental conditions under which any inferences are made are invariant.

(b) Causal analysis: Designed to infer the likelihood of events through dynamic changes induced by interventions or controlled manipulations of some system.

So given, for instance, a static probability distribution, causal analysis provides a framework for determining how the distribution would be affected if external conditions would change; it incorporates additional suppositions that can identify features of the distribution that remain invariant when controlled modifications take place.

As was just mentioned, an important component of causal analysis is the notion of an intervention. To perform a causal experiment, one must engage in a process of *external manipulation* or *intervention* in order to determine the comparative effects under different treatments on a group of subjects.[4] Under this sort of experimental policy, treatment is assigned to some subjects who would not ordinarily seek treatment, and a placebo (the control) is given to some subjects who would actually welcome treatment. Clearly, this approach is akin to short-circuiting one behavioral process and rewiring it to another such process via a splice formed by the act of randomization, for example, a coin flip determines who actually enters the treatment group.

Looked at from an alternative perspective, suppose we have variables X and Y and it is to be determined if X has an effect on Y. If we manipulate X and the resulting response Y does not depend on the value assigned to X, then we can conclude that X has no effect on Y. But if it is determined that the Y response depends on the level assigned to X, then our conclusion is that X does have an effect on Y. We thus need to consider the behavior of a system in a "disturbed state" that is precipitated by some external intervention.

This "intervention theme" will emerge quite often in our discussion of causality, especially when we get to counterfactuals. The next two sections house some important definitional material.

12.B.2 Rudiments of Experimental Design

In statistics, it is generally accepted that *observational studies* (e.g., ones that simply count the number of individuals with a specific condition or characteristic) can give clues about, but can never fully confirm, cause and effect. The standard procedure for testing and demonstrating causation is the *randomized experiment*.

[4] Readers unfamiliar with this terminology can quickly read the next section.

We may view an *experiment* as a directed study designed to determine the effect (if any) that varying at least one explanatory variable has on a *response variable*, the variable of interest. These explanatory variables are called *factors*, and any combination of the values of these factors is termed a *treatment*. An *experimental unit* can be a person (called a *subject*), an object, a place or location, and so on.

To conduct a randomized experiment, the first step is to divide the totality of, say, the subjects into two treatment groups: the *experimental group* (that receives the treatment) and the *control group* (that is given a *placebo*—an inert or innocuous substance such as a sugar tablet). It is the control group that bears the risk of being exposed in some fashion to an activity different from that reserved for the experimental group. Secondly, we must *randomize* the subjects to the treatment groups so that the effect(s) of extraneous factors, which are beyond the control of the experimenter, are minimized. Hence randomization "averages out" or tempers the impact of variation attributed to factors that are not subject to manipulation. Moreover, the randomized experiment must ideally be *double blind*—neither the subject nor the experimenter knows what treatment is being administered to the subject (it is not known by any party to the experiment whether a particular subject is in the experimental group or in the control group).

As far as the control is concerned, its level or status is set at some fixed value for the duration of the experiment. And regarding treatment levels for the experimental group, they are *manipulated* or set at prespecified levels. These treatment levels are the factors whose effect on the response variable is to be determined. Finally, *replication* occurs when each treatment level is applied to more than one subject or experimental unit.

For instance, suppose we are interested in measuring the impact of a new drug (in pill form) on reducing "bad" cholesterol. Our subjects (each having a cholesterol problem) can be randomly divided (by, say, the flip of a coin) into two groups: the control, which received the placebo; and the experimental group, which receives either 10, 20, or 40 mg doses of the drug (again under randomization). Neither the subjects nor the experimenter knows who gets an inert pill or a specific treatment level of the drug. Our goal is then to measure the effect of various treatment levels (relative to the control and to each other) on the response variable (bad cholesterol level).

12.B.3 Truth Sets, Propositions, and Logical Implications

Suppose the realm of all logical possibilities L constitutes a set whose elements correspond to propositions or statements. Think of a *proposition* as involving the content or substance of meaning of any declarative sentence, which is taken to be either true or false, but not both, for example, p can be the proposition "ice cream is delicious." Its *negation*, denoted p', is "ice cream is not delicious." So when p is true, p' is false; and when p is false, p' is true.) For statements p and q in L, let the sets P and Q represent the subsets of L for which these propositions are respectively true, that is, P and Q are *truth sets* for statements p and q.

Let us assign *truth values* to propositions p and q. Specifically, if proposition p is true, we write $p = 1$ ("1" denotes a proposition that is always true); and if proposition p

TABLE 12.B.1 Truth Values for p, q

p	q	pq (Conjunction)	$p+q$ (Disjunction)	$(pq)'$ (Negation)	$(p+q)'$ (Negation)
1	1	1	1	0	0
1	0	0	1	1	0
0	1	0	1	1	0
0	0	0	0	1	1

is false, we write $p=0$ ("0" is a proposition that is always false). Similar truth values as assigned to q. Given propositions p and q, the four possibilities for their truth values can be summarized in a *truth table* (Table 12.B.1): p and q both true; p true and q false; p false and q true; and p and q both false.

Some important logical operations are:

(a) *Conjunction* of p and q is "p and q" (both p and q) and denoted pq. Hence pq is true when both p and q are true; false when either one is false or both are false (Table 12.B.1).

(b) *Disjunction* of p and q is "p or q" (either p or q or both) and denoted $p+q$. Thus $p+q$ is true whenever either one of p and q or both are true; false only when both p and q are false (Table 12.B.1).

(c) *Negation:* $(pq)' = p' + q'$ (not p or not q); $(p+q)' = p'q'$ (not p and not q) (Table 12.B.1). (Hint: Think of DeMorgan's Laws.)

Consider the *material implication* "p implies q" or $p \rightarrow q = p' + q$ (either not p or q), where p is termed the *antecedent* and q is called the *consequent* (Table 12.B.2). (It is important to note that the word "implies" does not mean that q can logically be deduced from p.) This implication is equivalent to the *conditional statements:* "if p, then q" and "q if p." In this regard, the statement "if p, then q" is true if: p and q are true; p is false and q is true; and p is false and q is false. "If p, then q" is false if p is true and q is false.

Furthermore, given $p \rightarrow q$:

(a) Its *converse* is $q \rightarrow p = q' + p$.

(b) Its *inverse* is $p' \rightarrow q' = -p + q'$.

TABLE 12.B.2 Material Implication

P	q	$p' + q$ $p \rightarrow q =$	$q \rightarrow p =$ $q' + p$ (Converse)	$p' \rightarrow q' =$ $p + q'$ (Inverse)	$q' \rightarrow p' =$ $q + p'$ (Contrapositive)	$(p \rightarrow q)' =$ pq' (Negation)
1	1	1	1	1	1	0
1	0	0	1	1	0	1
0	1	1	0	0	1	0
0	0	1	1	1	1	0

(c) Its *contrapositive* is $q' \rightarrow p' = q + p'$.

(d) Its *negation* is $(p \rightarrow q)' = (p' + q)' = pq'$.
(It is false that "p implies q" is "p and not q").

The truth values for all these implications appear in Table 12.B.2.

12.B.4 Necessary and Sufficient Conditions

A statement $p \varepsilon P$ is a necessary condition for a statement $q \varepsilon Q$ to be true if $q \rightarrow p$ (q implies p or "if q, then p"). Here "q is true only if p is true" so that p is a prerequisite or requirement for q. So if we don't have p, then we don't have q. For instance,

> "purchasing a ticket" (p) is a necessary condition for "admission to a concert" (q).

(If a person is present at a concert, they must have purchased a ticket, that is, $q \rightarrow p$.)

Additionally, a statement $p \varepsilon P$ is a *sufficient condition* for a statement $q \varepsilon Q$ to be true if $p \rightarrow q$ (p implies q or "if p, then q"). Thus "q is true if p is true." Think of a sufficient condition as a condition that, if satisfied, guarantees that q obtains or assures the truth of q. That is,

> "not bringing a camcorder to a concert" (p) is a sufficient condition for "admission to the concert" (q).

(If a person does not bring a camcorder to a concert, they will be granted admission, that is, $p \rightarrow q$.)

Note that having a ticket is not sufficient for admission—it is possible to have a ticket and not be granted admission (a camcorder is discovered). And not having a camcorder is not a necessary condition for admission—it is possible to not have a camcorder and yet not be admitted (no ticket).

Let us consider another situation. The

> "presence of water" is a necessary condition for "life on earth."

(If life on earth exists, then that life has water.) However, the

> "presence of water" is not a sufficient condition for "life on earth."

(The presence of water does not, by itself, assure the existence of life since life also needs oxygen and generally a myriad of favorable conditions for its support.)

Also,

> "being male" is a necessary condition for "being a father."

(If a person is a father, then that individual must be male.) But

> "being male" is not a sufficient condition for "being a father."

(One can be male and not have any children.)

12.B.5 Causality Proper

Certainly everyone has, at one time or another, pondered the question "What caused a particular event to occur?" This is not an easy question to answer. In fact, it is a question that mankind has been addressing for hundreds of years. If one starts with the Western philosophical tradition, serious discussions of causation harken back to at least Aristotle [see his Metaphysics, Books V, XI (350BC), tr. by W. D. Ross (1941) and Posterior Analytics, Book II (350BC), tr. by E. S. Bouchier (1901)] and continue with David Hume (1740, 1748), John Stuart Mill (1843), and, more recently, with David Lewis (1973a, 1973b, 1973c, 1979, 1986), D. B. Rubin (1974, 1977, 1978, 1980), P. W. Holland (1986), and Judea Pearl (2009), among others.

We may view the notion of *causality* (or *causation*) as essentially a relationship between "cause and effect," that is, event A is said to cause event B if B is a consequence of A—the causal event A is the reason why effect B occurs.[5] Positing a cause and effect relationship requires the following:

1. The cause must occur before the effect.
2. The cause must be connected to the effect in a probabilistic sense, that is, the effect has a high probability of occurring.
3. There are no alternative or spurious causes due to so-called (unobserved) lurking variables (more on this point later on).

(While this might be considered a naïve portrayal of cause and effect, a more practicable version of this concept will be offered below.) Furthermore, we can distinguish between:

(a) Necessary cause: Event A is a *necessary cause* of event B if the presence of B implies the presence of A (the presence of A, however, does not imply that B will occur).
(b) Sufficient cause: Event A is a *sufficient cause* of event B if the presence of A implies the presence of B. However, another (confounding) event C may possibly cause B. (Thus the presence of B does not imply the presence of A.)

Let us consider the features of these two causes in greater detail.

Necessary causality emphasizes the absence of alternative causes or processes that are also capable of producing some effect. In this regard, this variety of causality is useful when focusing on a specific event. And since various external or background factors could also qualify as legitimate causal explanations, the isolation of a necessary cause may be impeded because of confounding. For example, consider the statement "A biking accident was the cause of Joe's concussion." Here we have a specific scenario—a single-event causal assertion.

Sufficient causality involves the operation of a functioning causal process capable of producing an effect. Here there exists a general or generic inclination for certain

[5] The phrase "cause and effect" should always be understood to mean "alleged cause and alleged effect."

types of events to generate other event types, where the particulars of event-specific information are lacking. For instance, the assertion "biking accidents cause concussions" is a generic causal statement.

12.B.6 Logical Implications and Causality

Do logical relations imply cause and effect? Suppose that for events A and B we have the material implication $A \rightarrow B$. That is, A is a sufficient condition for B or B if A. For concreteness, let events A and B appear as follows:

A: A person exhibits symptoms of Lyme disease.
B: A person was bitten by a deer tick.

For any individual under consideration, we can conclude that, if $A \rightarrow B$, then from column 3 of Table 12.B.2, "either not A or B" occurs (either the person does not exhibit symptoms of Lyme disease or the person was bitten by a deer tick).

Is there a cause-and-effect relationship between events A and B? No, event A cannot cause event B to occur (since symptoms of Lyme disease cannot cause a tick bite). In sum, we can infer that "A implies B," but certainly not that "A causes B;" we cannot infer causality from just the observation of individuals (or, more generally, from simply observing frequency data).

If conditional statements do not capture cause-and-effect relationships, then how should we view the notion of cause and effect? A cause-and-effect assertion is essentially a type of *predictive hypothesis* —a declaration that, if something changes (under a distribution perturbation or intervention), you cannot eliminate the effect without eliminating the cause. So given any adjustment that allows the cause to remain, we can expect the effect to remain as well. For our purposes, we shall recognize two forms of predictive hypotheses: a weak (or statistical) form; and a strong (or causal) form. Specifically,

(a) [Statistical] For propositions p and q, not rejecting the hypothesis that "p implies q" enables us to predict that, under repeated sampling from the "same distribution," we will continue to observe "either not p or q."
(b) [Causal] For events A and B, the assertion that "A causes B" enables us to predict that, if some intervention changes the distribution, then we will observe "either not A or B" under repeated sampling from the "new distribution."

Assertions about causality cannot be proved, only disproved. And causality is disproved if we can eliminate the effect without eliminating the cause. For instance, suppose that, day after day, the morning paper is delivered shortly after my morning pot of coffee is brewed. Can we conclude that brewing coffee caused the paper to be delivered? Obviously not. If the paper is not delivered because the carrier is taken ill and the coffee is brewed as usual, then we have disproved the notion that brewing coffee causes the paper to be delivered. (We have removed the effect but not the cause.) However, for events A and B given above, if the deer population is controlled (decreased) and, *ceteris paribus*, the incidence of Lyme disease symptoms concomitantly decreases (albeit with a lag), then, since both of these events move together, we have not disproved anything, that is, we have not eliminated the effect without eliminating the cause.

12.B.7 Correlation and Causality

In this chapter, we relied on the concept of correlation to determine if there exists a degree of linear association between the two random variables X and Y. To this end, we utilized the Pearson product–moment correlation coefficient r [Equation (12.29)] to reveal whether a relationship is strong or weak and whether it is positive or negative. If the relationship is found to be statistically significant [via the test statistic (12.31)], then regression analysis can subsequently be used to predict the value of one variable (say, Y) from the value of the other variable (X). It was also mentioned that even if two variables are significantly correlated, any such association does not imply causation.

Why is it the case that correlation does not imply causation? One possibility is that the observed degree of association between X and Y is actually *spurious* or due to *confounding*—we have *false correlation* due to the presence of an additional unobserved or *lurking variable* that causes movements in both X and Y. For instance:

(a) Suppose we find that there exists a significant positive linear association between the sale of bags of ice cubes and reported aggravated assaults. Will increased sales of bags of ice cubes actually precipitate an increase in aggravated assaults? Probably not. The explanation for this observed association is that high temperatures and oppressive dew points increase this type of aggressive behavior (people become much more irritable) as well as the sales of ice cubes (people enjoy cool drinks in hot weather). Thus ice cube sales and aggravated assaults are both affected by "temperature," a lurking (third) variable whose presence causes the observed correlation. Hence we are fooled into thinking that ice cube sales and aggravated assaults are significantly related.

(b) A study of school children found a significant positive linear association between reading comprehension and shoe size. Does having big feet lead to increased reading comprehension? It is not likely. It should be intuitively clear that as children grow older, their reading comprehension and shoe size concomitantly increase. Since a child's mental and physical prowess both develop as a child grows older, it is apparent that "age" is a lurking variable.

(c) A researcher found a high positive correlation between heavy drinking of alcoholic beverages and respiratory problems. Does heavy drinking cause respiratory difficulties? Not really. Since most heavy drinkers are also heavy smokers, it is the lurking variable "heavy smoking" that leads to respiratory problems. Thus heavy smoking is a confounder of heavy drinking and respiratory problems.

TABLE 12.B.3 Correlation Versus Causation

1. Correlation $\nrightarrow$ Causation
   (correlation is not a sufficient condition for causation)
2. Causation $\rightarrow$ Correlation
   (correlation is a necessary condition for causation—if a causal relationship between events *A* and *B* actually exists, then it is also highly likely that these events will be correlated)

The gist of the preceding discussion is that while "observed correlation does not imply causation," it is the case that correlation is a "necessary" but not a "sufficient" condition for causation. This is because, from a purely logical viewpoint, we know that the word "implies" means "is a sufficient circumstance for." Table 12.B.3 summarizes this line of argumentation.

It is interesting to note that the statement "correlation implies causation" is a logical fallacy of the *cum hoc, ergo propter hoc* ("with this, therefore because of this," *Latin*) variety and typically takes the form:

Events *A* and *B* happen at the same time.
Therefore *A* caused *B*.

This is clearly a logical fallacy since

1. The association between events *A* and *B* may be a pure coincidence (possibly because the sample size was too small).
2. In reality, *B* may actually be the cause of *A*.
3. Some *spurious* or *lurking factors* may be the cause of both *A* and *B*.

12.B.8 Causality from Counterfactuals

Within the counterfactual tradition of causal reasoning, the concept of causality is defined with respect to *counterfactual dependence*: for events *A* and *B*, *A* causes *B* if

1. *A* and *B* occur together.
2. If *A* had not occurred (this is the *counterfactual condition* since *A* did, in fact, occur), then, *ceteris paribus*, *B* would not have occurred.

It should be evident that this definition encompasses admissible background changes that can occur in the external world; it admits an implied modification that has predictive power when the *ceteris paribus* restriction holds. Under this approach, all statements about causality can be interpreted as counterfactual statements.

So what does a counterfactual statement look like? Consider the assertion:

"John's heavy drinking caused his liver failure."

Counterfactually, this statement is equivalent to

"Had John not been a heavy drinker, he would not have experienced liver failure."

So for an actual cause-and-effect relationship to exist between events A and B, we need something more than just a high degree of association between A and B (item #1)—we also need these events to display a counterfactual dependence (item #2).

For instance, suppose a student's performance on a statistics exam was less than stellar. After some reflection, he concludes that he did not solve enough practice problems. Counterfactually, now think of this same student taking the same exam under the same set of circumstances, but having solved a whole assortment of practice exercises the night before the exam. This "controlled intervention" changes only one small facet of the student's situation—solving more practice problems. In this circumstance, causation could be observed—we simply compare the exam scores obtained with and without the extra problem solving. However, since this type of experiment is impossible, causality can never be strictly proven, only inferred or suggested. This discussion leads to the specification of the

Fundamental Problem of Causal Inference: It is impossible to directly observe causal effects or outcomes.

Let us highlight the details of some of the subtleties underlying this fundamental problem by considering another (and more sharply focused) example. Suppose that each experimental unit (subject) in the population can be exposed to two alternative states of a cause: the treatment; and the control. That is, the effect of one cause is always specified relative to the effect of another cause. Hence the notion that "A causes B" will be taken to mean that A causes B relative to another cause that subsumes the proposition "not A." Hence one cause will be the treatment and the other will be the control. Clearly, the control includes the condition "not the treatment A."

For instance, high school seniors (the population of interest) are randomly assigned to take a new accelerated SAT review course (the treatment) or not (the control). Here we are implicitly assuming that each subject has a potential outcome under each causal state although, in reality, each subject can be observed in only one such state (either the treatment or the control) at a given point in time.

For this SAT review course (we are interested in the causal effect of taking or not taking the SAT review course on the student's SAT score), seniors who have taken the review have theoretical "what-if scores" under the control state, and seniors who have not taken the review have theoretical "what-if scores" under the treatment state. Clearly, these "what-if outcomes" are unobservable counterfactual outcomes.

Next, let Y be the response variable that measures the effect of a cause. (Here Y is the SAT score.) In this regard, let us define the *post-exposure values*:

$Y_T(s)$: The value of Y that would be observed if student s were exposed to the treatment (the SAT review course).

$Y_C(s)$: The value of Y that would be observed if student s were exposed to the control (did not take the SAT review course).

Then for student s, the effect of the treatment relative to the effect of the control is simply the difference

$$Y_T(s) - Y_C(s). \tag{12.B.1}$$

[The causal effect of the treatment relative to the control on student s as indexed by y.]

Here the treatment causes the net effect in (12.B.1)—the difference between observed and potential outcomes.

How may we determine the value of the causal effect provided by (12.B.1)? Note that this difference is defined for a particular student s. Can we actually observe the effects of causes for specific students? The answer is, no. We can never observe effect (12.B.1) for a particular student s. This is because one can never observe the "what-if outcome" under the treatment state for those subjects in the control state; and one can never observe the "what-if outcome" under the control state for those subjects in the treatment state. For a given student s, we cannot observe the values of both $Y_T(s)$ and $Y_C(s)$ at the same time. [Remember that Equation (12.B.1) is actually the "what-if" difference in achievement that could be calculated if student s could simultaneously be in the SAT review course and not in the SAT review course.] Thus one can never calculate subject-level causal states—the above-said fundamental problem of causal influence. Thus individuals contribute outcome information only from the causal state in which they are observed. Hence the typical response variable contains only a portion of the information that would be needed to compute causal effects for all subjects.

This said, the fundamental problem does, however, admit a solution. While individual causal effects (12.B.1) are unobserved, we can obtain the *average causal effect* (using expectations) as

$$E[Y_T(s) - Y_C(s)] = E(Y_T(s)) - E(Y_C(s)) = \alpha. \tag{12.B.2}$$

Clearly, this difference is a "statistical solution" to the fundamental problem: we can estimate an average causal effect α. And as long as the students are randomly assigned to the treatment and control groups ($Y_T(s)$ and $Y_C(s)$ must be random variables), Equation (12.B.2) is useful.

Given the above line of argumentation, it should by now be evident that causation encompasses the assessment of behavior under interventions or controlled manipulations. This observation is incorporated in the

General intervention principle: To infer a cause-and-effect relationship, we need to determine how a distribution or system responds to interventions or perturbations.

This principle now enables us to posit a workable definition of cause and effect. Specifically, we first ask: "What can serve as a cause?" A *cause* is anything that could conceivably be a treatment in a randomized experiment (e.g., various drug dosage levels, and so on) However, the attributes of a subject (sex, color, texture, and so on) cannot serve as a cause in an experiment. In fact, attributes can only be used when measuring association.

We next ask: "What can serve as a causal effect?" One may view an *effect* as an empirical outcome that would emerge in a controlled randomized experiment—it enables us to predict what circumstances would prevail under a hypothetical system intervention.

The stage is now set for addressing the issue of "testing causality." As will be explained in Section 12.B.9, the so-called "gold standard" for a test of causality is the implementation of a controlled randomized experiment. If performing this type of experiment is impractical, then, as a second best option, an observational study might be warranted.

12.B.9 Testing Causality

It is important to state at the outset that modeling for causal inferences emphasizes the measurement of the "effects of causes." Moreover, estimates of causal effects can be obtained by performing controlled randomized experiments in which the sample elements come under the direct control of, or are subject to manipulation by, the experimenter.

Remember that it is the "controlled" randomized experiment that constitutes a scientifically authenticated methodology for testing causality using sample data. Uncontrolled experiments are anathema. For instance, suppose we are interested in studying the ability of a new drug to lower blood pressure in women. The effectiveness of the drug depends upon both treatment and background factors (such as socioeconomic conditions, and so on). Without a control, the choice of the treatment is at the discretion of the subject—an obvious problem since we cannot then determine if response rates are due to the treatment or due to extraneous conditions.

Remember also that the role of "randomization" is to simulate equality between treatment and control groups. It is under randomization that the likelihood of these groups having similar characteristics, save for some suitably restricted manipulation or intervention, is rather high, that is, in the long run, randomization renders the treatment and control groups equivalent in terms of all other possible effects on the response so that any changes in response will reflect only the manipulation (treatment). In fact, this likelihood increases with the number of participants in the experiment. Now, if the measured difference between the effect of the treatment relative to that of the control is found to be statistically significant, then it is also the case that the likelihood of the treatment exerting a causal effect on some response variable is quite high.

In short, we need to compare subjects with similar backgrounds so that treatment effects can be isolated, and this is achieved under controlled randomized experimentation. It is only through well-designed experiments that we can uncover, as completely as possible, counterfactual dependencies between events.

While controlled randomized experiments are the cornerstone of causal analysis, it may be the case that such experiments are impractical, too expensive, and/or conflict with certain ethical or moral standards. For instance, suppose a subject with later-stage cancer ends up in the control group of some double-blind experiment designed to test the effect of a newly proposed cancer-curing wonder drug. And suppose further that they think that, by virtue of participating in the study, a cure might actually be forthcoming. What might be the reaction of these individuals if they found out that, being in the control, only a placebo was issued and "a cure was not in the cards?"

Additionally, what if an experiment was designed to assess the impact of weight gain on the speed of development of diabetes. Should subjects in the treatment group be forced to gain weight while those in the control group are required to diet until sufficient time passes so that their relative diabetes rates can be compared? How about the issue of smoking as a contributing factor to the development of lung cancer? Should subjects in the treatment group be required to smoke while subjects in the control are prohibited from smoking so that, after some period of time, we can determine if their relative cancer rates are significantly different? Clearly, most, if not all, practitioners would have a problem with these types of treatments. So under any moral or ethical restrictions, an alternative to a controlled randomized experiment is the performance of an observational study to possibly infer causal effects. Before exploring the efficacy of such studies, let us compare this form of inquiry with the process of experimental design. Specifically:

Experimental study: All aspects of the system under study are controlled by the experimenter (e.g., selection of experimental units, treatment(s), response measurement procedures, and so on).

Observational study: While this is an empirical investigation of treatments and the effects they cause, not all of the aforementioned particulars are under the control of the investigator (e.g., randomized designs are infeasible—no control over assigning treatments to subjects, and so on).

Can we rely on observational studies to infer causal effects? More formally, let us look at the

Fundamental Question of Observational Analysis: Under what circumstances can causal effects be inferred from statistical information obtained from passive, intervention-free behavior?

In a nutshell, the answer is—"it is very difficult." Conducting an observational study to uncover causation is not a trivial task. For instance,

(a) Some researchers [see, in particular, D. A. Freedman (1991, 1999)] have argued that relying primarily on statistical methodology can give questionable results. One needs to make "causal assumptions;" and the ability to make causal assumptions rests on subject matter expertise and good research design to alleviate confounding and eliminate competing explanations of reality. In fact, alternative explanations need to be tested.

(b) Modeling assumptions (normality, randomness, homogeneity, independence, and so on) must be tested. Only if the model itself passes muster, can we squeeze some useful information from it.

(c) To draw causal inferences from observational data, one needs to rely heavily on intuition, intelligence, perseverance, and luck (possible confounders must be identified, alternative paths of inquiry must be followed, a confluence of evidence must be analyzed, elements of natural variability between subjects must be identified, and so on).

(d) One must identify the process of causal exposure and how subjects end up in alternative treatment states. How is the choice of treatment made? To answer this, one must switch from a "treatment assignment" mindset to one focusing on "treatment selection."

(e) Large random-sample surveys are desirable (and it is assumed that within the underlying population the relative proportion and pattern of exposure of subjects to the causal process is fixed).

(f) If information about time is available, then a causal direction may possibly be inferred. In this regard, cross-sectional data provides less information about the direction and nature of causality than does longitudinal data.

It is important to note, however, that when it comes to observational studies, all is not lost. Observational or non-experimental information can at times be forthcoming from *natural experiments*, that is, experiments in which nature provides the data in a fashion that is tantamount to the performance of a randomized experiment. The classical case of a natural experiment is Snow's observational study of cholera [Snow (1855)]. Snow was able to conclude, via considerable effort (door-to-door canvassing, brute-force elimination of confounders, and so on) that nature itself had acted to mix subjects across treatments in a way that mirrors a randomized controlled experiment. Hence persistent leg-work and the process of elimination enabled Snow to eventually hit upon the notion that cholera is a water-borne disease.

What is the bottom-line conclusion regarding observational studies and causality? Observational studies can give hints or suggest directions, but can never fully establish a cause-and-effect relationship.

12.B.10 Suggestions for Further Reading

Readers interested in causal studies pertaining to epidemiology and the health effects of smoking should consult Friedman (1999), Freedman and Stark (1999), Greenland (1998), Greenland and Brumback (2002), and Glymour and Greenland (2008). A discussion pertaining to cause and effect in law and medicine can be found in Austin (2008), Croucher (2008), Cole (1997), Rubin (2007), Collier, et al. (2010), and Finkelstein and Levin (1990). Applications of causal inquiries in the sciences are treated in Psillos (1999), Humphreys (1989, 2004), and Godfrey-Smith (2003).

13

AN ASSORTMENT OF ADDITIONAL STATISTICAL TESTS

13.1 DISTRIBUTIONAL HYPOTHESES

In Chapter 10, we assumed that a population probability distribution had a particular functional form, for example, we assumed that a continuous random variable X was $N(\mu, \sigma)$. Under this "distributional assumption," we could then test a *parametric hypothesis* about, say, the unknown mean μ given that σ is known. For instance, we could test $H_0 : \mu = \mu_o$, against $H_1 : \mu > \mu_o$ using the standard normal distribution. However, if the distributional assumption of normality is not made, then we can actually test the *distributional hypothesis* H_0 : X is normally distributed, against H_1 : X is not normally distributed.

An important method of testing a distributional hypothesis is *Pearson's goodness-of-fit test*. Here we are interested in determining if a set of sample observations can be viewed as values of a population random variable having a given probability distribution. So given some distributional hypothesis, this test enables us to determine if the population follows a specific probability model.

13.2 THE MULTINOMIAL CHI-SQUARE STATISTIC

We found in Chapter 5 that the essential characteristics of a binomial random experiment are

Statistical Inference: A Short Course, First Edition. Michael J. Panik.

a. We have a discrete random variable X.
b. We have a simple alternative experiment.
c. The n trials are identical and independent.
d. p, the probability of a success, is constant from trial to trial.

With a simple alternative experiment, there are two possible outcomes: $X=0$ or $X=1$. Hence the elements of the population belong to one of two classes—success or failure.

Let us consider the more general case where the elements of the population are classified as belonging to one of k possible classes ($k \geq 2$). That is, we now perform what is called a *multinomial random experiment*. For this type of experiment, each of the n trials results in a k-fold alternative, that is, each trial results in any one of k mutually exclusive and collectively exhaustive outcomes $E_1, \ldots, E_k$ with respective probabilities $P(E_1) = p_1, \ldots, P(E_k) = p_k$, where $\Sigma_{i=1}^{k} p_i = 1$ and the p_i are constant from trial to trial. If the discrete random variable X_i, $i = 1, \ldots, k$, depicts the number of times the ith outcome type E_i occurs in the n independent and identical trials of the k-fold alternative, then the *multinomial random variable* $(X_1, \ldots, X_k)$ is said to follow a *multinomial probability distribution* (the joint probability distribution of the random variables $X_i = 1, \ldots, k$). If we view E_i as a particular outcome class or cell, then we may denote the number of trials in which E_i occurs or the number of observations falling into cell i as X_i, $i = 1, \ldots, k$. Hence $\Sigma_{i=1}^{k} X_i = n$.

For instance, suppose a population probability distribution consists of a set of k mutually exclusive and collectively exhaustive categories or cells (columns 1 and 2 of Table 13.1 depict such a probability distribution). Next, suppose we obtain a random sample of $n=100$ observations collected from some unknown source and each observation is placed within one and only one of the seven cells in Table 13.1. These *observed frequencies*, denoted o_i, appear in column 3 of Table 13.1. Does the *sample probability distribution* (consisting of columns 1 and 3) of 100 items differ from the *population* or *theoretical probability distribution* given in columns 1 and 2? Can we

TABLE 13.1 Observed and Expected Frequencies

Category or Cell	Population Relative Frequency or Theoretical Probability p_i	Observed Frequency o_i	Expected Frequency under $H_0, e_i(= np_i)$
1	0.14	10	14
2	0.17	18	17
3	0.33	24	33
4	0.21	18	21
5	0.10	15	10
6	0.03	10	3
7	0.02	5	2
	1.00	100	100

legitimately presume that the sample of 100 data points has been extracted from the population probability distribution?

To determine if the observed or sample distribution is the same as the given population distribution, let us test

H_0: The population distribution and the sample distribution are identical

against

H_1: The sample distribution differs from the population distribution. (13.1)

Here H_0 specifies the exact or theoretical distribution (our benchmark) and H_1 simply indicates some unspecified difference between these two distributions.

Given the relative frequencies listed for the population distribution (Table 13.1), an exact multinomial probability can be obtained via a multinomial probability calculation. However, since the direct calculation of multinomial probabilities is, to say the least, quite tedious, an alternative approach will be offered that enables us to easily address problems pertaining to goodness of fit. This alternative computational scheme is based upon the *Pearson multinomial chi-square distribution.*

In this regard, let the random variable $(X_1, \ldots, X_k)$follow a multinomial probability distribution with parameters $n, p_1, \ldots, p_k$. Then for fixed k, as the number of trials n increases without bound, the distribution of the random variable

$$U = \sum_{i=1}^{k} \frac{(X_i - np_i)^2}{np_i} \tag{13.2}$$

[*Pearson multinomial chi-square statistic*]

approaches a chi-square distribution with $k - 1$ degrees of freedom (denoted χ^2_{k-1}). So if the possible outcomes of a random experiment can be decomposed into k mutually exclusive and collectively exhaustive categories, with p_i denoting the probability that an outcome falls in category i, then, in n independent and identical trials of this k-fold alternative experiment, with X_i representing the number of outcomes favoring category i (so that $\Sigma_{i=1}^{k} X_i = n$), the distribution of U can be approximated by a chi-square distribution with $k - 1$ degrees of freedom as $n \rightarrow \infty$.

The importance of the quantity (Eq. 13.2) is that it can be used as a test statistic for conducting a multinomial goodness-of-fit test. So for $(X_1, \ldots, X_k)$a multinomial random variable, we may test a null hypothesis that specifies the parameters of a multinomial distribution as

$$H_0 : p_1 = p_1^o, \ldots, p_k = p_k^o \tag{13.3}$$

against

$$H_1 : p_i \neq p_i^o \text{ for at least one } i = 1, \ldots, k.$$

Then for a random sample of size n taken from a multinomial population, Equation (13.2) under H_0 becomes

$$U_o = \sum_{i=1}^{k} \frac{(X_i - np_i^o)^2}{np_i^o}. \tag{13.4}$$

Since U_o is non-decreasing as i goes from 1 to k, we will reject H_0 if U_o is an element of the (upper tail) critical region $R = \left\{U \middle| U > \chi^2_{1-\alpha,k-1}\right\}$, where $\chi^2_{1-\alpha,k-1}$ is the 100 $(1-\alpha)$ percentile of the chi-square distribution with $k-1$ degrees of freedom and $\alpha = P$ (Type I Error). Clearly, we reject H_0 for large U_o's. As Equation (13.4) reveals, a test of goodness of fit is essentially a test of a hypothesis concerning specified cell probabilities p_i, $i = 1, \ldots, k$. It is important to remember that the use of Equation (13.4) to approximate multinomial probabilities requires the following conditions:

a. Each outcome falls into one and only one cell or category.
b. The outcomes are independent.
c. n is large.

13.3 THE CHI-SQUARE DISTRIBUTION

Having rationalized the test statistic for conducting a multinomial goodness-of-fit test (Eq. 13.4), we next examine the properties of the chi-square distribution and its attendant probabilities. The *chi-square distribution* is a continuous distribution that represents the sampling distribution of a sum of squares of independent standard normal variables. That is, if the observations $X_1, \ldots, X_n$ constitute a random sample of size n taken from a normal population with mean μ and standard deviation σ, then the $Z_i = (X_i - \mu)/\sigma$, $i = 1, \ldots, n$, are independent $N(0,1)$ random variables and

$$Y = \sum_{i=1}^{n} Z_i^2 = \sum_{i=1}^{n} \frac{(X_i - \mu)^2}{\sigma^2} \text{ is } \chi^2_n.$$

Looking to the properties of the chi-square distribution:

1. The mean and standard deviation of a chi-square random variable X are $E(X) = v$ and $\sqrt{V(X)} = \sqrt{2v}$, respectively, where v denotes degrees of freedom.
2. The chi-square distribution is positively skewed and it has a peak that is sharper than that of a normal distribution.
3. Selected quantiles of the chi-square distribution can be determined from the chi-square table (Table A.3) for various values of the degrees of freedom parameter v. For various cumulative probabilities $1-\alpha$ (Fig. 13.1), the quantile $\chi^2_{1-\alpha,v}$ satisfies

$$P\left(X \leq \chi^2_{1-\alpha,v}\right) = 1 - \alpha$$

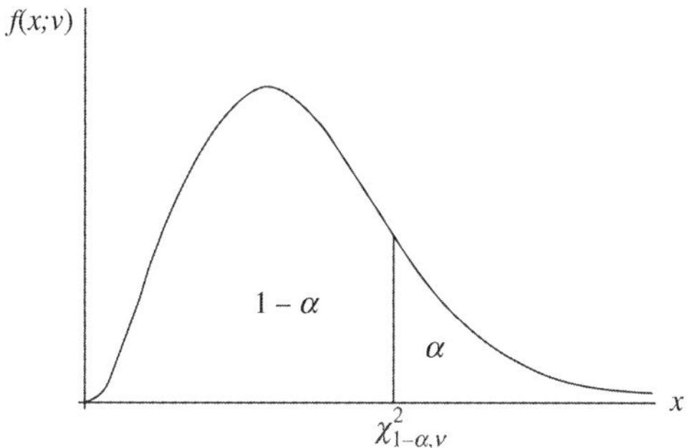

FIGURE 13.1 The χ^2_v distribution.

or, alternatively,

$$P\left(X > \chi^2_{1-\alpha,v}\right) = \alpha.$$

That is, for various degrees of freedom v, $\chi^2_{1-\alpha,v}$ gives the value of χ^2_v below which the proportion $1-\alpha$ of the χ^2_v distribution falls (or $\chi^2_{1-\alpha,v}$ is the value of χ^2_v above which the proportion α of the distribution is found).

Example 13.1

Suppose the random variable X is χ^2_{14}. Then for $1-\alpha=0.95$, we obtain, from Table A.3,

$$P\left(X \leq \chi^2_{0.95,14}\right) = P(X \leq 23.68) = 0.95.$$

Also, for $v=5$ and $\alpha=0.05$,

$$P\left(X \leq \chi^2_{0.95,5}\right) = P(X \leq 11.07) = 0.95.$$

And for $v=10$ and $1-\alpha=0.99$,

$$P\left(X > \chi^2_{0.99,10}\right) = P\left(X > 23.21\right) = 0.01.$$

13.4 TESTING GOODNESS OF FIT

Having discussed the particulars of the chi-square distribution, let us return to the use of the test statistic (Eq. 13.4). Why is this test statistic actually structured to reveal goodness of fit? It should be intuitively clear that the notion of goodness of fit should be assessed on the basis of the degree of disparity between the *sample* or *empirical distribution* and the *expected* or *theoretical distribution* given that the latter is specified by the null hypothesis. Under H_0, the *expected* or *theoretical frequencies* for the k categories or cells are simply $E(X_i) = np_i^o = e_i, i = 1, \ldots, k$. That is, e_i is the product between the sample size and the hypothesized relative frequency or theoretical probability p_i^o. So if H_0 is true, $e_i = np_i^o$ should be the expected number of occurrences for cell i under n repeated trials of our k-fold alternative experiment. The expected cell frequencies are given in column 4 of Table 13.1. Under this discussion, we may modify Equation (13.4) to read

$$U_0 = \sum_{i=1}^{k} \frac{(o_i - e_i)^2}{e_i}. \tag{13.5}$$

Here U_0 serves as an index of goodness of fit given that H_0 specifies the theoretical distribution that is fitted to sample data. As this expression reveals, each squared difference between observed and expected frequencies is "weighted inversely" by the expected frequency for cell i; any departure from expectation receives relatively more weight if we expect only a few outcomes in that cell than if we expect many such outcomes.

Equation (13.5) enables us to determine how well the sample distribution and the expected or theoretical distribution agree. The difference $o_i - e_i$ serves as a measure of the agreement between the observed and expected frequencies belonging to cell i. Under H_0 or perfect agreement, $E(o_i - e_i) = 0$. To circumvent this problem, we use the squares $(o_i - e_i)^2$, with $(o_i - e_i)^2$ weighted inversely by e_i to gauge its relative importance so that $(o_i - e_i)^2/e_i$ is the individual contribution of cell i to U_0, with the latter thus serving as an overall measure of goodness to fit.

When there is a high degree of agreement between the sample and theoretical distribution, U_0 tends to be relatively small and thus it is highly unlikely that it would lie within the critical region and that H_0 would be rejected. But if there is considerable disagreement between the observed and expected frequencies, then U_0 tends to be relatively large, thus favoring the terms upon which it is likely to fall within the critical region and H_0 is rejected. As this discussion reveals, the region of rejection must be located in the upper tail of the chi-square distribution.

The goodness-of-fit test just described can be executed for "any" theoretical population distribution we choose to specify under H_0. All that is required is that the theoretical population distribution be discrete or its domain can be partitioned into a collection of class intervals and all of its parameters are *completely specified* (e.g., X being $N(50,10)$ is completely specified whereas X being $N(50,\sigma)$ is not). And as indicated above, this chi-square goodness-of-fit test is based upon the difference between the observed cell frequencies obtained from the sample and the cell frequencies we would expect to obtain if the random variable $(X_1, \ldots, X_k)$ conformed to the theoretical distribution (given by H_0).

For the distribution appearing in Table 13.1, let the null and alternative hypotheses be given by Equation (13.1). And for $\alpha = 0.05$, the critical region $R = \left\{U \middle| U > \chi^2_{0.95,6} = 12.59\right\}$ (see Table A.3 of the appendix). From Equation (13.5),

$$U_0 = \sum_{i=1}^{7} \frac{(o_i - e_i)^2}{e_i} = \frac{(10-14)^2}{14} + \frac{(18-17)^2}{17} + \ldots + \frac{(5-2)^2}{2} = 27.42.$$

Since U_0 lies well within R, we can reject H_0 at the 5% level—the sample and theoretical distributions are not identical. Thus we can safely say that the sample of size $n = 100$ has not been drawn from the population probability distribution. Here the p-value or observed level of significance for this test is less than 0.1%. [In general, $p\text{-value} = P(\chi^2 > U_0|k - 1)$ degrees of freedom.]

Example 13.2

Ten years ago, the residents of a certain city were represented by four major ethnic groups according to the 8:2:2:1 ratio. A recent random sample of size $n = 1000$ resulted in the observed numbers 410, 218, 297, and 75 for the four ethnic groups. Do these recent observed values agree with the previous ratio for $\alpha = 0.05$? The null hypothesis is

$$H_0 : p_1 = p_1^o = \frac{8}{13}, p_2 = p_2^o = \frac{2}{13}, p_3 = p_3^o = \frac{2}{13}, p_4 = p_4^o = \frac{1}{13}$$

and the alternative hypothesis is

$$H_1 : p_i \neq p_i^o \text{ for at least one } i = 1, 2, 3, 4.$$

Our calculations supporting the use of Equation (13.5) appear in Table 13.2. Then from Equation (13.5),

$$U_0 = \sum_{i=1}^{4} \frac{(o_i - e_i)^2}{e_i} = \frac{(410 - 615)^2}{615} + \frac{(218 - 154)^2}{154} + \frac{(297 - 154)^2}{154} + \frac{(75 - 77)^2}{77} = 227.77.$$

TABLE 13.2 Observed and Expected Frequencies

Ethnic Group	Observed Frequency o_i	Expected Frequency Under H_0, $e_i = np_i^o$
1	410	$1000\left(\frac{8}{13}\right) = 615$
2	218	$1000\left(\frac{2}{13}\right) = 154$
3	297	$1000\left(\frac{2}{13}\right) = 154$
4	75	$1000\left(\frac{1}{13}\right) = 77$
	1000	1000

Since $U_0 = 227.77$ is a member of $R = \left\{U \middle| U > \chi^2_{0.95,3} = 7.81\right\}$, we can readily reject H_0 at the 5% level—there has been a statistically significant change in the ethnic composition of this city.

Example 13.3

If we have a binomial population, then the number of cells is $k=2$ (we have two possible outcomes for a binomial experiment—success or failure). We can then test H_0: $p = p_o$, against H_1: $p \neq p_o$, where p is the proportion of successes in the population. Then the proportion of failures must be $1-p$ and thus the expected numbers of successes and failures in a sample of size n are np and $n(1-p)$, respectively. If we have X successes in a sample of size n, then there must be $n-X$ failures. Under H_0, the expected number of successes is np_o and thus the expected number of failures is $n(1-p_o)$. Then for a binomial population, the *Pearson multinomial chi-square statistic* (Eq. 13.2) with $k-1=1$ degrees of freedom appears as

$$U_0 = \frac{(X - np_o)^2}{np_o} + \frac{[n - X - n(1-p_o)]^2}{n(1-p_o)}. \tag{13.2.1}$$

For instance, 10 years ago in a particular city the single to married ratio was known to be 5 to 1 in favor of singles (there were five single people for every married individual). In a recent survey, a random sample of 650 people contained 100 married individuals. Is this sufficient evidence for us to conclude that this ratio has changed? For a ratio of 5 to 1, one out of every six people is married. So if the ratio has not changed, the proportion of married in the population is $p=\frac{1}{6}$. Hence we test H_0: $p = p_o = \frac{1}{6}$, against H_1: $p \neq \frac{1}{6}$. Let $\alpha = 0.05$. Given the above information, $X = 100$, $n - X = 650 - 100 = 550$, $np_o = 650\ \left(\frac{1}{6}\right) = 108$, and $n(1-p_o) = 650\ \left(\frac{5}{6}\right) = 542$. Then from Equation (13.2.1),

$$U_0 = \frac{(100 - 108)^2}{108} + \frac{(550 - 542)^2}{542} = 0.70.$$

From Table A.3 of the appendix, $R = \left\{U \middle| U > \chi^2_{0.95,1} = 3.84\right\}$. Clearly, U_0 is not a member of R so that, at the 5% level, we cannot reject H_0—there has not been a statistically significant change in the ratio of singles to married.

An important application of the Pearson multinomial chi-square statistic (Eq. 13.5) is a test to determine if a sample distribution can be thought of as having been drawn from a normal population. For instance, given the absolute frequency distribution presented in Table 13.3, we would like to determine if this sample of size $n = 200$ came from a normal distribution. Here we test H_0: X is normally distributed versus H_1: X is not normally distributed. Let us use $\alpha = 0.05$. Since the null hypothesis is not completely specified, let us estimate μ and σ from the sample distribution. It is easily demonstrated that $\bar{X} = 115.25$ and $s = 62.18$ (See Appendix 3.A).

TABLE 13.3 Observed and Expected Frequencies

Classes of X	Observed Frequency o_i	Estimated Expected Frequency Under H_0, $\hat{e}_i = n\hat{p}_i$
0–49	28	28.92
50–99	68	51.34
100–149	41	61.50
150–199	39	40.86
200–249	24	17.38
	200	200.00

Next, let us employ Table A.1 of the appendix to find the probability that a N (115.25, 62.18) random variable would assume a value in each of the five cells appearing in column 1 of Table 13.3. To this end, the cell probabilities are estimated by using the class boundaries (–0.5, 49.5, 99.5, 149.5, 199.5, 249.5) and, under the restriction that these cell probabilities must total unity, the first and last cells are replaced by $-\infty < X \leq 49.5$ and $199.5 \leq X < +\infty$, respectively. Then for $Z = (X - \bar{X})/s$, the estimated cell probabilities $\hat{p}_i, i = 1, \ldots, 5$, are

$$\begin{aligned}
\hat{p}_1 &= P(-\infty < X \leq 49.5) = P(-\infty < Z \leq -1.057) = 0.1446 \\
\hat{p}_2 &= P(49.5 \leq X \leq 99.5) = P(-1.057 \leq Z \leq -0.253) = 0.2567 \\
\hat{p}_3 &= P(99.5 \leq X \leq 149.5) = P(-0.253 \leq Z \leq 0.551) = 0.3075 \\
\hat{p}_4 &= P(149.5 \leq X \leq 199.5) = P(0.551 \leq Z \leq 1.355) = 0.2043 \\
\hat{p}_5 &= P(199.5 \leq X \leq +\infty) = P(1.355 \leq Z < +\infty) = 0.0869.
\end{aligned}$$

Now, if H_0 is true, then the estimated expected cell frequencies are $\hat{e}_i = n\hat{p}_i = 200\,\hat{p}_i, i = 1, \ldots, 5$ (see Table 13.3). Then using a modification of Equation (13.5) that utilizes the estimated expected cell frequencies,

$$U_0 = \sum_{i=1}^{k} \frac{(o_i - \hat{e}_i)^2}{\hat{e}_i} \tag{13.5.1}$$

and we can readily determine that

$$U_0 = \sum_{i=1}^{5} \frac{(o_i - \hat{e}_i)^2}{\hat{e}_i} = \frac{(28 - 28.92)^2}{28.92} + \frac{(68 - 51.34)^2}{51.34} + \ldots + \frac{(24 - 17.38)^2}{17.38} = 14.875.$$

For degrees of freedom $k - r - 1 = 5 - 2 - 1 = 2$ (since μ and σ were unknown, $r = 2$ parameters had to be estimated), the critical region is $R = \left\{U \middle| U > \chi^2_{0.95,2} = 5.99\right\}$.

Since U_0 exceeds 5.99, we can reject the null hypothesis that the sample was drawn from a normal population; at the 5% level the normal distribution does not provide a good fit to the sample data.

It is important to note that if X is a continuous random variable, then to use this chi-square goodness-of-fit test, the data must be grouped into k classes or cells of uniform length, where the number of cells typically ranges between 5 and 10 and each cell has an expected frequency of at least 5. If it should happen that the expected frequency in any cell is less than 5, then adjacent cells or classes should be combined so that all the expected frequencies will be at least 5, since otherwise the approximation will not be adequate. Note also that degrees of freedom will be $k-1$ less the number of parameters (r) we have to estimate in order to determine the $\hat{e}_i$'s, $i=1, \ldots, k$.

13.5 TESTING INDEPENDENCE

In the Pearson chi-square tests covered in the preceding section, the random sample data were classified via a single criterion into a set of mutually exclusive and collectively exhaustive categories or cells. However, as will be demonstrated shortly, this approach can be applied to situations other than these resulting from multinomial experiments. For example, suppose our observations are classified according to two or more attributes. Then the question naturally arises: "Are these attributes or categories of classification statistically independent?" ("Do they lack statistical association?") To answer this, we shall look to *Pearson's chi-square test for independence* of attributes or *Pearson's chi-square test of association.* As will be evidenced below, this technique can be applied to either quantitative or qualitative data.

Given a specific criterion of classification, suppose we cross-classify our data set according to a second criterion or attribute, where the population proportion within each resulting cell is unknown, for example, given a set of categories on the factor "educational attainment" (high school diploma, college degree, graduate school degree), we may cross-classify according to a second factor such as "sex" or "income level" (low, medium, high). We shall refer to the resulting two-way classification scheme as a *two-way contingency table.*

As stated above, we are interested in determining whether there is any dependency relationship or association between two population characteristics. We thus test

H_0: The two attributes are independent (or the rows and columns represent independent classifications)

against

H_1: The two attributes are not independent (or the rows and columns are not independent classifications). (13.6)

Note that while the alternative hypothesis states that the attributes are dependent or related in some way, the extent or nature of the dependency is not specified.

Suppose that n independent sample outcomes have been obtained from a population of size N (N is assumed to be large) and each such outcome is classified according to two attributes or criteria (simply called A and B). Suppose further that there are r mutually exclusive and collectively exhaustive categories for attribute $A(A_1, \ldots, A_r)$ and c such categories for attribute $B(B_1, \ldots, B_c)$. All n sample outcomes can then be classified into a two-way table, with A_i, $i = 1, \ldots, r$, categories for factor A constituting the rows and the B_j, $j = 1, \ldots, c$, categories for factor B making up the columns. The resulting table is termed an $r \times c$ *contingency table*, where each possible combination of factors A and B, (A_i, B_j), $i = 1, \ldots, r; j = 1, \ldots, c$, represents a distinct cell within the table (Table 13.4). Additionally, all n sample outcomes are to be placed within the $r \times c$ contingency table on the basis of their specific A and B attribute categories. In this regard, let n_{ij} denote the number of outcomes falling into row i and column j, that is, n_{ij} is the observed number of items possessing attribute A_i and B_j in combination.

Let us denote the total number of sample outcomes having attribute A_i as

$$n_{i\cdot} = \sum_{j=1}^{c} n_{ij} = ith \text{ row total} \qquad (13.7)$$
$$(i \text{ is held fixed and we sum over columns})$$

(the dot "." indicates the subscript we sum over); and the total number of sample outcomes possessing attribute B_j is

$$n_{\cdot j} = \sum_{i=1}^{r} n_{ij} = jth \text{ column total} \qquad (13.8)$$
$$(j \text{ is held fixed and we sum over rows}).$$

Here both Equations (13.7) and (13.8) are termed *marginal totals*. Obviously $n_{i.} + n_{2.} + \cdots + n_{r.} = n = n_{.1} + n_{.2} + \cdots + n_{.c}$.

TABLE 13.4 $r \times c$ **Contingency Table**

	Columns				
Rows	B_1	B_2	...	B_c	Row Totals
A_1	n_{11}	n_{12}	...	n_{1c}	$n_{1\cdot}$
A_2	n_{21}	n_{22}	...	n_{2c}	$n_{2\cdot}$
.	.	.	...	.	.
.	.	.	...	.	.
.	.	.	...	.	.
A_r	n_{r1}	n_{r2}	...	n_{rc}	$n_{r\cdot}$
Column totals	$n_{\cdot 1}$	$n_{\cdot 2}$	...	$n_{\cdot c}$	n

As stated in Equation (13.6), the null hypothesis is that factors A and B are independent. Under H_0, these factors for the population at large are independent if and only if

$$P(A_i \cap B_j) = P(A_i) \cdot P(B_j), i = 1, \ldots, r;\ j = 1, \ldots, c \tag{13.9}$$

for all possible joint events $A_i \cap B_j$ [see Section 4.4 and Equation (4.7)] or

$$p_{ij} = (p_{i\cdot})(p_{\cdot j}), i = 1, \ldots, r;\ j = 1, \ldots, c. \tag{13.9.1}$$

Here p_{ij} is the unknown joint probability that an item selected at random from the population will be from cell (i, j), $p_{i.}$ is the unknown marginal probability that an item drawn at random from the population is from category i of characteristic A, and $p_{.j}$ is the unknown marginal probability that a randomly selected item from the population is from category j of characteristic B. Then under the independence assumption, Equation (13.6) can be rewritten as

$$\begin{aligned} &H_0 : p_{ij} = (p_{i\cdot})(p_{\cdot j}) \text{ for all } i \text{ and } j, \text{ and} \\ &H_1 : p_{ij} \neq (p_{i\cdot})(p_{\cdot j}) \text{ for at least one cell } (i,j), i = 1, \ldots, r;\ j = 1, \ldots, c. \end{aligned} \tag{13.6.1}$$

Given that the population marginal probabilities $p_{i.}$ and $p_{.j}$ are unknown, the *estimated sample marginal probabilities* can be obtained as

$$\hat{p}_{i\cdot} = \frac{n_{i\cdot}}{n}, i = 1, \ldots, r, \tag{13.10a}$$

$$\hat{p}_{\cdot j} = \frac{n_{\cdot j}}{n}, j = 1, \ldots, c \tag{13.10b}$$

respectively. Hence the *estimated population probability of the joint event* $A_i \cap B_j$ is

$$\hat{p}_{ij} = (\hat{p}_{i\cdot})(\hat{p}_{\cdot j}) = \frac{(n_{i\cdot})(n_{\cdot j})}{n^2} \text{ for all } i, j \tag{13.11}$$

and thus the expected frequency of the joint event $A_i \cap B_j$ must be

$$e_{ij} = n\hat{p}_{ij} = \frac{(n_{i\cdot})(n_{\cdot j})}{n} \text{ for all } i, j. \tag{13.12}$$

So in a test of independence of attributes, the expected frequency in cell (i, j) is simply the product between the marginal total for row i and the marginal total for column j divided by the sample size n.

On the basis of Equation (13.12), we can extend Pearson's multinomial chi-square statistic (Eq. 13.2) to the case where we test for independence among two factors A and B in an $r \times c$ contingency table. To this end, for n independent trials of a random experiment classified by the categories of attributes A and B, if the random variable X_{ij} is the frequency of event $A_i \cap B_j$, then the statistic

$$U = \sum_{i=1}^{r}\sum_{j=1}^{c}\frac{\left(X_{ij} - np_{ij}\right)^2}{np_{ij}} \tag{13.13}$$

is, for large n, approximately chi-square distributed with $rc - 1$ degrees of freedom. Under Equation (13.6.1), with Equations (13.10)–(13.12) holding, Equation (13.13) becomes the *Pearson chi-square statistic for a test of association*

$$\begin{aligned} U_0 &= \sum_{i=1}^{r}\sum_{j=1}^{c}\frac{\left(n_{ij} - n\hat{p}_{ij}\right)^2}{n\hat{p}_{ij}} \\ &= \sum_{i=1}^{r}\sum_{j=1}^{c}\frac{\left(o_{ij} - e_{ij}\right)^2}{e_{ij}}, \end{aligned} \tag{13.14}$$

where $o_{ij}\,(=n_{ij})$ is the *observed frequency for cell* (i, j). For large n, Equation (13.14) is approximately chi-square distributed with $(r - 1)(c - 1)$ degrees of freedom. How have we determined degrees of freedom for Equation (13.14)? Since we estimate marginal probabilities from sample outcomes, we subtract 1 degree of freedom for each parameter estimated. Since $\Sigma_{i=1}^{r} p_{i\cdot} = 1 = \Sigma_{j=1}^{c} p_{\cdot j}$, there are, respectively, $r - 1$ row parameters and $c - 1$ column parameters to be estimated. Hence it follows that degrees of freedom $= rc - 1 - (r - 1) - (c - 1) = (r - 1)(c - 1)$.

Note that the finite double sum in Equation (13.14) has us calculate the grand total over all $r \times c$ cells, with the contribution of the (i, j)th cell to the Pearson chi-square statistic being $(o_{ij} - e_{ij})^2/e_{ij}$. As was the case with Equation (13.2), we compare the observed cell frequencies with the expected cell frequencies under H_0. If the deviations between the observed and expected cell frequencies are larger than that which could be attributed to sampling variation alone, then we are inclined to reject H_0 for large values of U_0. Hence the critical region for this upper tail test is $R = \left\{U \middle| U > \chi^2_{1-\alpha,(r-1)(c-1)}\right\}$, where $\chi^2_{1-\alpha,(r-1)(c-1)}$ is the $100(1 - \alpha)$ percentile of the chi-square distribution with $(r - 1)(c - 1)$ degrees of freedom and $\alpha = P(\text{TIE})$. It is important to mention that in order for the Pearson chi-square statistic to provide a good approximation to the chi-square distribution with $(r - 1)(c - 1)$ degrees of freedom, each expected cell frequency should be at least 5. If this requirement is not met, then adjacent cells can be combined as is appropriate.

Example 13.4

Suppose the principal (Miss Apple) of the ABC Elementary School is contemplating the imposition of a new dress code for children in grades 1 to 6. She has decided to sample the opinion of all "interested parties" in order to determine whether opinion on this issue is independent of school affiliation. A random sample of $n = 300$ interested

TABLE 13.5 Observed and Expected Frequencies

	Opinion			
Affiliation	In Favor	Neutral	Opposed	Row Totals
Students	80 (51.66)	10 (11.66)	10 (70)	100 $(=n_{1.})$
Parents	45 (36.16)	15 (8.16)	10 (49)	70 $(=n_{2.})$
Teachers	20 (41.33)	0 (9.33)	60 (56)	80 $(=n_{3.})$
Staff	10 (25.83)	10 (5.83)	30 (18.33)	50 $(=n_{4.})$
Column totals	155 $(=n_{.1})$	35 $(=n_{.2})$	110 $(=n_{.3})$	300 $(=n)$

parties yielded the collection of observed frequencies displayed in Table 13.5, for example, 80 students are in favor of the new dress code, 60 teachers are opposed to the same, 10 staff members are neutral, and so on.

Let H_0: $p_{ij}=(p_{i\cdot})(p_{.j})$, with H_1: $p_{ij} \neq (p_{i.})(p_{.j})$ for at least one cell (i, j), $i=1, \ldots, 4$; $j=1, \ldots, 3$, where p_{ij} is the joint probability that an interested party selected at random will fall into category (i,j); $p_{i.}$ is the marginal probability that a randomly chosen person belongs to affiliation category i; and $p_{.j}$ is the marginal probability that a randomly selected person is of opinion j. By Equation (13.12), the expected frequency of cell (i,j) is $(n_{i.})(n_{.j})/n$ (appearing in parentheses to the right of the observed cell frequency o_{ij} for cell (i,j)). For instance, the expected cell frequency of a staff member opposed to the new dress code is $(n_{4.})(n_{.3})/n=(50)(110)/300=18.33$. Then the final value of the Pearson chi-square statistic (Eq. 13.14) is

$$U_0=\frac{(80-51.66)^2}{51.66}+\frac{(10-11.66)^2}{11.66}+\frac{(10-70)^2}{70}+\frac{(45-36.16)^2}{36.16}$$
$$+\cdots+\frac{(10-5.83)^2}{5.83}+\frac{(30-18.33)^2}{18.33}=146.85.$$

With $r=4$ and $c=3$, degrees of freedom is $(r-1)(c-1)=3\cdot 2=6$. Using Table A.3 of the Appendix with $\alpha=0.05$, we find that $R=\left\{U \middle| U>\chi^2_{0.95,6}=12.59\right\}$. Since U_0 lies well within this critical region, we can safely reject the null hypothesis of independence between opinion and school affiliation at the 5% level. In fact, the p-value is less than 0.005.

Example 13.5

At times, one is faced with performing a chi-square test of independence when there are only two categories of classification for each of the A and B attributes. This obviously involves a 2×2 contingency table. To streamline the calculations in this instance, let us display the 2×2 table as Table 13.6.

Then a particularization of Equation (13.14) to the 2×2 case, which incorporates the *Yates continuity correction*, is

TABLE 13.6 2 × 2 Contingency Table

		Columns		
		B_1	B_2	Row Totals
	A_1	a	b	e
Rows	A_2	c	d	f
	Column Totals	g	h	

$$U_0' = \sum_{i=1}^{2}\sum_{j=1}^{2}\frac{\left(\left|o_{ij}-e_{ij}\right|-0.5\right)^2}{e_{ij}} = \frac{n\left(|ad-bc|-\frac{n}{2}\right)^2}{efgh}, \tag{13.14.1}$$

where $R = \left\{U \middle| U > \chi^2_{1-\alpha,1}\right\}$.

For instance, suppose the results of a random survey of employees of ABC Corp. pertaining to a proposed modification of the leave program appears in Table 13.7. Are opinion and sex statistically independent at the $\alpha = 0.05$ level? Here we test H_0: $p_{ij} = (p_{i.})(p_{.j})$, against H_1: $p_{ij} \neq (p_{i.})(p_{.j})$ for at least one cell (i, j), $i = 1, 2$; $j = 1, 2$. Using Equation (13.14.1),

$$U_0' = \frac{300(|(157)(15)-(30)(98)|-150)^2}{(187)(113)(255)(45)} = 0.234.$$

With $R = \left\{U \middle| U > \chi^2_{0.95,1} = 3.84\right\}$, we cannot reject the null hypothesis of independence between sex and opinion at the 5% level of significance.

13.6 TESTING k PROPORTIONS

We now turn to a special case of Equation (13.14) in that we examine the instance in which we want to test for the significance of the difference among k population proportions p_i, $i = 1, \ldots, k$. In this regard, suppose we have k independent random samples and that $X_1, \ldots, X_k$ comprise a set of independent binomial random variables with the parameters p_1 and n_1; p_2 and n_2; …; and p_k and n_k, respectively, where p_i, $i = 1, \ldots, k$, is the proportion of successes in the ith population. Here X_i depicts the number of successes obtained in a sample of size n_i, $i = 1, \ldots, k$.

TABLE 13.7 Observed and Expected Frequencies

	Opinion		
Sex	Modify	Do Not Modify	Row Totals
Male	157 $(=a)$	30 $(=b)$	187 $(=e)$
Female	98 $(=c)$	15 $(=d)$	113 $(=f)$
Column totals	255 $(=g)$	45 $(=h)$	300 $(=n)$

TABLE 13.8 Successes and Failures for k Independent Random Samples

	Successes	Failures	n_i
Sample 1	o_{11}	$o_{12}=n_1-o_{11}$	$n_1=o_{11}+o_{12}$
Sample 2	o_{21}	$o_{22}=n_2-o_{21}$	$n_2=o_{21}+o_{22}$
⋮	⋮	⋮	⋮
Sample k	o_{k1}	$o_{k2}=n_k-o_{k1}$	$n_k=o_{k1}+o_{k2}$

Let us arrange the observed number of successes and failures for the k independent random samples in the following $k \times 2$ table (Table 13.8). Here the $2k$ entries within this table are the *observed cell frequencies* o_{ij}, $i=1,\ldots,k; j=1,2$. Our objective is to test H_0: $p_1=p_2=\ldots=p_k=p_o$, against H_1: $p_i \neq p_o$ for at least one $i=1,\ldots,k$, where p_o is the null value of p_i. Under H_0, the expected number of successes for sample i is n_ip_o, $i=1,\ldots k$ (since $p_o=X_i/n_i$); and the expected number of failures for sample i is $n_i(1-p_o)$, $i=1,\ldots,k$ (since $1-p_o=\frac{n_i-X_i}{n_i}$). In this regard, the expected cell frequencies for columns 1 and 2 are, respectively, $e_{i1}^o=n_ip_o$ and $e_{i2}^o=n_i(1-p_o), i=1,\ldots,k$. So given p_o, Equation (13.14) becomes

$$U_{p_o}=\sum_{i=1}^{k}\sum_{j=1}^{2}\frac{\left(o_{ij}-e_{ij}^o\right)^2}{e_{ij}^o}, \tag{13.15}$$

where o_{i1} is the collection of observed cell frequencies for the success column; o_{i2} is the set of observed cell frequencies for the failure column; the expected cell frequencies for the success category are $e_{i1}^o=n_ip_o$; and the expected cell frequencies for the failure category are $e_{i2}=n_i(1-p_o), i=1,\ldots,k$. For $\alpha=P(\text{TIE})$, the critical region is $R=\left\{U\middle|U>\chi^2_{1-\alpha,k-1}\right\}$.

When p_o is not specified, then we can use a *pooled estimate of the common proportion of success* $\hat{p}$ as

$$\hat{p}=\frac{X_1+X_2+\cdots+X_k}{n_1+n_2+\cdots+n_k}. \tag{13.16}$$

Then from this expression, the *estimated expected cell frequencies* are

$$\hat{e}_{i1}=n_i\hat{p} \text{ and } \hat{e}_{i2}=n_i(1-\hat{p}), i=1,\ldots,k. \tag{13.17}$$

So when Equation (13.16) is required, we can test H_0: $p_1=p_2=\ldots=p_k$, against H_1: the p_i's are not equal using the test statistic:

$$U_{\hat{p}}=\sum_{i=1}^{k}\sum_{j=1}^{2}\frac{\left(o_{ij}-\hat{e}_{ij}\right)^2}{\hat{e}_{ij}}. \tag{13.18}$$

TABLE 13.9 Voter Preference for a Bond Issue (Observed Cell Frequencies)

	In Favor	Opposed	n_i
District 1	60	40	$n_1=100$
District 2	70	80	$n_2=150$
District 3	40	40	$n_3=80$
District 4	50	80	$n_4=130$

Example 13.6

Given the sample outcomes appearing in Table 13.9, determine if the population of voters favoring a bond issue (defined as a success) is the same across the four districts polled just before a special referendum on school financing. Let $\alpha=0.05$.

Here we test H_0: $p_1=p_2=p_3=p_4$, against H_1: the p_i's are not all equal, $i=1,\ldots,4$.

For these observed cell frequencies, the pooled estimate of the common proportion of successes is, via Equation (13.16),

$$\hat{p}=\frac{60+70+40+50}{100+150+80+130}=0.478.$$

Then from Equation (13.7), the estimated expected cell frequencies are

$$\begin{aligned}
\hat{e}_{11}&=n_1\hat{p}=100(0.478)=47.80 & \hat{e}_{12}&=n_1(1-\hat{p})=100(0.522)=52.20\\
\hat{e}_{21}&=n_2\hat{p}=150(0.478)=71.70 & \hat{e}_{22}&=n_2(1-\hat{p})=150(0.522)=78.30\\
\hat{e}_{31}&=n_3\hat{p}=80\ (0.478)=38.24 & \hat{e}_{32}&=n_3(1-\hat{p})=80\ (0.522)=41.76\\
\hat{e}_{41}&=n_4\hat{p}=130(0.478)=62.14 & \hat{e}_{42}&=n_4(1-\hat{p})=130(0.522)=67.86
\end{aligned}$$

Upon substituting these observed and estimated cell frequencies into Equation (13.18) renders

$$U_{\hat{p}}=\frac{(60-47.80)^2}{47.80}+\frac{(70-71.70)^2}{71.70}+\cdots+\frac{(40-41.76)^2}{41.76}+\frac{(80-67.86)^2}{67.86}=10.75.$$

Since $U_{\hat{p}}$ is an element of $R=\left\{U\middle|U\geq\chi^2_{1-\alpha,k-1}=\chi^2_{0.95,3}=7.81\right\}$, we reject the null hypothesis of equal district proportions at the 5% level; the proportion of voters favoring the bond issue is not the same across the four districts polled.

13.7 A MEASURE OF STRENGTH OF ASSOCIATION IN A CONTINGENCY TABLE

We noted in Section 13.5 that the Pearson chi-square statistic is used to test for association between two attributes A and B. To conduct the test, we hypothesize independence among the A and B factors and then, for a given level of significance α, we see if we can reject the null hypothesis of independence in favor of the alternative hypothesis that the A and B characteristics are not independent but, instead, exhibit statistical association. If we detect statistical association between A and B, then the next logical step in our analysis of these attributes is to assess how strong the association between the A and B categories happens to be. That is, we need a measure of the strength of association between factors A and B.

One device for depicting statistical association in an $r \times c$ contingency table is *Cramer's phi-squared statistic*

$$\Phi^2 = \frac{U}{n(q-1)}, 0 \leq \Phi^2 \leq 1, \tag{13.19}$$

obtained by dividing Pearson's chi-square statistic U by its theoretical maximum value $n\ (q-1)$, where n is the sample size and $q = min\{r,c\}$. Here Φ^2 measures the overall strength of association, that is, for $\Phi^2 = 0$, the sample exhibits complete independence between the A and B attributes; and when $\Phi^2 = 1$, there exists complete dependence between these factors.

Example 13.7

Given the chi-square test results determined in Example 13.5 (we had a 4×3 contingency table involving opinion and school affiliation and we obtained $U_0 = 146.85$), let us calculate Cramer's phi-squared statistic (Eq. 13.19) as

$$\Phi^2 = \frac{146.85}{300(3)} = 0.1632.$$

So while we found a significant lack of independence between opinion and school affiliation for $\alpha = 0.05$, the strength of association between these factors is rather modest.

13.8 A CONFIDENCE INTERVAL FOR σ^2 UNDER RANDOM SAMPLING FROM A NORMAL POPULATION

Let us first discuss the properties of the sampling distribution of the sample variance $s^2 = \Sigma_{i=1}^{n} (X_i - \bar{X})^2 / (n-1)$ when sampling from a population that is normally distributed. Specifically, if s^2 is determined from a sample of size n taken from a normal population with mean μ and variance σ^2, then the quantity

$$Y = \frac{(n-1)s^2}{\sigma^2} \tag{13.20}$$

is distributed as χ^2_{n-1}. Moreover, it can be shown that

$$E(Y) = n - 1,$$
$$V(Y) = \sqrt{2(n-1)}.$$

Then given Equation (13.20), it follows that $s^2 = \sigma^2 Y/(n-1)$ and that

$$E(s^2) = \frac{\sigma^2}{n-1} E(Y) = \sigma^2,$$
$$V(s^2) = \left(\frac{\sigma^2}{n-1}\right)^2 V(Y) = \frac{2\sigma^4}{n-1}.$$

For Y defined above, we seek to determine quantiles $\chi^2_{\alpha/2,n-1}$ and $\chi^2_{1-(\alpha/2),n-1}$ such that

$$\begin{aligned} &P\left(\chi^2_{\alpha/2,n-1} < Y < \chi^2_{1-(\alpha/2),n-1}\right) = \\ &P\left(\chi^2_{\alpha/2,n-1} < \frac{(n-1)s^2}{\sigma^2} < \chi^2_{1-(\alpha/2),n-1}\right) = 1 - \alpha \end{aligned} \tag{13.21}$$

(Fig. 13.2). Upon transforming Equation (13.21) to the equivalent probability statement

$$P\left(\underbrace{\frac{(n-1)s^2}{\chi^2_{1-(\alpha/2),n-1}}}_{L_1} < \sigma^2 < \underbrace{\frac{(n-1)s^2}{\chi^2_{\alpha/2,n-1}}}_{L_2}\right) = 1 - \alpha,$$

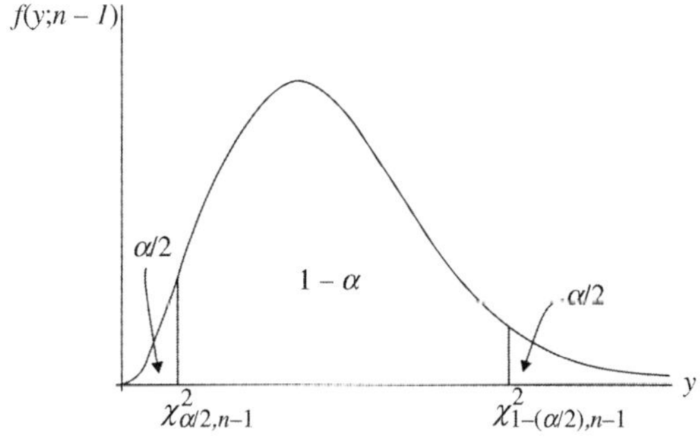

FIGURE 13.2 $P\left(\chi^2_{\alpha/2,n-1} < Y < \chi^2_{1-(\alpha/2),n-1}\right) = 1 - \alpha$.

we see that the random interval (L_1, L_2) with confidence probability $1-\alpha$ becomes a *100(1 − α)% confidence interval for* σ^2 once s^2 is determined from the sample values, that is, we may be $100(1-\alpha)\%$ confident that the interval

$$\left(\frac{(n-1)s^2}{\chi^2_{1-(\alpha/2),n-1}}, \frac{(n-1)s^2}{\chi^2_{\alpha/2,n-1}}\right) \tag{13.22}$$

contains the population variance σ^2.

In addition, a *100(1 − α)% confidence interval for* σ is readily obtained from Equation (13.22) as

$$\left(\sqrt{\frac{(n-1)s^2}{\chi^2_{1-(\alpha/2),n-1}}}, \sqrt{\frac{(n-1)s^2}{\chi^2_{\alpha/2,n-1}}}\right). \tag{13.23}$$

Hence we may be $100(1-\alpha)\%$ confident that this interval contains σ.

Example 13.8

Suppose we extract a random sample of size $n=25$ from a normal population and we find $s^2=61.67$. Find a 95% confidence interval for the population variance σ^2. With $1-\alpha=0.95$, Table A.3 of the appendix Yields $\chi^2_{\alpha/2,n-1}=\chi^2_{0.025,24}=12.40$ and $\chi^2_{1-(\alpha/2),n-1}=\chi^2_{0.975.24}=39.36$. Then from Equation (13.22), a 95% confidence interval for σ^2 is (37.60, 119.36). That is, we may be 95% confident that the population variance lies between 37.60 and 119.36. If we take the positive square root of the confidence limits for σ^2, then we obtain a 95% confidence interval for σ [see Equation (13.23)] or (6.13, 10.93). Hence we may be 95% confident that the population standard deviation lies between 6.13 and 10.93.

13.9 THE *F* DISTRIBUTION

Suppose we conduct a random sampling experiment whose outcome is a set of observations on two independent chi-square random variables X and Y with degrees of freedom v_1 and v_2, respectively. If we define a new variable

$$F=\frac{X/v_1}{Y/v_2} \tag{13.24}$$

as the ratio of two independent chi-square random variables, each divided by its degrees of freedom, then F is said to follow an F distribution with v_1 and v_2 degrees of freedom and will be denoted as F_{v_1,v_2}. As this notation reveals, the F distribution is a two-parameter family of distributions; and as we vary the parameters v_1 and v_2, we generate a whole assortment of different F distributions. It can be demonstrated that the F distribution is positively skewed for any values of v_1 and v_2, with

$$E(F) = \frac{v_2}{v_2 - 2}, v_2 > 2,$$

$$V(F) = \frac{2v_2(v_1 + v_2 - 2)}{v_1(v_2 - 2)^2(v_2 - 4)}, v_2 > 4.$$

Moreover, unlike the *t* or *Z* distributions, the *F* distribution can only assume positive values.

What is the connection between Equation (13.24) and sampling from a normal population? Suppose a random sample of size n_1 with variance s_1^2 is drawn from a $N(\mu_1,\sigma_1)$ population and a second random sample size n_2 with variance s_2^2 is extracted from a population which is $N(\mu_2,\sigma_2)$. If these two random samples are independent, then, via Equation (13.20), $X = (n_1 - 1)s_1^2/\sigma_1^2$ and $Y = (n_2 - 1)s_2^2/\sigma_2^2$ are independent chi-square random variables with $v_1 = n_1 - 1$ and $v_2 = n_2 - 1$ degrees of freedom, respectively. Then from Equation (13.24), the random variable

$$F_{n_1-1,n_2-1} = \frac{s_1^2/\sigma_1^2}{s_2^2/\sigma_2^2} \tag{13.25}$$

is *F* distributed with $n_1 - 1$ and $n_2 - 1$ degrees of freedom.

How do we obtain the percentage points or quantiles of the *F* distribution? The *upper tail* α *percentage point* F_{α,v_1,v_2} *for F* appears in Table A.4 of the appendix and is defined by

$$P(F \geq F_{\alpha,v_1,v_2}) = \alpha \tag{13.26}$$

(Fig. 13.3). Only the upper tail α percentage points are given since the *lower tail* $1 - \alpha$ *percentage point* $F_{1-\alpha,v_1,v_2}$ *for F* is the reciprocal of the upper tail α percentage point with degrees of freedom reversed or

$$F_{1-\alpha,v_1,v_2} = \frac{1}{F_{\alpha,v_2,v_1}}. \tag{13.27}$$

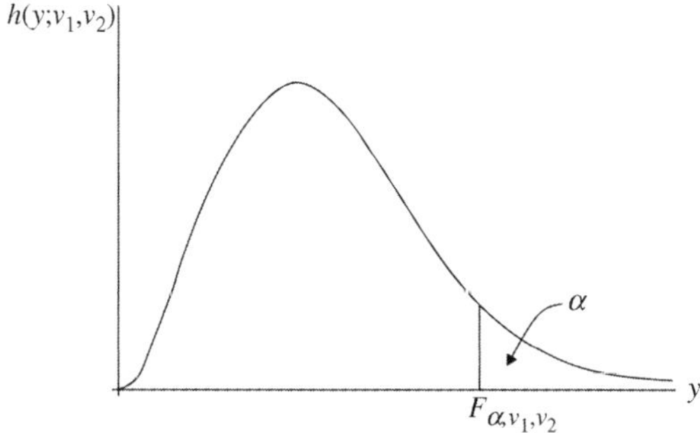

FIGURE 13.3 The F_{α,v_1,v_2} distribution.

Example 13.9

From Table A.4 of the appendix it is easily verified that:

a. $F_{\alpha,\nu_1,\nu_2} = F_{0.05,5,10} = 3.33$
b. $F_{\alpha,\nu_1,\nu_2} = F_{0.01,4,2} = 99.25$
c. $F_{1-\alpha,\nu_1,\nu_2} = F_{0.95,5,10} = \dfrac{1}{F_{0.05,10,5}} = \dfrac{1}{4.74} = 0.2110$

13.10 APPLICATIONS OF THE *F* STATISTIC TO REGRESSION ANALYSIS

13.10.1 Testing the Significance of the Regression Relationship Between *X* and *Y*

We previously used a t test to determine if X contributes significantly to the variation in Y. If H_0: $\beta_1 = 0$ is true (we posit "no linear relationship" between X and Y), then the sole source of variation in Y is the random disturbance term ε since the population regression sum of squares is zero. Now, it can be demonstrated that the statistic

$$F = \frac{\left(\hat{\beta}_1 - \beta_1\right)^2 \sum x_i^2}{\sum e_i^2/(n-2)} \text{ is } F_{1,n-2}. \tag{13.28}$$

Under H_0: $\beta_1 = 0$, Equation (13.28) becomes

$$F_o = \frac{\hat{\beta}_1 \sum x_i y_i}{s_\varepsilon^2} = \frac{\text{Regression mean square}}{\text{Error mean square}} = \frac{\text{SSR}/1}{\text{SSE}/(n-2)} \text{ is } F_{1,n-2}. \tag{13.28.1}$$

Here the appropriate alternative hypothesis is H_1: $\beta_1 \neq 0$ so that the critical region is $R = \left\{F \middle| F > F_{\alpha,1,n-2}\right\}$(we have a one-tail alternative using the upper tail of the F distribution). So if we reject H_0 in favor of H_1, then we can safely conclude that there exists a statistically significant linear relationship between X and Y at the 100 α% level.

Example 13.10

Using the information provided in Table 12.4, let us conduct a significance test of the linear relationship between X and Y for $\alpha = 0.05$. Here we test H_0: $\beta_1 = 0$, against H_1: $\beta_1 \neq 0$ with $R = \left\{F \middle| F > F_{\alpha,1,n-2} = F_{0.05,1,10} = 4.96\right\}$. Using Equation (13.28.1), we obtain the sample F value

$$F = \frac{\text{Regression MS}}{\text{Error MS}} = \frac{46.71}{2.33} = 20.05.$$

Since this sample F value lies well with R, we may conclude that the linear relationship between X and Y is statistically significant at the 5% level—"the model matters."

It is interesting to note that if we square the t statistic $t = \hat{\beta}_1/s_{\hat{\beta}_1}$, we obtain the F statistic (Eq. 13.28.1). That is, under H_0: $\beta_1 = 0$, $t^2 = F$. To see this, we need only glance back at Example 12.5 to see that $|t| = 4.48$ and $R = \{t||t| > 2.228\}$. Clearly, this calculated two-tail t value squared is $|t|^2 = (4.48)^2 = 20.07$, which equals the above calculated F value of 20.05 (the difference is due to rounding); likewise, the tabular t of $t_{0.025,10} = 2.228$ squared $(= 4.963)$ equals the tabular F value of $F_{0.05,1,10} = 4.96$. Hence we may conclude that the preceding one-sided F test of H_0: $\beta_1 = 0$ versus H_1: $\beta_1 \neq 0$ is equivalent to the two-sided t test of H_0: $\beta_1 = 0$ versus H_1: $\beta_1 \neq 0$. So if for a given α the statistical significance of the regression slope emerges under the t test, then the F test must also reveal a significant slope and vice versa.

13.10.2 A Joint Test of the Regression Intercept and Slope

Up to now, our regression diagnostics have consisted mainly of performing hypothesis tests on β_0 and β_1 separately. However, we can actually test β_0 and β_1 jointly in order to determine if the entire population regression line itself is statistically significant. Here we address the question: "Are β_0 and β_1 jointly significantly different from zero?" That is, we test the *joint null hypothesis* H_0: $\beta_0 = \beta_1 = 0$, against the *joint alternative hypothesis* H_1: $\beta_0 \neq 0$, $\beta_1 \neq 0$ at the 100α% level. Now, it can be shown that the quantity

$$F = \frac{\sum\left[\left(\hat{\beta}_0 - \beta_0\right) + \left(\hat{\beta}_1 - \beta_1\right)X_i\right]^2/2}{s_\varepsilon^2} \text{ is } F_{2,n-2}, \tag{13.29}$$

where the term $\Sigma[\cdot]^2$ serves as a measure of the overall discrepancy between the estimated values of β_0 and β_1 and their actual population values for given X_i's. Clearly, this sum varies directly with $\left|\hat{\beta}_0 - \beta_0\right|$ and $\left|\hat{\beta}_1 - \beta_1\right|$. Under H_0,

$$F_o = \frac{\sum\left(\hat{\beta}_0 - \hat{\beta}_1 X_i\right)^2/2}{s_\varepsilon^2} = \frac{\left(n\bar{Y}^2 + \hat{\beta}_1 x_i y_i\right)/2}{s_\varepsilon^2} \tag{13.29.1}$$

with $R = \{F|F > F_{\alpha,2,n-2}\}$. Here, we have a one-tail alternative on the upper tail of the F distribution; and we reject H_0 when F_o is a member of R.

Example 13.11

Do the estimates for β_0 and β_1 obtained earlier in Example 12.1 warrant the conclusion that the entire population regression line is significant at the 5% level? Here we aim to test H_0: $\beta_0 = \beta_1 = 0$, against H_1: $\beta_0 \neq 0$, $\beta_1 \neq 0$ (the entire population regression equation is significant). With $R = \{F | F > F_{\alpha,2,n-2} = F_{0.05,2,10} = 4.10\}$ and, from Equation (13.29.1),

$$F_o = \frac{[12(25) + 0.6869(68)]/2}{2.329} = 74.43,$$

we immediately see that F_o falls into R, thus prompting us to reject H_0 and conclude that the entire population regression line is statistically significant at the 5% level.

EXERCISES

1 For the given ν and α values, find the implied quantiles of the chi-square distribution:

a. $\alpha = 0.05,\ \nu = 12$ **c.** $\alpha = 0.01,\ \nu = 20$

b. $\alpha = 0.01,\ \nu = 5$ **d.** $\alpha = 0.05,\ \nu = 8$

2 Ten years ago the non-athlete versus athlete graduation ratio at a local liberal arts college was known to be 8 to 1 in favor of non-athletes (there were eight non-athletes graduating for every athlete who graduated). Suppose that in a recent survey conducted by the Registrar's Office it was found that in a random sample of 450 degree applicants, 68 were athletes. Has the ratio changed significantly? Use $\alpha = 0.05$.

3 If four variants of a certain plant species are believed to be present in a South American rain forest in the 9:3:3:1 ratio, what are the null and alternative hypotheses if this belief is to be tested?

4 Billings Motors surveyed 250 of its customers regarding their preferred time for servicing their vehicles: morning (M), afternoon (A), or evening (E). The results were M = 101, A = 60, and E = 89. The service department thinks that the proportions for the three groups should be $\frac{1}{2}$, $\frac{1}{4}$, and $\frac{1}{4}$, respectively. Is there any evidence that these proportions differ from 2:1:1? Use $\alpha = 0.01$.

5 SMART Publications has four salesmen (denoted 1, 2, 3, and 4) in its advertising department. These salesmen have similar territories. Sales calls for last month were:

Salesman	1	2	3	4
Number of calls	57	65	60	66

For $\alpha = 0.05$, is the number of sales calls made uniformly distributed over the sales force?

6 Given the following absolute frequency distribution, can we conclude, for $\alpha = 0.05$, that it was drawn from a normal distribution?

Classes of X	Absolute Frequency (o_i)
30–39	4
40–49	6
50–59	8
60–69	12
70–79	9
80–89	7
90–99	4
	50

7 Invitations to homecoming were mailed to the class of 2000 graduates from a small Midwestern liberal arts college. The numbers from the class of 2000 who attended the festivities are shown below. Test the hypothesis that the proportion attending is independent of the highest degree earned.

	Degree		
Homecoming	B.A.	M.A.	Ph.D.
Attended	47	42	46
Did not attend	13	13	8

8 Two fertilizers (A and B) were applied to random samples of azalea seeds. After treatment, germination tests were conducted, with the results given below. Do these data indicate the fertilizers differ in their effects on germination? Use $\alpha = 0.05$.

Fertilizer	Number of seeds	% Germination
A	150	80
B	125	88

9 A random sample of employees at XYZ Corp. were asked to indicate their preference for one of three health plans (1, 2, or 3). Is there reason to believe that preferences are dependent upon job classification? Use $\alpha = 0.01$.

	Health plans		
Job classification	1	2	3
Clerical	39	50	19
Labor	52	57	41
Manager	9	13	20

10 The manager of a retail outlet store is interested in the relationship between days of sales training and customer satisfaction with the service. Customer responses to a brief survey taken randomly at the time of purchase are indicated below. Are days of sales training and customer satisfaction independent? Use $\alpha = 0.05$.

		Customer satisfaction		
		Poor	Good	Excellent
Days	1 day	10	15	2
of	2 days	7	10	9
training	3 days	5	15	8
	4 days	3	20	10

11 Given the information contained in the following table, determine if the population of voters favoring candidate A over candidate B is the same across the four wards polled. Use $\alpha = 0.05$. What assumption must be made?

	Candidate	
Ward	A	B
1	65	40
2	80	80
3	40	40
4	50	90

12 Five groups were independently and randomly polled regarding the issue of building more nuclear power plants. Are the population proportions of individuals favoring the building of more such plants the same across the groups polled. Set $\alpha = 0.01$.

	Outcome	
Group	In favor	Opposed
Investment bankers	37	13
Environmentalists	10	40
High school teachers	25	25
Construction industry	40	10
Small business assn.	30	20

13 Find the value of Cramer's phi-squared statistic for the contingency tables appearing in Exercises 13.8, 13.9, and 13.10.

14 Suppose a random sample of size $n = 15$ is taken from a normally distributed population. It is determined that $\Sigma_{i=1}^{15}(X_i - \bar{X})^2 = 131.43$. Find a 95% confidence interval for σ.

15 From a random sample of size $n = 20$ from an approximately normal population it was determined that $s^2 = 5.36$. Find a 99% confidence interval for σ.

16 Given α, v_1, and v_2, find the implied quantiles of the F distribution:

a. $F_{0.01,8,10}$
b. $F_{0.05,6,9}$
c. $F_{0.99,7,5}$
d. $F_{0.99,4,12}$

17 For the data set appearing in Exercise 12.1 (see also the result for Exercise 12.4), conduct a significance test of the linear relationship between X and Y for $\alpha = 0.05$. [Hint: Use the F test.]

18 For the Exercise 12.2 data set (see the result of Exercise 12.5), determine if there exists a statistically significant linear relationship between X and Y. Use $\alpha = 0.05$. [Hint: Use the F test.]

19 For the Exercise 12.7 data set, conduct a joint test of the regression intercept and slope. Use $\alpha = 0.05$.

APPENDIX A

TABLE A.1 Standard Normal Areas [Z is $N(0,1)$]

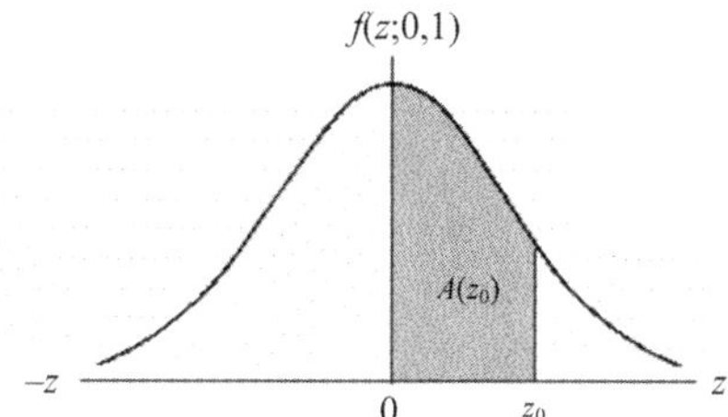

FIGURE A.1 $A(z_0) = \frac{1}{\sqrt{2\pi}}\int_0^{z_0} e^{-z^2/2}dz$ gives the total area under the standard normal distribution between 0 and any point z_0 on the positive z-axis (e.g., for $z_0 = 1.96$, $A(z_0) = 0.475$).

z	0.00	0.01	0.02	0.03	0.04	0.05	0.06	0.07	0.08	0.09
0.0	0.0000	0.0040	0.0080	0.0120	0.0150	0.0199	0.0239	0.0279	0.0319	0.0359
0.1	0.0398	0.0438	0.0478	0.0517	0.0557	0.0596	0.0636	0.0675	0.0714	0.0754
0.2	0.0793	0.0832	0.0871	0.0910	0.0948	0.0987	0.1026	0.1064	0.1103	0.1141
0.3	0.1179	0.1217	0.1253	0.1293	0.1331	0.1368	0.1406	0.1443	0.1480	0.1517
0.4	0.1554	0.1591	0.1628	0.1664	0.1700	0.1736	0.1772	0.1808	0.1844	0.1879
0.5	0.1915	0.1950	0.1985	0.2019	0.2054	0.2088	0.2123	0.2157	0.2190	0.2224
0.6	0.2258	0.2291	0.2324	0.2357	0.2389	0.2422	0.2454	0.2486	0.2518	0.2549
0.7	0.2580	0.2612	0.2642	0.2673	0.2704	0.2734	0.2764	0.2794	0.2823	0.2852

Statistical Inference: A Short Course, First Edition. Michael J. Panik.

TABLE A.1 (*Continued*)

z	0.00	0.01	0.02	0.03	0.04	0.05	0.06	0.07	0.08	0.09
0.8	0.2881	0.2910	0.2939	0.2967	0.2996	0.3023	0.3051	0.3078	0.3106	0.3133
0.9	0.3159	0.3186	0.3212	0.3288	0.3264	0.3289	0.3315	0.3340	0.3365	0.3389
1.0	0.3413	0.3438	0.3461	0.3485	0.3508	0.3531	0.3554	0.3557	0.3559	0.3621
1.1	0.3642	0.3665	0.3686	0.3708	0.3729	0.3749	0.3770	0.3790	0.3810	0.3830
1.2	0.3849	0.3869	0.3888	0.3907	0.3925	0.3944	0.3962	0.3980	0.3997	0.4015
1.3	0.4032	0.4049	0.4066	0.4082	0.4099	0.4115	0.4131	0.4147	0.4162	0.4177
1.4	0.4192	0.4207	0.4222	0.4236	0.4251	0.4265	0.4279	0.4292	0.4306	0.4319
1.5	0.4332	0.4345	0.4357	0.4370	0.4382	0.4394	0.4406	0.4418	0.4429	0.4441
1.6	0.4452	0.4463	0.4474	0.4484	0.4495	0.4505	0.4515	0.4525	0.4535	0.4545
1.7	0.4554	0.4564	0.4573	0.4582	0.4591	0.4599	0.4608	0.4616	0.4625	0.4633
1.8	0.4641	0.4649	0.4656	0.4664	0.4671	0.4678	0.4686	0.4693	0.4699	0.4706
1.9	0.4713	0.4719	0.4726	0.4732	0.4738	0.4744	0.4750	0.4756	0.4761	0.4767
2.0	0.4772	0.4778	0.4783	0.4788	0.4793	0.4798	0.4803	0.4808	0.4812	0.4817
2.1	0.4821	0.4826	0.4830	0.4834	0.4838	0.4842	0.4846	0.4850	0.4854	0.4857
2.2	0.4861	0.4864	0.4868	0.4871	0.4875	0.4878	0.4881	0.4884	0.4887	0.4890
2.3	0.4893	0.4896	0.4898	0.4901	0.4904	0.4906	0.4909	0.4911	0.4913	0.4916
2.4	0.4918	0.4920	0.4922	0.4925	0.4927	0.4929	0.4931	0.4932	0.4934	0.4936
2.5	0.4938	0.4940	0.4941	0.4943	0.4945	0.4946	0.4948	0.4949	0.4951	0.4952
2.6	0.4953	0.4955	0.4956	0.4957	0.4959	0.4960	0.4961	0.4962	0.4963	0.4964
2.7	0.4965	0.4966	0.4967	0.4968	0.4969	0.4970	0.4971	0.4972	0.4973	0.4974
2.8	0.4974	0.4975	0.4976	0.4977	0.4977	0.4978	0.4979	0.4979	0.4980	0.4981
2.9	0.4981	0.4982	0.4982	0.4983	0.4984	0.4984	0.4985	0.4985	0.4986	0.4986
3.0	0.4987	0.4987	0.4987	0.4988	0.4988	0.4989	0.4989	0.4989	0.4990	0.4990
3.1	0.4990	0.4990	0.4991	0.4991	0.4991	0.4991	0.4992	0.4992	0.4992	0.4992
3.2	0.4993	0.4993	0.4993	0.4993	0.4994	0.4994	0.4994	0.4994	0.4994	0.4995
3.3	0.4995	0.4995	0.4995	0.4995	0.4995	0.4996	0.4996	0.4996	0.4996	0.4996
3.4	0.4996	0.4996	0.4996	0.4997	0.4997	0.4997	0.4997	0.4997	0.4997	0.4997
3.5	0.4997	0.4997	0.4997	0.4997	0.4998	0.4998	0.4998	0.4998	0.4998	0.4998
3.6	0.4998	0.4998	0.4998	0.4998	0.4998	0.4998	0.4998	0.4998	0.4998	0.4998
3.7	0.4998	0.4999	0.4999	0.4999	0.4999	0.4999	0.4999	0.4999	0.4999	0.4999

TABLE A.2 Quantiles of the *t* Distribution (T is t_v)

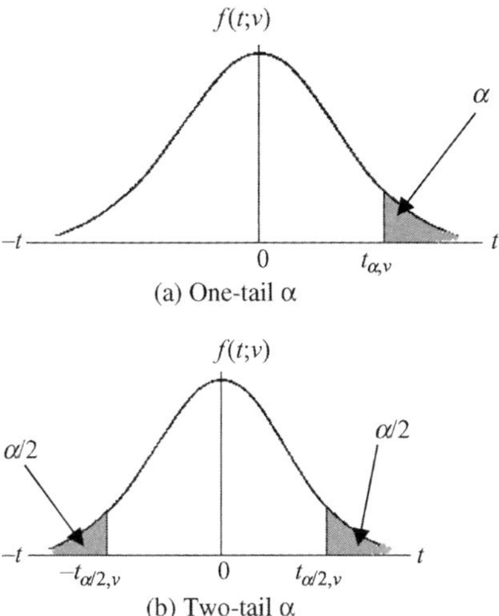

FIGURE A.2 Given degrees of freedom v, the table gives either: (a) the one-tail $t_{\alpha,v}$ value such that $P(T \geq t_{\alpha,v}) = \alpha$; or (b) the two-tail $\pm t_{\alpha/2,v}$ values for which $P(T \leq -t_{\alpha/2,v}) + P(T \geq t_{\alpha/2,v}) = \alpha/2 + \alpha/2 = \alpha$ (e.g., for $v = 15$ and $\alpha = 0.05$, $t_{0.05,\ 15} = 1.753$ while $t_{0.025,15} = 2.131$).

			One-tail α			
	0.10	0.05	0.025	0.01	0.005	0.001
			Two-tail α			
v	0.20	0.10	0.05	0.02	0.01	0.002
1	3.078	6.314	12.706	31.821	63.657	318.309
2	1.886	2.920	4.303	6.965	9.925	22.327
3	1.638	2.353	3.182	4.541	5.841	10.215
4	1.533	2.132	2.776	3.747	4.604	7.173
5	1.476	2.015	2.571	3.365	4.032	5.893
6	1.440	1.943	2.447	3.143	3.707	5.208
7	1.415	1.895	2.365	2.998	3.499	4.785
8	1.397	1.860	2.306	2.896	3.355	4.501
9	1.383	1.833	2.262	2.821	3.250	4.297
10	1.372	1.812	2.228	2.764	3.169	4.144
11	1.363	1.796	2.201	2.718	3.106	4.025
12	1.356	1.782	2.179	2.681	3.055	3.930
13	1.350	1.771	2.160	2.650	3.012	3.852
14	1.345	1.761	2.145	2.624	2.977	3.787
15	1.341	1.753	2.131	2.602	2.947	3.733
16	1.337	1.746	2.120	2.583	2.921	3.686
17	1.333	1.740	2.110	2.567	2.898	3.646
18	1.330	1.734	2.101	2.552	2.878	3.610
19	1.328	1.729	2.093	2.539	2.861	3.579
20	1.325	1.725	2.086	2.528	2.845	3.552

TABLE A.2 (*Continued*)

	One-tail α					
	0.10	0.05	0.025	0.01	0.005	0.001
	Two-tail α					
v	0.20	0.10	0.05	0.02	0.01	0.002
21	1.323	1.721	2.080	2.518	2.831	3.527
22	1.321	1.717	2.074	2.508	2.819	3.505
23	1.319	1.714	2.069	2.500	2.807	3.485
24	1.318	1.711	2.064	2.492	2.797	3.467
25	1.316	1.708	2.060	2.485	2.787	3.450
29	1.311	1.699	2.045	2.462	2.756	3.396
30	1.310	1.697	2.042	2.457	2.750	3.385
31	1.309	1.696	2.040	2.453	2.744	3.375
32	1.309	1.694	2.037	2.449	2.738	3.365
33	1.308	1.692	2.035	2.445	2.733	3.356
34	1.307	1.691	2.032	2.441	2.728	3.348
35	1.306	1.690	2.030	2.438	2.724	3.340
36	1.306	1.688	2.028	2.435	2.719	3.333
37	1.305	1.687	2.026	2.431	2.715	3.326
38	1.304	1.686	2.024	2.429	2.712	3.319
39	1.304	1.685	2.023	2.426	2.708	3.313
40	1.303	1.684	2.021	2.423	2.704	3.307
60	1.296	1.671	2.000	2.390	2.660	3.232
80	1.292	1.664	1.990	2.374	2.639	3.195
100	1.290	1.660	1.984	2.364	2.626	3.174
∞	1.282	1.645	1.960	2.326	2.576	3.090

TABLE A.3 Quantiles of the Chi-Square Distribution (X is χ_v^2)

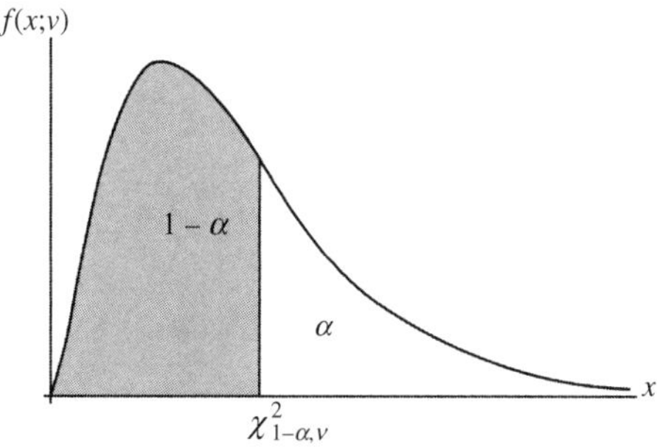

FIGURE A.3 For the cumulative probability $1-\alpha$ and degrees of freedom v, the quantile $\chi^2_{1-\alpha,v}$ satisfies $F(\chi^2_{1-\alpha}; v) = P(X \leq \chi^2_{1-\alpha,v}) = 1-\alpha$ or, alternatively, $P(X > \chi^2_{1-\alpha,v}) = 1 - P(X \leq \chi^2_{1-\alpha,v}) = \alpha$ (e.g., for $v = 10$ and $\alpha = 0.05$, $1-\alpha = 0.95$ and thus $\chi^2_{0.95,10} = 18.31$).

			$1-\alpha$					
ν	0.005	0.01	0.025	0.05	0.95	0.975	0.99	0.995
1	—	—	—	0.004	3.84	5.02	6.63	7.88
2	0.01	0.02	0.05	0.10	5.99	7.38	9.21	10.60
3	0.07	0.11	0.22	0.35	7.81	9.35	11.34	12.84
4	0.21	0.30	0.48	0.71	9.49	11.14	13.28	14.86
5	0.41	0.55	0.83	1.15	11.07	12.83	15.09	16.75
6	0.68	0.87	1.24	1.64	12.59	14.45	16.81	18.55
7	0.99	1.24	1.69	2.17	14.07	16.01	18.48	20.28
8	1.34	1.65	2.18	2.73	15.51	17.53	20.09	21.96
9	1.73	2.09	2.70	3.33	16.92	19.02	21.67	23.59
10	2.16	2.56	3.25	3.94	18.31	20.48	23.21	25.19
11	2.60	3.05	3.82	4.57	19.68	21.92	24.72	26.76
12	3.07	3.57	4.40	5.23	21.03	23.34	26.22	28.30
13	3.57	4.11	5.01	5.89	22.36	24.74	27.69	29.82
14	4.07	4.66	5.63	6.57	23.68	26.12	29.14	31.32
15	4.60	5.23	6.26	7.26	25.00	27.490	30.58	32.80
16	5.14	5.81	6.91	7.96	26.30	28.80	32.00	34.27
17	5.70	6.41	7.56	8.67	27.59	30.19	33.41	35.72
18	6.26	7.01	8.23	9.39	28.87	31.53	34.81	37.16
19	6.84	7.63	8.91	10.12	30.14	32.85	36.19	38.58
20	7.43	8.26	9.59	10.85	31.41	34.17	37.57	40.00
21	8.03	8.90	10.28	11.59	32.67	35.48	38.93	41.40
22	8.64	9.54	10.98	12.34	33.92	36.78	40.29	42.80
23	9.26	10.20	11.69	13.09	35.17	38.08	41.64	44.18
24	9.89	10.86	12.40	13.85	36.42	39.36	42.98	45.56
25	10.52	11.52	13.12	14.61	37.65	40.65	44.31	46.93
26	11.16	12.20	13.84	15.38	38.89	41.92	45.64	48.29
27	11.81	12.88	14.57	16.15	40.11	43.19	46.96	49.64
28	12.46	13.56	15.31	16.93	41.34	44.46	48.28	50.99

TABLE A.3 (*Continued*)

				$1-\alpha$				
ν	0.005	0.01	0.025	0.05	0.95	0.975	0.99	0.995
29	13.12	14.26	16.05	17.71	42.56	45.72	49.59	52.34
30	13.79	14.95	16.79	18.49	43.77	46.98	50.89	53.67
40	20.71	22.16	24.43	26.51	55.76	59.34	63.69	66.77
50	27.99	29.71	32.36	34.76	67.50	71.42	76.15	79.49
60	35.53	37.48	40.48	43.19	79.08	83.30	88.38	91.95
70	43.28	45.44	48.76	51.74	90.53	95.02	100.43	104.22
80	51.17	53.54	57.15	60.39	101.88	106.63	112.33	116.32
90	59.20	61.75	65.65	69.13	113.14	118.14	124.12	128.30
100	67.33	70.06	74.22	77.93	124.34	129.56	135.81	140.17

TABLE A.4 Quantiles of the F Distribution (F is F_{ν_1,ν_2})

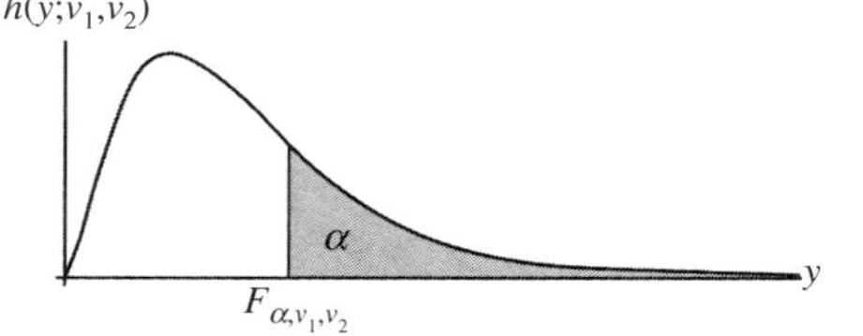

FIGURE A.4 Given the cumulative probability $1-\alpha$ and numerator and denominator degrees of freedom ν_1 and ν_2, respectively, the table gives the upper α-quantile F_{α,ν_1,ν_2} such that $P(F \geq F_{\alpha,\nu_1,\nu_2}) = \alpha$ (e.g., for $\alpha = 0.05$, $\nu_1 = 6$, and $\nu_2 = 10$, $F_{0.05,6,10} = 3.22$).

$\alpha = 0.10$ (Upper 10% Fractile)

ν_2 \ ν_1	1	2	3	4	5	6	7	8	9	10	12	15	20	24	30	40	60	120	∞
1	39.86	49.50	53.59	55.83	57.24	58.20	58.91	59.44	59.86	60.19	60.71	61.22	61.74	62.00	62.26	62.53	62.79	63.06	63.33
2	8.53	9.00	9.16	9.24	9.29	9.33	9.35	9.37	9.38	9.39	9.41	9.42	9.44	9.45	9.46	9.47	9.47	9.48	9.49
3	5.54	5.46	5.39	5.34	5.31	5.28	5.27	5.25	5.24	5.23	5.22	5.20	5.18	5.18	5.17	5.16	5.15	5.14	5.13
4	4.54	4.32	4.19	4.11	4.05	4.01	3.98	3.95	3.94	3.92	3.90	3.87	3.84	3.83	3.82	3.80	3.79	3.78	3.76
5	4.06	3.78	3.62	3.52	3.45	3.40	3.37	3.34	3.32	3.30	3.27	3.24	3.21	3.19	3.17	3.16	3.14	3.12	3.10
6	3.78	3.46	3.29	3.18	3.11	3.05	3.01	2.98	2.96	2.94	2.90	2.87	2.84	2.82	2.80	2.78	2.76	2.74	2.72
7	3.59	3.26	3.07	2.96	2.88	2.83	2.78	2.75	2.72	2.70	2.67	2.63	2.59	2.58	2.56	2.54	2.51	2.49	2.47
8	3.46	3.11	2.92	2.81	2.73	2.67	2.62	2.59	2.56	2.54	2.50	2.46	2.42	2.40	2.38	2.36	2.34	2.32	2.29
9	3.36	3.01	2.81	2.69	2.61	2.55	2.51	2.47	2.44	2.42	2.38	2.34	2.30	2.28	2.25	2.23	2.21	2.18	2.16
10	3.29	2.92	2.73	2.61	2.52	2.46	2.41	2.38	2.35	2.32	2.28	2.24	2.20	2.18	2.16	2.13	2.11	2.08	2.06
11	3.23	2.86	2.66	2.54	2.45	2.39	2.34	2.30	2.27	2.25	2.21	2.17	2.12	2.10	2.08	2.05	2.03	2.00	1.97
12	3.18	2.81	2.61	2.48	2.39	2.33	2.28	2.24	2.21	2.19	2.15	2.10	2.06	2.04	2.01	1.99	1.96	1.93	1.90
13	3.14	2.76	2.56	2.43	2.35	2.28	2.23	2.20	2.16	2.14	2.10	2.05	2.01	1.98	1.96	1.93	1.90	1.88	1.85

TABLE A.4 (*Continued*)

$\alpha = 0.10$ (Upper 10% Fractile)

ν_2 \ ν_1	1	2	3	4	5	6	7	8	9	10	12	15	20	24	30	40	60	120	∞
14	3.10	2.73	2.52	2.39	2.31	2.24	2.19	2.15	2.12	2.10	2.05	2.01	1.96	1.94	1.91	1.89	1.86	1.83	1.80
15	3.07	2.70	2.49	2.36	2.27	2.21	2.16	2.12	2.09	2.06	2.02	1.97	1.92	1.90	1.87	1.85	1.82	1.79	1.76
16	3.05	2.67	2.46	2.33	2.24	2.18	2.13	2.09	2.06	2.03	1.99	1.94	1.89	1.87	1.84	1.81	1.78	1.75	1.72
17	3.03	2.64	2.44	2.31	2.22	2.15	2.10	2.06	2.03	2.00	1.96	1.91	1.86	1.84	1.81	1.78	1.75	1.72	1.69
18	3.01	2.62	2.42	2.29	2.20	2.13	2.08	2.04	2.00	1.98	1.93	1.89	1.84	1.81	1.78	1.75	1.72	1.69	1.66
19	2.99	2.61	2.40	2.27	2.18	2.11	2.06	2.02	1.98	1.96	1.91	1.86	1.81	1.79	1.76	1.73	1.70	1.67	1.63
20	2.97	2.59	2.38	2.25	2.16	2.09	2.04	2.00	1.96	1.94	1.89	1.84	1.79	1.77	1.74	1.71	1.68	1.64	1.61
21	2.96	2.57	2.36	2.23	2.14	2.08	2.02	1.98	1.95	1.92	1.87	1.83	1.78	1.75	1.72	1.69	1.66	1.62	1.59
22	2.95	2.56	2.35	2.22	2.13	2.06	2.01	1.97	1.93	1.90	1.86	1.81	1.76	1.73	1.70	1.67	1.64	1.60	1.57
23	2.94	2.55	2.34	2.21	2.11	2.05	1.99	1.95	1.92	1.89	1.84	1.80	1.74	1.72	1.69	1.66	1.62	1.59	1.55
24	2.93	2.54	2.33	2.19	2.10	2.04	1.98	1.94	1.91	1.88	1.83	1.78	1.73	1.70	1.67	1.64	1.61	1.57	1.53
25	2.92	2.53	2.32	2.18	2.09	2.02	1.97	1.93	1.89	1.87	1.82	1.77	1.72	1.69	1.66	1.63	1.59	1.56	1.52
26	2.91	2.52	2.31	2.17	2.08	2.01	1.96	1.92	1.88	1.86	1.81	1.76	1.71	1.68	1.65	1.61	1.58	1.54	1.50
27	2.90	2.51	2.30	2.17	2.07	2.00	1.95	1.91	1.87	1.85	1.80	1.75	1.70	1.67	1.64	1.60	1.57	1.53	1.49
28	2.89	2.50	2.29	2.16	2.06	2.00	1.94	1.90	1.87	1.84	1.79	1.74	1.69	1.66	1.63	1.59	1.56	1.52	1.48
29	2.89	2.50	2.28	2.15	2.06	1.99	1.93	1.89	1.86	1.83	1.78	1.73	1.68	1.65	1.62	1.58	1.55	1.51	1.47
30	2.88	2.49	2.28	2.14	2.05	1.98	1.93	1.88	1.85	1.82	1.77	1.72	1.67	1.64	1.61	1.57	1.54	1.50	1.46
40	2.84	2.44	2.23	2.09	2.00	1.93	1.87	1.83	1.79	1.76	1.71	1.66	1.61	1.57	1.54	1.51	1.47	1.42	1.38
60	2.79	2.39	2.18	2.04	1.95	1.87	1.82	1.77	1.74	1.71	1.66	1.60	1.54	1.51	1.48	1.44	1.40	1.35	1.29
120	2.75	2.35	2.13	1.99	1.90	1.82	1.77	1.72	1.68	1.65	1.60	1.55	1.48	1.45	1.41	1.37	1.32	1.26	1.19
∞	2.71	2.30	2.08	1.94	1.85	1.77	1.72	1.67	1.63	1.60	1.55	1.49	1.42	1.38	1.34	1.30	1.24	1.17	1.00

$\alpha = 0.05$ (Upper 5% Fractile)

v_2 \ v_1	1	2	3	4	5	6	7	8	9	10	12	15	20	24	30	40	60	120	∞
1	161.4	199.5	215.7	224.6	230.2	234	236.8	238.9	240.5	241.9	243.9	245.9	248	249.1	250.1	251.1	252.2	253.3	254.3
2	18.51	19	19.16	19.25	19.3	19.33	19.35	19.37	19.38	19.4	19.41	19.43	19.45	19.45	19.46	19.47	19.48	19.49	19.5
3	10.13	9.55	9.28	9.12	9.01	8.94	8.89	8.85	8.81	8.79	8.74	8.7	8.66	8.64	8.62	8.59	8.57	8.55	8.53
4	7.71	6.94	6.59	6.39	6.26	6.16	6.09	6.04	6	5.96	5.91	5.86	5.8	5.77	5.75	5.72	5.69	5.66	5.63
5	6.61	5.79	5.41	5.19	5.05	4.95	4.88	4.82	4.77	4.74	4.68	4.62	4.56	4.53	4.5	4.46	4.43	4.4	4.36
6	5.99	5.14	4.76	4.53	4.39	4.28	4.21	4.15	4.1	4.06	4	3.94	3.87	3.84	3.81	3.77	3.74	3.7	3.67
7	5.59	4.74	4.35	4.12	3.97	3.87	3.79	3.73	3.68	3.64	3.57	3.51	3.44	3.41	3.38	3.34	3.3	3.27	3.23
8	5.32	4.46	4.07	3.84	3.69	3.58	3.5	3.44	3.39	3.35	3.28	3.22	3.15	3.12	3.08	3.04	3.01	2.97	2.93
9	5.12	4.26	3.86	3.63	3.48	3.37	3.29	3.23	3.18	3.14	3.07	3.01	2.94	2.9	2.86	2.83	2.79	2.75	2.71
10	4.96	4.1	3.71	3.48	3.33	3.22	3.14	3.07	3.02	2.98	2.91	2.85	2.77	2.74	2.7	2.66	2.62	2.58	2.54
11	4.84	3.98	3.59	3.36	3.2	3.09	3.01	2.95	2.9	2.85	2.79	2.72	2.65	2.61	2.57	2.53	2.49	2.45	2.4
12	4.75	3.89	3.49	3.26	3.11	3	2.91	2.85	2.8	2.75	2.69	2.62	2.54	2.51	2.47	2.43	2.38	2.34	2.3
13	4.67	3.81	3.41	3.18	3.03	2.92	2.83	2.77	2.71	2.67	2.6	2.53	2.46	2.42	2.38	2.34	2.3	2.25	2.21
14	4.6	3.74	3.34	3.11	2.96	2.85	2.76	2.7	2.65	2.6	2.53	2.46	2.39	2.35	2.31	2.27	2.22	2.18	2.13
15	4.54	3.68	3.29	3.06	2.9	2.79	2.71	2.64	2.59	2.54	2.48	2.4	2.33	2.29	2.25	2.2	2.16	2.11	2.07
16	4.49	3.63	3.24	3.01	2.85	2.74	2.66	2.59	2.54	2.49	2.42	2.35	2.28	2.24	2.19	2.15	2.11	2.06	2.01
17	4.45	3.59	3.2	2.96	2.81	2.7	2.61	2.55	2.49	2.45	2.38	2.31	2.23	2.19	2.15	2.1	2.06	2.01	1.96
18	4.41	3.55	3.16	2.93	2.77	2.66	2.58	2.51	2.46	2.41	2.34	2.27	2.19	2.15	2.11	2.06	2.02	1.97	1.92
19	4.38	3.52	3.13	2.9	2.74	2.63	2.54	2.48	2.42	2.38	2.31	2.23	2.16	2.11	2.07	2.03	1.98	1.93	1.88
20	4.35	3.49	3.1	2.87	2.71	2.6	2.51	2.45	2.39	2.35	2.28	2.2	2.12	2.08	2.04	1.99	1.95	1.9	1.84
21	4.32	3.47	3.07	2.84	2.68	2.57	2.49	2.42	2.37	2.32	2.25	2.18	2.1	2.05	2.01	1.96	1.92	1.87	1.81
22	4.3	3.44	3.05	2.82	2.66	2.55	2.46	2.4	2.34	2.3	2.23	2.15	2.07	2.03	1.98	1.94	1.89	1.84	1.78
23	4.28	3.42	3.03	2.8	2.64	2.53	2.44	2.37	2.32	2.27	2.2	2.13	2.05	2.01	1.96	1.91	1.86	1.81	1.76
24	4.26	3.4	3.01	2.78	2.62	2.51	2.42	2.36	2.3	2.25	2.18	2.11	2.03	1.98	1.94	1.89	1.84	1.79	1.73
25	4.24	3.39	2.99	2.76	2.6	2.49	2.4	2.34	2.28	2.24	2.16	2.09	2.01	1.96	1.92	1.87	1.82	1.77	1.71
26	4.23	3.37	2.98	2.74	2.59	2.47	2.39	2.32	2.27	2.22	2.15	2.07	1.99	1.95	1.9	1.85	1.8	1.75	1.69
27	4.21	3.35	2.96	2.73	2.57	2.46	2.37	2.31	2.25	2.2	2.13	2.06	1.97	1.93	1.88	1.84	1.79	1.73	1.67

TABLE A.4 (*Continued*)

$\alpha = 0.05$ (Upper 5% Fractile)

v_2 \ v_1	1	2	3	4	5	6	7	8	9	10	12	15	20	24	30	40	60	120	∞
28	4.2	3.34	2.95	2.71	2.56	2.45	2.36	2.29	2.24	2.19	2.12	2.04	1.96	1.91	1.87	1.82	1.77	1.71	1.65
29	4.18	3.33	2.93	2.7	2.55	2.43	2.35	2.28	2.22	2.18	2.1	2.03	1.94	1.9	1.85	1.81	1.75	1.7	1.64
30	4.17	3.32	2.92	2.69	2.53	2.42	2.33	2.27	2.21	2.16	2.09	2.01	1.93	1.89	1.84	1.79	1.74	1.68	1.62
40	4.08	3.23	2.84	2.61	2.45	2.34	2.25	2.18	2.12	2.08	2	1.92	1.84	1.79	1.74	1.69	1.64	1.58	1.51
60	4	3.15	2.76	2.53	2.37	2.25	2.17	2.1	2.04	1.99	1.92	1.84	1.75	1.7	1.65	1.59	1.53	1.47	1.39
120	3.92	3.07	2.68	2.45	2.29	2.17	2.09	2.02	1.96	1.91	1.83	1.75	1.66	1.61	1.55	1.5	1.43	1.35	1.25
∞	3.84	3	2.6	2.37	2.21	2.1	2.01	1.94	1.88	1.83	1.75	1.67	1.57	1.52	1.46	1.39	1.32	1.22	1.00

$\alpha = 0.01$ (Upper 1% Fractile)

v_2 \ v_1	1	2	3	4	5	6	7	8	9	10	12	15	20	24	30	40	60	120	∞
1	4052	4999.5	5403	5625	5764	5859	5928	5982	6022	6056	6106	6157	6209	6235	6261	6287	6313	6339	6366
2	98.50	99.00	99.17	99.25	99.30	99.33	99.36	99.37	99.39	99.40	99.42	99.43	99.45	99.46	99.47	99.47	99.48	99.49	99.50
3	34.12	30.82	29.46	28.71	28.24	27.91	27.67	27.49	27.35	27.23	27.05	26.87	26.69	26.60	26.50	26.41	26.32	26.22	26.13
4	21.20	18.00	16.69	15.98	15.52	15.21	14.98	14.80	14.66	14.55	14.37	14.20	14.02	13.93	13.84	13.75	13.65	13.56	13.46
5	16.26	13.27	12.06	11.39	10.97	10.67	10.46	10.29	10.16	10.05	9.89	9.72	9.55	9.47	9.38	9.29	9.20	9.11	9.02
6	13.75	10.92	9.78	9.15	8.75	8.47	8.26	8.10	7.98	7.87	7.72	7.56	7.40	7.31	7.23	7.14	7.06	6.97	6.88
7	12.25	9.55	8.45	7.85	7.46	7.19	6.99	6.84	6.72	6.62	6.47	6.31	6.16	6.07	5.99	5.91	5.82	5.74	5.65
8	11.26	8.65	7.59	7.01	6.63	6.37	6.18	6.03	5.91	5.81	5.67	5.52	5.36	5.28	5.20	5.12	5.03	4.95	4.86
9	10.56	8.02	6.99	6.42	6.06	5.80	5.61	5.47	5.35	5.26	5.11	4.96	4.81	4.73	4.65	4.57	4.48	4.40	4.31
10	10.04	7.56	6.55	5.99	5.64	5.39	5.20	5.06	4.94	4.85	4.71	4.56	4.41	4.33	4.25	4.17	4.08	4.00	3.91
11	9.65	7.21	6.22	5.67	5.32	5.07	4.89	4.74	4.63	4.54	4.40	4.25	4.10	4.02	3.94	3.86	3.78	3.69	3.60
12	9.33	6.93	5.95	5.41	5.06	4.82	4.64	4.50	4.39	4.30	4.16	4.01	3.86	3.78	3.70	3.62	3.54	3.45	3.36
13	9.07	6.70	5.74	5.21	4.86	4.62	4.44	4.30	4.19	4.10	3.96	3.82	3.66	3.59	3.51	3.43	3.34	3.25	3.17
14	8.86	6.51	5.56	5.04	4.69	4.46	4.28	4.14	4.03	3.94	3.80	3.66	3.51	3.43	3.35	3.27	3.18	3.09	3.00

15	8.68	6.36	5.42	4.89	4.56	4.32	4.14	4.00	3.89	3.80	3.67	3.52	3.37	3.29	3.21	3.13	3.05	2.96	2.87
16	8.53	6.23	5.29	4.77	4.44	4.20	4.03	3.89	3.78	3.69	3.55	3.41	3.26	3.18	3.10	3.02	2.93	2.84	2.75
17	8.40	6.11	5.18	4.67	4.34	4.10	3.93	3.79	3.68	3.59	3.46	3.31	3.16	3.08	3.00	2.92	2.83	2.75	2.65
18	8.29	6.01	5.09	4.58	4.25	4.01	3.84	3.71	3.60	3.51	3.37	3.23	3.08	3.00	2.92	2.84	2.75	2.66	2.57
19	8.18	5.93	5.01	4.50	4.17	3.94	3.77	3.63	3.52	3.43	3.30	3.15	3.00	2.92	2.84	2.76	2.67	2.58	2.49
20	8.10	5.85	4.94	4.43	4.10	3.87	3.70	3.56	3.46	3.37	3.23	3.09	2.94	2.86	2.78	2.69	2.61	2.52	2.42
21	8.02	5.78	4.87	4.37	4.04	3.81	3.64	3.51	3.40	3.31	3.17	3.03	2.88	2.80	2.72	2.64	2.55	2.46	2.36
22	7.95	5.72	7.82	4.31	3.99	3.76	3.59	3.45	3.35	3.26	3.12	2.98	2.83	2.75	2.67	2.58	2.50	2.40	2.31
23	7.88	5.66	4.76	4.26	3.94	3.71	3.54	3.41	3.30	3.21	3.07	2.93	2.78	2.70	2.62	2.54	2.45	2.35	2.26
24	7.82	5.61	7.72	4.22	3.90	3.67	3.50	3.36	3.26	3.17	3.03	2.89	2.74	2.66	2.58	2.49	2.40	2.31	2.21
25	7.77	5.57	4.68	4.18	3.85	3.63	3.46	3.32	3.22	3.13	2.99	2.85	2.70	2.62	2.54	2.45	2.36	2.27	2.17
26	7.72	5.53	4.64	4.14	3.82	3.59	3.42	3.29	3.18	3.09	2.96	2.81	2.66	2.58	2.50	2.42	2.33	2.23	2.13
27	7.68	5.49	4.60	4.11	3.78	3.56	3.39	3.26	3.15	3.06	2.93	2.78	2.63	2.55	2.47	2.38	2.29	2.20	2.10
28	7.64	5.45	4.57	4.07	3.75	3.53	3.36	3.23	3.12	3.03	2.90	2.75	2.60	2.52	2.44	2.35	2.26	2.17	2.06
29	7.60	5.42	4.54	4.04	3.73	3.50	3.33	3.20	3.09	3.00	2.87	2.73	2.57	2.49	2.41	2.33	2.23	2.14	2.03
30	7.56	5.39	4.51	4.02	3.70	3.47	3.30	3.17	3.07	2.98	2.84	2.70	2.55	2.47	2.39	2.30	2.21	2.11	2.01
40	7.31	5.18	4.31	3.83	3.51	3.29	3.12	2.99	2.89	2.80	2.66	2.52	2.37	2.29	2.20	2.11	2.02	1.92	1.80
60	7.08	4.98	4.13	3.65	3.34	3.12	2.95	2.82	2.72	2.63	2.50	2.35	2.20	2.12	2.03	1.94	1.84	1.73	1.60
120	6.85	4.79	3.95	3.48	3.17	2.96	2.79	2.66	2.56	2.47	2.34	2.19	2.03	1.95	1.86	1.76	2.66	1.53	1.38
∞	6.63	4.61	3.78	3.32	3.02	2.80	2.64	2.51	2.41	2.32	2.18	2.04	1.88	1.79	1.70	1.59	1.47	1.32	1.00

TABLE A.5 Binomial Probabilities $P(X;n,p)$

Given n and p, the table gives the binomial probability of x successes in n independent trials or $P(x;n,p) = \binom{n}{x} p^x (1-p)^{n-x}$, $x = 0, 1, \ldots, n$.

		p																			
n	x	0.01	0.05	0.10	0.15	0.20	0.25	0.30	0.35	0.40	0.45	0.50	0.55	0.60	0.65	0.70	0.75	0.80	0.85	0.90	0.95
2	0	0.9801	0.9025	0.8100	0.7225	0.6400	0.5625	0.4900	0.4225	0.3600	0.3025	0.2500	0.2025	0.1600	0.1225	0.0900	0.0625	0.0400	0.0225	0.0100	0.0025
	1	0.0198	0.0950	0.1800	0.2550	0.3200	0.3750	0.4200	0.4550	0.4800	0.4950	0.5000	0.4950	0.4800	0.4550	0.4200	0.3750	0.3200	0.2550	0.1800	0.0950
	2	0.0001	0.0025	0.0100	0.0225	0.0400	0.0625	0.0900	0.1225	0.1600	0.2025	0.2500	0.3025	0.3600	0.4225	0.4900	0.5625	0.8400	0.7225	0.8100	0.9025
3	0	0.9703	0.8574	0.7290	0.6141	0.5120	0.4219	0.3430	0.2746	0.2160	0.1664	0.1250	0.0911	0.0640	0.0429	0.0270	0.0156	0.0080	0.0034	0.0010	0.0001
	1	0.0294	0.1354	0.2430	0.3251	0.3840	0.4219	0.4410	0.4436	0.4320	0.4084	0.3750	0.3341	0.2880	0.2389	0.1890	0.1406	0.0960	0.0574	0.0270	0.0071
	2	0.0003	0.0071	0.0270	0.0574	0.0960	0.1406	0.1890	0.2389	0.2880	0.3341	0.3750	0.4084	0.4320	0.4436	0.4410	0.4219	0.3840	0.3251	0.2430	0.1354
	3	0.0000+	0.0001	0.0010	0.0034	0.0080	0.0156	0.0270	0.0429	0.0640	0.0911	0.1250	0.1664	0.2160	0.2746	0.3430	0.4219	0.5120	0.6141	0.7290	0.8574
4	0	0.9606	0.8145	0.6561	0.5220	0.4096	0.3164	0.2401	0.1785	0.1296	0.0915	0.0625	0.0410	0.0256	0.0150	0.0081	0.0039	0.0016	0.0005	0.0001	0.0000+
	1	0.0388	0.1715	0.2916	0.3585	0.4096	0.4219	0.4116	0.3845	0.3456	0.2995	0.2500	0.2005	0.1536	0.1115	0.0756	0.0469	0.0256	0.0116	0.0036	0.0005
	2	0.0006	0.0135	0.0486	0.0975	0.1536	0.2109	0.2646	0.3105	0.3456	0.3675	0.3760	0.3675	0.3456	0.3105	0.2646	0.2109	0.1536	0.0975	0.0486	0.0135
	3	0.0000+	0.0005	0.0036	0.0115	0.0256	0.0469	0.0756	0.1115	0.1536	0.2005	0.2500	0.2995	0.3456	0.3845	0.4116	0.4219	0.4096	0.3685	0.2916	0.1715
	4	0.0000+	0.0000+	0.0001	0.0005	0.0016	0.0039	0.0081	0.0160	0.0256	0.0410	0.0625	0.0915	0.1296	0.1785	0.2401	0.3164	0.4096	0.5220	0.6561	0.8145
5	0	0.9510	0.7738	0.5905	0.4437	0.3277	0.2373	0.1681	0.1160	0.0778	0.0503	0.0313	0.0185	0.0102	0.0053	0.0024	0.0010	0.0003	0.0001	0.0000+	0.0000+
	1	0.0480	0.2036	0.3281	0.3915	0.4096	0.3955	0.3602	0.3124	0.2592	0.2059	0.1563	0.1128	0.0768	0.0488	0.0284	0.0146	0.0064	0.0022	0.0005	0.0000+
	2	0.0010	0.0214	0.0729	0.1382	0.2048	0.2637	0.3087	0.3364	0.3456	0.3369	0.3125	0.2757	0.2304	0.1811	0.1323	0.0879	0.0512	0.0244	0.0081	0.0011
	3	0.0000+	0.0011	0.0081	0.0244	0.0512	0.0879	0.1323	0.1811	0.2304	0.2757	0.3125	0.3369	0.3456	0.3364	0.3087	0.2637	0.2048	0.1382	0.0729	0.0214
	4	0.0000+	0.0000+	0.0005	0.0022	0.0064	0.0146	0.0284	0.0488	0.0768	0.1128	0.1563	0.2059	0.2592	0.3124	0.3602	0.3955	0.4096	0.3915	0.3281	0.2036
	5	0.0000+	0.0000+	0.0000+	0.0001	0.0003	0.0010	0.0024	0.0053	0.0102	0.0185	0.0313	0.0503	0.0778	0.1160	0.1681	0.2373	0.3277	0.4437	0.5905	0.7738
6	0	0.9415	0.7351	0.5314	0.3771	0.2621	0.1780	0.1176	0.0754	0.0467	0.0277	0.0156	0.0083	0.0041	0.0018	0.0007	0.0002	0.0001	0.0000+	0.0000+	0.0000+
	1	0.0571	0.2321	0.3543	0.3993	0.3932	0.3560	0.3025	0.2437	0.1866	0.1359	0.0938	0.0609	0.0369	0.0205	0.0102	0.0044	0.0015	0.0004	0.0001	0.0000+
	2	0.0014	0.0305	0.0984	0.1762	0.2458	0.2966	0.3241	0.3280	0.3110	0.2780	0.2344	0.1861	0.1382	0.0951	0.0595	0.0330	0.0154	0.0056	0.0012	0.0001
	3	0.0000+	0.0021	0.0146	0.0415	0.0819	0.1318	0.1852	0.2355	0.2765	0.3032	0.3125	0.3032	0.2765	0.2355	0.1852	0.1318	0.0819	0.0415	0.0146	0.0021
	4	0.0000+	0.0001	0.0012	0.0056	0.0154	0.0330	0.0595	0.0951	0.1382	0.1861	0.2344	0.2780	0.3110	0.3280	0.3241	0.2966	0.2458	0.1762	0.0984	0.0305
	5	0.0000+	0.0000+	0.0001	0.0004	0.0015	0.0044	0.0102	0.0205	0.0369	0.0609	0.0938	0.1359	0.1866	0.2437	0.3025	0.3560	0.3932	0.3993	0.3543	0.2321
	6	0.0000+	0.0000+	0.0000+	0.0000+	0.0001	0.0002	0.0007	0.0018	0.0041	0.0083	0.0156	0.0277	0.0467	0.0754	0.1176	0.1780	0.2621	0.3771	0.5314	0.7351
7	0	0.9321	0.6983	0.4783	0.3206	0.2097	0.1335	0.0824	0.0490	0.0280	0.0152	0.0078	0.0037	0.0016	0.0006	0.0002	0.0001	0.0000+	0.0000+	0.0000+	0.0000+
	1	0.0659	0.2573	0.3720	0.3960	0.3670	0.3115	0.2471	0.1848	0.1306	0.0872	0.0547	0.0320	0.0172	0.0084	0.0036	0.0013	0.0004	0.0001	0.0000+	0.0000+
	2	0.0020	0.0406	0.1240	0.2097	0.2753	0.3115	0.3177	0.2985	0.2613	0.2140	0.1641	0.1172	0.0774	0.0466	0.0250	0.0115	0.0043	0.0012	0.0002	0.0000+
	3	0.0000+	0.0036	0.0230	0.0617	0.1147	0.1730	0.2269	0.2679	0.2903	0.2918	0.2734	0.2388	0.1935	0.1442	0.0972	0.0577	0.0287	0.0109	0.0026	0.0002
	4	0.0000+	0.0002	0.0026	0.0109	0.0287	0.0577	0.0972	0.1442	0.1935	0.2388	0.2734	0.2918	0.2903	0.2679	0.2269	0.1730	0.0617	0.0617	0.0230	0.0036
	5	0.0000+	0.0000+	0.0002	0.0012	0.0043	0.0115	0.0250	0.0466	0.0774	0.1172	0.1641	0.2140	0.2613	0.2985	0.3177	0.3115	0.1147	0.2097	0.1240	0.0406
	6	0.0000+	0.0000+	0.0000+	0.0001	0.0004	0.0013	0.0036	0.0084	0.0172	0.0320	0.0547	0.0872	0.1306	0.1848	0.2471	0.3115	0.3670	0.3960	0.3720	0.2573
	7	0.0000+	0.0000+	0.0000+	0.0000+	0.0000+	0.0001	0.0002	0.0006	0.0016	0.0037	0.0078	0.0152	0.0280	0.0490	0.0824	0.1335	0.2097	0.3206	0.4783	0.6983

8	0	0.9227	0.6634	0.4305	0.2725	0.1678	0.1001	0.0576	0.0319	0.0168	0.0084	0.0039	0.0017	0.0007	0.0002	0.0001	0.0000+	0.0000+	0.0000+	0.0000+	0.0000+
	1	0.0746	0.2793	0.3826	0.3847	0.3355	0.2670	0.1977	0.1373	0.0896	0.0548	0.0313	0.0164	0.0079	0.0033	0.0012	0.0004	0.0001	0.0000+	0.0000+	0.0000+
	2	0.0026	0.0515	0.1488	0.2376	0.2936	0.3115	0.2965	0.2587	0.2090	0.1569	0.1094	0.0703	0.0413	0.0217	0.0100	0.0038	0.0011	0.0002	0.0000+	0.0000+
	3	0.0001	0.0054	0.0331	0.0839	0.1468	0.2076	0.2541	0.2786	0.2787	0.2568	0.2188	0.1719	0.1239	0.0808	0.0467	0.0231	0.0092	0.0026	0.0004	0.0000+
	4	0.0000+	0.0004	0.0045	0.0185	0.0459	0.0865	0.1361	0.1875	0.2322	0.2627	0.2734	0.2627	0.2322	0.1675	0.1361	0.0865	0.0459	0.0185	0.0046	0.0004
	5	0.0000+	0.0000+	0.0004	0.0026	0.0092	0.0231	0.0467	0.0808	0.1239	0.1719	0.2188	0.2568	0.2767	0.2786	0.2541	0.2076	0.1468	0.0839	0.0331	0.0054
	6	0.0000+	0.0000+	0.0000+	0.0002	0.0011	0.0038	0.0100	0.0217	0.0413	0.0703	0.1094	0.1569	0.2090	0.2587	0.2965	0.3115	0.2936	0.2376	0.1488	0.0515
	7	0.0000+	0.0000+	0.0000+	0.0000+	0.0001	0.0004	0.0012	0.0033	0.0079	0.0164	0.0313	0.0548	0.0896	0.1373	0.1977	0.2670	0.3355	0.3647	0.3826	0.2793
	8	0.0000+	0.0000+	0.0000+	0.0000+	0.0000+	0.0000+	0.0001	0.0002	0.0007	0.0017	0.0039	0.0084	0.0168	0.0319	0.0576	0.1001	0.1678	0.2725	0.4305	0.6634
9	0	0.9135	0.6302	0.3874	0.2316	0.1342	0.0751	0.0404	0.0207	0.0101	0.0046	0.0020	0.0008	0.0003	0.0001	0.0000+	0.0000+	0.0000+	0.0000+	0.0000+	0.0000+
	1	0.0830	0.2985	0.3874	0.3679	0.3020	0.2253	0.1556	0.1004	0.0605	0.0339	0.0176	0.0083	0.0035	0.0013	0.0004	0.0001	0.0000+	0.0000+	0.0000+	0.0000+
	2	0.0034	0.0629	0.1722	0.2597	0.3020	0.3003	0.2668	0.2162	0.1612	0.1110	0.0703	0.0407	0.0212	0.0098	0.0039	0.0012	0.0003	0.0000+	0.0000+	0.0000+
	3	0.0001	0.0077	0.0446	0.1069	0.1762	0.2336	0.2668	0.2716	0.2508	0.2119	0.1641	0.1160	0.0743	0.0424	0.0210	0.0087	0.0028	0.0006	0.0001	0.0000+
	4	0.0000+	0.0006	0.0074	0.0283	0.0661	0.1168	0.1715	0.2194	0.2508	0.2600	0.2461	0.2128	0.1672	0.1181	0.0735	0.0389	0.0165	0.0050	0.0008	0.0000+
	5	0.0000+	0.0000+	0.0008	0.0050	0.0165	0.0389	0.0735	0.1181	0.1672	0.2128	0.2461	0.2600	0.2508	0.2194	0.1715	0.1168	0.0661	0.0283	0.0074	0.0006
	6	0.0000+	0.0000+	0.0001	0.0006	0.0028	0.0087	0.0210	0.0424	0.0743	0.1160	0.1641	0.2119	0.2508	0.2716	0.2668	0.2336	0.1762	0.1069	0.0446	0.0077
	7	0.0000+	0.0000+	0.0000+	0.0000+	0.0003	0.0012	0.0039	0.0098	0.0212	0.0407	0.0703	0.1110	0.1612	0.2162	0.2668	0.3003	0.3020	0.2597	0.1722	0.0629
	8	0.0000+	0.0000+	0.0000+	0.0000+	0.0000+	0.0001	0.0004	0.0013	0.0035	0.0083	0.0176	0.0339	0.0605	0.1004	0.1556	0.2253	0.3020	0.3679	0.3874	0.2985
	9	0.0000+	0.0000+	0.0000+	0.0000+	0.0000+	0.0000+	0.0000+	0.0001	0.0003	0.0008	0.0020	0.0046	0.0101	0.0207	0.0404	0.0751	0.1342	0.2316	0.3874	0.6302
10	0	0.9044	0.5987	0.3487	0.1969	0.1074	0.0563	0.0282	0.0135	0.0060	0.0025	0.0010	0.0003	0.0001	0.0000+	0.0000+	0.0000+	0.0000+	0.0000+	0.0000+	0.0000+
	1	0.0914	0.3151	0.3874	0.3474	0.2684	0.1877	0.1211	0.0725	0.0403	0.0207	0.0098	0.0042	0.0016	0.0005	0.0001	0.0000+	0.0000+	0.0000+	0.0000+	0.0000+
	2	0.0042	0.0748	0.1937	0.2759	0.3020	0.2816	0.2335	0.1757	0.1209	0.0763	0.0439	0.0229	0.0106	0.0043	0.0014	0.0004	0.0001	0.0000+	0.0000+	0.0000+
	3	0.0001	0.0105	0.0574	0.1298	0.2013	0.2503	0.2668	0.2522	0.2150	0.1665	0.1172	0.0746	0.0425	0.0212	0.0090	0.0031	0.0008	0.0001	0.0000+	0.0000+
	4	0.0000+	0.0010	0.0112	0.0401	0.0881	0.1460	0.2001	0.2377	0.2508	0.2384	0.2051	0.1596	0.1115	0.0689	0.0368	0.0162	0.0055	0.0012	0.0001	0.0000+
	5	0.0000+	0.0001	0.0015	0.0085	0.0264	0.0584	0.1029	0.1536	0.2007	0.2340	0.2461	0.2340	0.2007	0.1536	0.1029	0.0584	0.0264	0.0085	0.0015	0.0001
	6	0.0000+	0.0000+	0.0001	0.0012	0.0055	0.0162	0.0368	0.0689	0.1115	0.1596	0.2051	0.2384	0.2508	0.2377	0.2001	0.1460	0.0881	0.0401	0.0112	0.0010
	7	0.0000+	0.0000+	0.0000+	0.0001	0.0008	0.0031	0.0090	0.0212	0.0425	0.0746	0.1172	0.1665	0.2150	0.2522	0.2668	0.2503	0.2013	0.1298	0.0574	0.0105
	8	0.0000+	0.0000+	0.0000+	0.0000+	0.0001	0.0004	0.0014	0.0043	0.0106	0.0229	0.0439	0.0763	0.1209	0.1757	0.2335	0.2818	0.3020	0.2759	0.1937	0.0746
	9	0.0000+	0.0000+	0.0000+	0.0000+	0.0000+	0.0000+	0.0001	0.0005	0.0016	0.0042	0.0098	0.0207	0.0403	0.0725	0.1211	0.1877	0.2684	0.3474	0.3874	0.3151
	10	0.0000+	0.0000+	0.0000+	0.0000+	0.0000+	0.0000+	0.0000+	0.0000+	0.0001	0.0003	0.0010	0.0025	0.0060	0.0135	0.0282	0.0563	0.1074	0.1969	0.3487	0.5987
11	0	0.8953	0.5688	0.3136	0.1673	0.0859	0.0422	0.0196	0.0088	0.0036	0.0014	0.0005	0.0002	0.0000+	0.0000+	0.0000+	0.0000+	0.0000+	0.0000+	0.0000+	0.0000+
	1	0.0995	0.3293	0.3835	0.3246	0.2362	0.1549	0.0932	0.0518	0.0266	0.0125	0.0054	0.0021	0.0007	0.0002	0.0000+	0.0000+	0.0000+	0.0000+	0.0000+	0.0000+
	2	0.0050	0.0867	0.2131	0.2866	0.2953	0.2581	0.1998	0.1395	0.0887	0.0513	0.0269	0.0126	0.0052	0.0018	0.0005	0.0001	0.0000+	0.0000+	0.0000+	0.0000+
	3	0.0002	0.0137	0.0710	0.1517	0.2215	0.2581	0.2568	0.2254	0.1774	0.1259	0.0806	0.0462	0.0234	0.0102	0.0037	0.0011	0.0002	0.0000+	0.0000+	0.0000+
	4	0.0000+	0.0014	0.0158	0.0536	0.1107	0.1721	0.2201	0.2426	0.2365	0.2060	0.1611	0.1128	0.0701	0.0379	0.0173	0.0064	0.0017	0.0003	0.0000+	0.0000+
	5	0.0000+	0.0001	0.0025	0.0132	0.0388	0.0803	0.1321	0.1830	0.2207	0.2360	0.2256	0.1931	0.1471	0.0985	0.0568	0.0268	0.0097	0.0023	0.0003	0.0000+
	6	0.0000+	0.0000+	0.0003	0.0023	0.0097	0.0268	0.0566	0.0985	0.1471	0.1931	0.2256	0.2360	0.2207	0.1830	0.1321	0.0803	0.0388	0.0132	0.0025	0.0001
	7	0.0000+	0.0000+	0.0000+	0.0003	0.0017	0.0064	0.0173	0.0379	0.0701	0.1128	0.1611	0.2060	0.2365	0.2428	0.2201	0.1721	0.1107	0.0536	0.0158	0.0014
	8	0.0000+	0.0000+	0.0000+	0.0000+	0.0002	0.0011	0.0037	0.0102	0.0234	0.0462	0.0806	0.1259	0.1774	0.2254	0.2568	0.2581	0.2215	0.1517	0.0710	0.0137
	9	0.0000+	0.0000+	0.0000+	0.0000+	0.0000+	0.0001	0.0005	0.0016	0.0052	0.0126	0.0269	0.0513	0.0887	0.1395	0.1998	0.2561	0.2953	0.2866	0.2131	0.0867
	10	0.0000+	0.0000+	0.0000+	0.0000+	0.0000+	0.0000+	0.0000+	0.0002	0.0007	0.0021	0.0054	0.0125	0.0266	0.0516	0.0932	0.1549	0.2362	0.3248	0.3835	0.3293
	11	0.0000+	0.0000+	0.0000+	0.0000+	0.0000+	0.0000+	0.0000+	0.0000+	0.0000+	0.0002	0.0005	0.0014	0.0036	0.0088	0.0196	0.0422	0.0659	0.1673	0.3138	0.5888

TABLE A.5 *(Continued)*

Given n and p, the table gives the binomial probability of x successes in n independent trials or $P(x; n, p) = \binom{n}{x} p^x (1-p)^{n-x}$, $x = 0, 1, \ldots, n$.

		p																			
n	x	0.01	0.05	0.10	0.15	0.20	0.25	0.30	0.35	0.40	0.45	0.50	0.55	0.60	0.65	0.70	0.75	0.80	0.85	0.90	0.95
12	0	0.8864	0.5404	0.2824	0.1422	0.0687	0.0317	0.0138	0.0057	0.0022	0.0008	0.0002	0.0001	0.0000	0.0000	0.0000	0.0000+	0.0000+	0.0000+	0.0000+	0.0000+
	1	0.1074	0.3413	0.3766	0.3012	0.2062	0.1267	0.0712	0.0368	0.0174	0.0075	0.0029	0.0010	0.0003	0.0001	0.0000	0.0000+	0.0000+	0.0000+	0.0000+	0.0000+
	2	0.0060	0.0988	0.2301	0.2924	0.2835	0.2323	0.1678	0.1088	0.0639	0.0339	0.0161	0.0068	0.0025	0.0008	0.0002	0.0000+	0.0000+	0.0000+	0.0000+	0.0000+
	3	0.0002	0.0173	0.0852	0.1720	0.2362	0.2581	0.2397	0.1954	0.1419	0.0923	0.0537	0.0277	0.0125	0.0048	0.0015	0.0004	0.0001	0.0000+	0.0000+	0.0000+
	4	0.0000+	0.0021	0.0213	0.0683	0.1329	0.1936	0.2311	0.2367	0.2128	0.1700	0.1208	0.0762	0.0420	0.0199	0.0078	0.0024	0.0005	0.0001	0.0000+	0.0000+
	5	0.0000+	0.0002	0.0038	0.0193	0.0532	0.1032	0.1585	0.2039	0.2270	0.2225	0.1934	0.1489	0.1009	0.0591	0.0291	0.0115	0.0033	0.0006	0.0000+	0.0000+
	6	0.0000+	0.0000+	0.0005	0.0040	0.0155	0.0401	0.0792	0.1281	0.1766	0.2124	0.2256	0.2124	0.1766	0.1281	0.0792	0.0401	0.0155	0.0040	0.0005	0.0000+
	7	0.0000+	0.0000+	0.0000+	0.0006	0.0033	0.0115	0.0291	0.0591	0.1009	0.1489	0.1934	0.2225	0.2270	0.2039	0.1585	0.1032	0.0532	0.0193	0.0038	0.0002
	8	0.0000+	0.0000+	0.0000+	0.0001	0.0005	0.0024	0.0078	0.0199	0.0420	0.0762	0.1208	0.1700	0.2128	0.2367	0.2311	0.1936	0.1329	0.0683	0.0213	0.0021
	9	0.0000+	0.0000+	0.0000+	0.0000+	0.0001	0.0004	0.0015	0.0048	0.0125	0.0277	0.0537	0.0923	0.1419	0.1954	0.2397	0.2581	0.2362	0.1720	0.0852	0.0173
	10	0.0000+	0.0000+	0.0000+	0.0000+	0.0000+	0.0000+	0.0002	0.0008	0.0025	0.0068	0.0161	0.0339	0.0639	0.1088	0.1678	0.2323	0.2835	0.2924	0.2301	0.0988
	11	0.0000+	0.0000+	0.0000+	0.0000+	0.0000+	0.0000+	0.0000+	0.0001	0.0003	0.0010	0.0029	0.0075	0.0174	0.0368	0.0712	0.1267	0.2062	0.3012	0.3766	0.3413
	12	0.0000+	0.0000+	0.0000+	0.0000+	0.0000+	0.0000+	0.0000+	0.0000+	0.0000+	0.0001	0.0002	0.0008	0.0022	0.0057	0.0138	0.0317	0.0687	0.1422	0.2824	0.5404
15	0	0.8601	0.4633	0.2059	0.0874	0.0352	0.0134	0.0047	0.0016	0.0005	0.0001	0.0000+	0.0000+	0.0000+	0.0000+	0.0000+	0.0000+	0.0000+	0.0000+	0.0000+	0.0000+
	1	0.1303	0.3658	0.3432	0.2312	0.1319	0.0668	0.0305	0.0126	0.0047	0.0016	0.0005	0.0001	0.0000+	0.0000+	0.0000+	0.0000+	0.0000+	0.0000+	0.0000+	0.0000+
	2	0.0092	0.1348	0.2669	0.2856	0.2309	0.1559	0.0916	0.0476	0.0219	0.0090	0.0032	0.0010	0.0003	0.0001	0.0000+	0.0000+	0.0000+	0.0000+	0.0000+	0.0000+
	3	0.0004	0.0307	0.1285	0.2184	0.2501	0.2262	0.1700	0.1110	0.0634	0.0318	0.0139	0.0052	0.0016	0.0004	0.0001	0.0000+	0.0000+	0.0000+	0.0000+	0.0000+
	4	0.0000+	0.0049	0.0428	0.1156	0.1876	0.2252	0.2186	0.1792	0.1268	0.0780	0.0417	0.0191	0.0074	0.0024	0.0006	0.0001	0.0000	0.0000	0.0000	0.0000
	5	0.0000+	0.0006	0.0105	0.0449	0.1032	0.1651	0.2081	0.2123	0.1859	0.1404	0.0916	0.0515	0.0245	0.0096	0.0030	0.0007	0.0001	0.0000+	0.0000+	0.0000+
	6	0.0000+	0.0000+	0.0019	0.0132	0.0430	0.0917	0.1472	0.1906	0.2066	0.1914	0.1527	0.1048	0.0612	0.0298	0.0116	0.0034	0.0007	0.0001	0.0000+	0.0000+
	7	0.0000+	0.0000+	0.0003	0.0030	0.0138	0.0393	0.0811	0.1319	0.1771	0.2013	0.1964	0.1647	0.1181	0.0710	0.0348	0.0131	0.0035	0.0005	0.0000+	0.0000+
	8	0.0000+	0.0000+	0.0000+	0.0005	0.0035	0.0131	0.0348	0.0710	0.1181	0.1647	0.1964	0.2013	0.1771	0.1319	0.0811	0.0393	0.0138	0.0030	0.0003	0.0000+
	9	0.0000+	0.0000+	0.0000+	0.0001	0.0007	0.0034	0.0116	0.0298	0.0612	0.1048	0.1527	0.1914	0.2066	0.1906	0.1472	0.0917	0.0430	0.0132	0.0019	0.0000+
	10	0.0000+	0.0000+	0.0000+	0.0000+	0.0001	0.0007	0.0030	0.0096	0.0245	0.0515	0.0916	0.1404	0.1859	0.2123	0.2061	0.1651	0.1032	0.0449	0.0105	0.0006
	11	0.0000+	0.0000+	0.0000+	0.0000+	0.0000+	0.0001	0.0006	0.0024	0.0074	0.0191	0.0417	0.0780	0.1268	0.1792	0.2186	0.2252	0.1876	0.1156	0.0428	0.0049
	12	0.0000+	0.0000+	0.0000+	0.0000+	0.0000+	0.0000+	0.0001	0.0004	0.0016	0.0052	0.0139	0.0318	0.0634	0.1110	0.1700	0.2252	0.2501	0.2184	0.1285	0.0307
	13	0.0000+	0.0000+	0.0000+	0.0000+	0.0000+	0.0000+	0.0000+	0.0001	0.0003	0.0010	0.0032	0.0090	0.0219	0.0476	0.0916	0.1559	0.2309	0.2856	0.2669	0.1348
	14	0.0000+	0.0000+	0.0000+	0.0000+	0.0000+	0.0000+	0.0000+	0.0000+	0.0000+	0.0001	0.0005	0.0016	0.0047	0.0126	0.0305	0.0668	0.1319	0.2312	0.3432	0.3658
	15	0.0000+	0.0000+	0.0000+	0.0000+	0.0000+	0.0000+	0.0000+	0.0000+	0.0000+	0.0000+	0.0000+	0.0001	0.0005	0.0016	0.0047	0.0134	0.0352	0.0874	0.2059	0.4633
20	0	0.8179	0.3585	0.1216	0.0388	0.0115	0.0032	0.0008	0.0002	0.0000+	0.0000+	0.0000+	0.0000+	0.0000+	0.0000+	0.0000+	0.0000+	0.0000+	0.0000+	0.0000+	0.0000+
	1	0.1652	0.3774	0.2702	0.1368	0.0576	0.0211	0.0068	0.0020	0.0005	0.0001	0.0000+	0.0000+	0.0000+	0.0000+	0.0000+	0.0000+	0.0000+	0.0000+	0.0000+	0.0000+
	2	0.0159	0.1887	0.2852	0.2293	0.1369	0.0669	0.0278	0.0100	0.0031	0.0008	0.0002	0.0000+	0.0000+	0.0000+	0.0000+	0.0000+	0.0000+	0.0000+	0.0000+	0.0000+
	3	0.0010	0.0596	0.1901	0.2428	0.2054	0.1339	0.0716	0.0323	0.0123	0.0040	0.0011	0.0002	0.0000+	0.0000+	0.0000+	0.0000+	0.0000+	0.0000+	0.0000+	0.0000+
	4	0.0000+	0.0133	0.0898	0.1821	0.2182	0.1897	0.1304	0.0738	0.0350	0.0139	0.0046	0.0013	0.0003	0.0000+	0.0000+	0.0000+	0.0000+	0.0000+	0.0000+	0.0000+
	5	0.0000+	0.0022	0.0319	0.1028	0.1746	0.2023	0.1789	0.1272	0.0746	0.0365	0.0148	0.0049	0.0013	0.0003	0.0000+	0.0000+	0.0000+	0.0000+	0.0000+	0.0000+
	6	0.0000+	0.0003	0.0089	0.0454	0.1091	0.1686	0.1916	0.1712	0.1244	0.0746	0.0370	0.0150	0.0049	0.0012	0.0002	0.0000+	0.0000+	0.0000+	0.0000+	0.0000+

7	0.0000+	0.0000+	0.0020	0.0160	0.0545	0.1124	0.1643	0.1844	0.1659	0.1221	0.0739	0.0366	0.0146	0.0045	0.0010	0.0002	0.0000+	0.0000+	0.0000+	0.0000+
8	0.0000+	0.0000+	0.0004	0.0046	0.0222	0.0609	0.1144	0.1614	0.1797	0.1623	0.1201	0.0727	0.0355	0.0136	0.0039	0.0008	0.0001	0.0000+	0.0000+	0.0000+
9	0.0000+	0.0000+	0.0001	0.0011	0.0074	0.0271	0.0654	0.1158	0.1597	0.1771	0.1602	0.1185	0.0710	0.0336	0.0120	0.0030	0.0005	0.0000+	0.0000+	0.0000+
10	0.0000+	0.0000+	0.0000+	0.0002	0.0020	0.0099	0.0308	0.0686	0.1171	0.1593	0.1762	0.1593	0.1171	0.0686	0.0308	0.0099	0.0020	0.0002	0.0000+	0.0000+
11	0.0000+	0.0000+	0.0000+	0.0000+	0.0005	0.0030	0.0120	0.0336	0.0710	0.1185	0.1602	0.1771	0.1597	0.1158	0.0654	0.0271	0.0074	0.0011	0.0001	0.0000+
12	0.0000+	0.0000+	0.0000+	0.0000+	0.0001	0.0008	0.0039	0.0136	0.0355	0.0727	0.1201	0.1623	0.1797	0.1614	0.1144	0.0609	0.0222	0.0046	0.0004	0.0000+
13	0.0000+	0.0000+	0.0000+	0.0000+	0.0000+	0.0002	0.0010	0.0045	0.0145	0.0366	0.0739	0.1221	0.1659	0.1844	0.1643	0.1124	0.0545	0.0160	0.0020	0.0000+
14	0.0000+	0.0000+	0.0000+	0.0000+	0.0000+	0.0000+	0.0002	0.0012	0.0049	0.0150	0.0370	0.0746	0.1244	0.1712	0.1916	0.1686	0.1091	0.0454	0.0089	0.0003
15	0.0000+	0.0000+	0.0000+	0.0000+	0.0000+	0.0000+	0.0000+	0.0003	0.0013	0.0049	0.0148	0.0365	0.0746	0.1272	0.1789	0.2023	0.1746	0.1028	0.0319	0.0022
16	0.0000+	0.0000+	0.0000+	0.0000+	0.0000+	0.0000+	0.0000+	0.0000+	0.0003	0.0013	0.0046	0.0139	0.0350	0.0738	0.1304	0.1897	0.2182	0.1821	0.0898	0.0133
17	0.0000+	0.0000+	0.0000+	0.0000+	0.0000+	0.0000+	0.0000+	0.0000+	0.0000+	0.0002	0.0011	0.0040	0.0123	0.0323	0.0716	0.1339	0.2054	0.2428	0.1901	0.0596
18	0.0000+	0.0000+	0.0000+	0.0000+	0.0000+	0.0000+	0.0000+	0.0000+	0.0000+	0.0000+	0.0002	0.0008	0.0031	0.0100	0.0278	0.0669	01369	0.2293	0.2852	0.1887
19	0.0000+	0.0000+	0.0000+	0.0000+	0.0000+	0.0000+	0.0000+	0.0000+	0.0000+	0.0000+	0.0000+	0.0001	0.0005	0.0020	0.0068	0.0211	0.0576	0.1368	0.2702	0.3774
20	0.0000+	0.0000+	0.0000+	0.0000+	0.0000+	0.0000+	0.0000+	0.0000+	0.0000+	0.0000+	0.0000+	0.0000+	0.0000+	0.0002	0.0008	0.0032	0.0115	0.0388	0.1216	0.3585

Note: 0.0000 + means the probability is 0.0000 rounded to four decimal places. However, the probability is not zero.

TABLE A.6 Cumulative Binomial Probabilities

Given n and p, the table gives the probability that the binomial random variable X assumes a value $\leq x$.

		p																			
n	x	0.01	0.05	0.10	0.15	0.20	0.25	0.30	0.35	0.40	0.45	0.50	0.55	0.60	0.65	0.70	0.75	0.80	0.85	0.90	0.95
2	0	0.9801	0.9025	0.8100	0.7225	0.6400	0.5625	0.4900	0.4226	0.3600	0.3025	0.2500	0.2025	0.1600	0.1225	0.0900	0.0625	0.0400	0.0225	0.0100	0.0025
	1	0.8989	0.9025	0.8100	0.7225	0.6400	0.5525	0.4900	0.4225	0.3600	0.3025	0.2500	0.2025	0.1600	0.1225	0.0900	0.0625	0.0400	0.0225	0.0100	0.0025
	2	1.0000	1.0000	1.0000	1.0000	1.0000	1.0000	1.0000	1.0000	1.0000	1.0000	1.0000	1.0000	1.0000	1.0000	1.0000	1.0000	1.0000	1.0000	1.0000	1.0000
3	0	0.9703	0.8574	0.7290	0.6141	0.5120	0.4219	0.3430	0.2746	0.2160	0.1664	0.1250	0.0911	0.0640	0.0429	0.0270	0.0156	0.0080	0.0034	0.0010	0.0001
	1	0.9997	0.9928	0.9720	0.9393	0.8960	0.8438	0.7840	0.7183	0.6480	0.5748	0.5000	0.4253	0.3520	0.2818	0.2160	0.1563	0.1040	0.0608	0.0280	0.0073
	2	1.0000	0.9999	0.9990	0.9966	0.9920	0.9844	0.9730	0.9571	0.9360	0.9089	0.8750	0.8336	0.7840	0.7254	0.6570	0.5781	0.4880	0.3859	0.2710	0.1426
	3	1.0000	1.0000	1.0000	1.0000	1.0000	1.0000	1.0000	1.0000	1.0000	1.0000	1.0000	1.0000	1.0000	1.0000	1.0000	1.0000	1.0000	1.0000	1.0000	1.0000
4	0	0.9606	0.8145	0.6561	0.5220	0.4096	0.3164	0.2401	0.1785	0.1296	0.0915	0.0625	0.0410	0.0258	0.0150	0.0081	0.0039	0.0016	0.0005	0.0001	0.0000+
	1	0.9994	0.9860	0.9477	0.8905	0.8192	0.7383	0.6517	0.5830	0.4752	0.3910	0.3125	0.2415	0.1792	0.1265	0.0837	0.0508	0.0272	0.0120	0.0037	0.0005
	2	1.0000	0.9995	0.9863	0.9880	0.9728	0.9492	0.9163	0.8735	0.8208	0.7585	0.6875	0.6090	0.5248	0.4370	0.3483	0.2617	0.1808	0.1095	0.0523	0.0140
	3	1.0000	1.0000	0.9999	0.9995	0.9984	0.9961	0.9819	0.9850	0.9744	0.9500	0.9376	0.9085	0.8704	0.8215	0.7599	0.6836	0.5904	0.4780	0.3439	0.1855
	4	1.0000	1.0000	1.0000	1.0000	1.0000	1.0000	1.0000	1.0000	1.0000	1.0000	1.0000	1.0000	1.0000	1.0000	1.0000	1.0000	1.0000	1.0000	1.0000	1.0000
5	0	0.9510	0.7738	0.5905	0.4437	0.3277	0.2373	0.1681	0.1160	0.0778	0.0503	0.0313	0.0185	0.0102	0.0053	0.0024	0.0010	0.0003	0.0001	0.0000+	0.0000+
	1	0.9990	0.9774	0.9185	0.8352	0.7373	0.6328	0.5282	0.4284	0.3370	0.2562	0.1875	0.1312	0.0870	0.0540	0.0308	0.0156	0.0067	0.0022	0.0005	0.0000
	2	1.0000	0.9988	0.9914	0.9734	0.9421	0.8965	0.8369	0.7648	0.6826	0.5931	0.5000	0.4069	0.3174	0.2352	0.1631	0.1035	0.0579	0.0266	0.0086	0.0012
	3	1.0000	1.0000	0.9995	0.9978	0.9933	0.9844	0.9692	0.9460	0.9130	0.8688	0.8125	0.7438	0.6630	0.5716	0.4718	0.3672	0.2627	0.1648	0.0815	0.0226
	4	1.0000	1.0000	1.0000	0.9999	0.9997	0.9990	0.9976	0.9947	0.9898	0.9815	0.9688	0.9497	0.9222	0.8840	0.8319	0.7627	0.6723	0.5563	0.4095	0.2262
	5	1.0000	1.0000	1.0000	1.0000	1.0000	1.0000	1.0000	1.0000	1.0000	1.0000	1.0000	1.0000	1.0000	1.0000	1.0000	1.0000	1.0000	1.0000	1.0000	1.0000
6	0	0.9415	0.7351	0.5314	0.3771	0.2621	0.1780	0.1176	0.0754	0.0467	0.0277	0.0156	0.0083	0.0043	0.0018	0.0007	0.0002	0.0001	0.0000+	0.0000+	0.0000+
	1	0.9985	0.9672	0.8857	0.7765	0.6554	0.5339	0.4202	0.3191	0.2333	0.1636	0.1094	0.0692	0.0410	0.0223	0.0109	0.0046	0.0016	0.0004	0.0001	0.0000+
	2	1.0000–	0.9978	0.9642	0.9527	0.9011	0.8306	0.7443	0.6471	0.5443	0.4415	0.3438	0.2553	0.1792	0.1174	0.0705	0.0376	0.0170	0.0059	0.0013	0.0001
	3	1.0000–	0.9899	0.9987	0.9941	0.9830	0.9624	0.9295	0.8826	0.8208	0.7447	0.6563	0.5585	0.4557	0.3529	0.2557	0.1694	0.0989	0.0473	0.0159	0.0022
	4	1.0000–	1.0000–	0.9999	0.9996	0.9984	0.9954	0.9891	0.9777	0.9590	0.9308	0.8906	0.8364	0.7667	0.6809	0.6798	0.4661	0.3446	0.2235	0.1143	0.0328
	5	1.0000–	1.0000–	1.0000–	1.0000–	0.9989	0.9998	0.9993	0.9982	0.9959	0.8917	0.9844	0.9723	0.9533	0.9246	0.8824	0.8220	0.7379	0.6229	0.4686	0.2649
	6	1.0000–	1.0000–	1.0000–	1.0000–	1.0000–	1.0000–	1.0000–	1.0000–	1.0000–	1.0000–	1.0000–	1.0000–	1.0000–	1.0000–	1.0000–	1.0000–	1.0000–	1.0000–	1.0000–	1.0000–
7	0	0.9321	0.6983	0.4783	0.3206	0.2097	0.1335	0.0824	0.0490	0.0280	0.0152	0.0078	0.0037	0.0016	0.0006	0.0002	0.0001	0.0000+	0.0000+	0.0000+	0.0000+
	1	0.9980	0.9556	0.8503	0.7166	0.5767	0.4449	0.3294	0.2338	0.1586	0.1024	0.0625	0.0357	0.0188	0.0090	0.0038	0.0013	0.0004	0.0001	0.0000+	0.0000+
	2	1.0000–	0.9962	0.9743	0.9262	0.8520	0.7564	0.6471	0.5323	0.4199	0.3164	0.2266	0.1529	0.0963	0.0556	0.0288	0.0129	0.0047	0.0012	0.0002	0.0000+
	3	1.0000–	0.9998	0.9973	0.9879	0.9667	0.9294	0.8740	0.8002	0.7102	0.6083	0.5000	0.3917	0.2898	0.1998	0.1260	0.0706	0.0333	0.0121	0.0027	0.0002

	4	1.0000–	1.0000–	0.9998	0.9988	0.9953	0.9871	0.9712	0.9444	0.9037	0.8471	0.7734	0.6836	0.5801	0.4677	0.3529	0.2436	0.1480	0.0738	0.0257	0.0038
	5	1.0000–	1.0000–	1.0000–	0.9999	0.9996	0.9987	0.9962	0.9910	0.9812	0.9643	0.9375	0.8976	0.8414	0.7662	0.6706	0.5551	0.4233	0.2834	0.1497	0.0444
	6	1.0000–	1.0000–	1.0000–	1.0000–	1.0000–	0.9999	0.9998	0.9994	0.9984	0.9963	0.9922	0.9848	0.9720	0.9510	0.9176	0.8665	0.7903	0.6794	0.5217	0.3017
	7	1.0000–	1.0000–	1.0000–	1.0000–	1.0000–	1.0000–	1.0000–	1.0000–	1.0000–	1.0000–	1.0000–	1.0000–	1.0000–	1.0000–	1.0000–	1.0000–	1.0000–	1.0000–	1.0000–	1.0000–
8	0	0.9227	0.6634	0.4305	0.2725	0.1678	0.1001	0.0576	0.0319	0.0168	0.0084	0.0039	0.0017	0.0007	0.0002	0.0001	0.0000+	0.0000+	0.0000+	0.0000+	0.0000+
	1	0.9973	0.9428	0.8131	0.6572	0.5033	0.3671	0.2553	0.1691	0.1064	0.0632	0.0352	0.0181	0.0085	0.0036	0.0013	0.0004	0.0001	0.0000+	0.0000+	0.0000+
	2	0.9999	0.9942	0.9619	0.8948	0.7969	0.6785	0.5518	0.4278	0.3154	0.2201	0.1445	0.0885	0.0498	0.0253	0.0113	0.0042	0.0012	0.0002	0.0000+	0.0000+
	3	1.0000–	0.9996	0.9950	0.9786	0.9437	0.8862	0.8059	0.7064	0.5941	0.4770	0.3633	0.2604	0.1737	0.1061	0.0580	0.0273	0.0104	0.0029	0.0004	0.0000+
	4	1.0000–	1.0000–	0.9996	0.9971	0.9896	0.9727	0.9420	0.5939	0.8263	0.7396	0.6367	0.5230	0.4059	0.2936	0.1941	0.1138	0.0563	0.0214	0.0050	0.0004
	5	1.0000–	1.0000–	1.0000–	0.9998	0.9988	0.9958	0.9887	0.9747	0.9502	0.9115	0.8556	0.7799	0.8846	0.5722	0.4482	0.3215	0.2031	0.1052	0.0381	0.0058
	6	1.0000–	1.0000–	1.0000–	1.0000–	0.9999	0.9996	0.9987	0.9964	0.9915	0.9819	0.9648	0.9368	0.8936	0.8309	0.7447	0.6329	0.4967	0.3428	0.1869	0.0572
	7	1.0000–	1.0000–	1.0000–	1.0000–	1.0000–	1.0000–	0.9999	0.9998	0.9993	0.9983	0.9961	0.9916	0.9832	0.9681	0.9424	0.8999	0.8322	0.7275	0.5695	0.3366
	8	1.0000–	1.0000–	1.0000–	1.0000–	1.0000–	1.0000–	1.0000–	1.0000–	1.0000–	1.0000–	1.0000–	1.0000–	1.0000–	1.0000–	1.0000–	1.0000–	1.0000–	1.0000–	1.0000–	1.0000–
9	0	0.9135	0.6302	0.3874	0.2316	0.1342	0.0751	0.0404	0.0207	0.0101	0.0046	0.0020	0.0008	0.0003	0.0001	0.0000+	0.0000+	0.0000+	0.0000+	0.0000+	0.0000+
	1	0.9966	0.9288	0.7748	0.5995	0.4362	0.3003	0.1960	0.1211	0.0705	0.0385	0.0195	0.0091	0.0038	0.0014	0.0004	0.0001	0.0000+	0.0000+	0.0000+	0.0000+
	2	0.9999	0.9916	0.9470	0.8591	0.7382	0.6007	0.4628	0.3373	0.2318	0.1495	0.0898	0.0498	0.0250	0.0112	0.0043	0.0013	0.0003	0.0000+	0.0000+	0.0000+
	3	1.0000–	0.9994	0.9917	0.9661	0.9144	0.8343	0.7297	0.6089	0.4826	0.3614	0.2539	0.1658	0.0994	0.0536	0.0253	0.0100	0.0031	0.0006	0.0001	0.0000+
	4	1.0000–	1.0000–	0.9991	0.9944	0.9804	0.9511	0.9012	0.8283	0.7334	0.6214	0.5000	0.3786	0.2666	0.1717	0.0988	0.0489	0.0196	0.0056	0.0009	0.0000+
	5	1.0000–	1.0000–	0.9999	0.9944	0.9969	0.9900	0.9747	0.9464	0.9006	0.8342	0.7461	0.6386	0.5174	0.3911	0.2703	0.1657	0.0856	0.0339	0.0083	0.0006
	6	1.0000–	1.0000–	1.0000–	1.0000–	0.9997	0.9987	0.9957	0.9888	0.9750	0.9502	0.9102	0.8505	0.7682	0.6627	0.5372	0.3993	0.2618	0.1409	0.0530	0.0077
	7	1.0000–	1.0000–	1.0000–	1.0000–	1.0000–	0.9999	0.9996	0.9986	0.9962	0.9909	0.9805	0.9615	0.9295	0.8789	0.8040	0.6997	0.5638	0.4005	0.2252	0.0629
	8	1.0000–	1.0000–	1.0000–	1.0000–	1.0000–	1.0000–	1.0000–	0.9999	0.9997	0.9992	0.9980	0.9954	0.9899	0.9793	0.9596	0.9249	0.8658	0.7684	0.6126	0.2985
	9	1.0000–	1.0000–	1.0000–	1.0000–	1.0000–	1.0000–	1.0000–	1.0000–	1.0000–	1.0000–	1.0000–	1.0000–	1.0000–	1.0000–	1.0000–	1.0000–	1.0000–	1.0000–	1.0000–	1.0000–
10	0	0.8044	0.5987	0.3487	0.1969	0.1074	0.0563	0.0282	0.0135	0.0060	0.0025	0.0010	0.0003	0.0001	0.0000+	0.0000+	0.0000+	0.0000+	0.0000+	0.0000+	0.0000+
	1	0.9957	0.9139	0.7361	0.5443	0.3758	0.2440	0.1493	0.0860	0.0464	0.0233	0.0107	0.0046	0.0017	0.0005	0.0001	0.0000+	0.0000+	0.0000+	0.0000+	0.0000+
	2	0.9999	0.9885	0.9288	0.8202	0.6778	0.5256	0.3828	0.2616	0.1673	0.0996	0.0547	0.0274	0.0123	0.0048	0.0016	0.0004	0.0001	0.0000+	0.0000+	0.0000+
	3	1.0000–	0.9990	0.9872	0.9500	0.8791	0.7759	0.6496	0.5138	0.3823	0.2660	0.1719	0.1020	0.0548	0.0280	0.0106	0.0035	0.0009	0.0001	0.0000+	0.0000+
	4	1.0000–	0.9999	0.9884	0.9901	0.9672	0.9219	0.8497	0.7515	0.6331	0.5044	0.3770	0.2616	0.1662	0.0949	0.0473	0.0197	0.0064	0.0014	0.0001	0.0000+
	5	1.0000–	1.0000–	0.9999	0.9956	0.9936	0.9803	0.9527	0.9051	0.8338	0.7384	0.6230	0.4956	0.3669	0.2485	0.1503	0.0781	0.0328	0.0099	0.0016	0.0001
	6	1.0000–	1.0000–	1.0000–	0.9999	0.9991	0.9965	0.9894	0.9740	0.9452	0.8980	0.8281	0.7340	0.6177	0.4862	0.3504	0.2241	0.1209	0.0500	0.0128	0.0010
	7	1.0000–	1.0000–	1.0000–	1.0000–	0.9999	0.9996	0.9964	0.9052	0.9877	0.8726	0.8453	0.9004	0.8327	0.7384	0.6172	0.4744	0.3222	0.1798	0.0702	0.0115
	8	1.0000–	1.0000–	1.0000–	1.0000–	1.0000–	1.0000–	0.9999	0.9995	0.9983	0.9955	0.9893	0.9767	0.9536	0.9140	0.6507	0.7560	0.6242	0.4557	0.2639	0.0861
	9	1.0000–	1.0000–	1.0000–	1.0000–	1.0000–	1.0000–	1.0000–	1.0000–	0.8999	0.8887	0.9990	0.8975	0.8940	0.9865	0.9718	0.9437	0.8926	0.8031	0.6513	0.4013
	10	1.0000–	1.0000–	1.0000–	1.0000–	1.0000–	1.0000–	1.0000–	1.0000–	1.0000–	1.0000–	1.0000–	1.0000–	1.0000–	1.0000–	1.0000–	1.0000–	1.0000–	1.0000–	1.0000–	1.0000–
11	0	0.8953	0.5688	0.3138	0.1673	0.0859	0.0422	0.0198	0.0088	0.0036	0.0014	0.0005	0.0002	0.0000+	0.0000+	0.0000+	0.0000+	0.0000+	0.0000+	0.0000+	0.0000+
	1	0.9948	0.8981	0.6974	0.4922	0.3221	0.1971	0.1130	0.0606	0.0302	0.0139	0.0059	0.0022	0.0007	0.0002	0.0000+	0.0000+	0.0000+	0.0000+	0.0000+	0.0000+
	2	0.9998	0.9848	0.9104	0.7788	0.6174	0.4552	0.3127	0.2001	0.1189	0.0652	0.0327	0.0148	0.0059	0.0020	0.0006	0.0001	0.0000+	0.0000+	0.0000+	0.0000+
	3	1.0000–	0.9984	0.9815	0.9306	0.8389	0.7133	0.5696	0.4256	0.2963	0.1911	0.1133	0.0610	0.0293	0.0122	0.0043	0.0012	0.0002	0.0000+	0.0000+	0.0000+

TABLE A.6 (*Continued*)

Given n and p, the table gives the probability that the binomial random variable X assumes a value $\leq x$.

n	x	0.01	0.05	0.10	0.15	0.20	0.25	0.30	0.35	0.40	0.45	0.50	0.55	0.60	0.65	0.70	0.75	0.80	0.85	0.90	0.95
	4	1.0000−	0.9999	0.9972	0.9841	0.9496	0.8854	0.7897	0.6683	0.5326	0.3971	0.2744	0.1738	0.0994	0.0501	0.0216	0.0076	0.0020	0.0003	0.0000+	0.0000+
	5	1.0000−	1.0000	0.9997	0.9973	0.9883	0.9657	0.9218	0.8513	0.7535	0.6331	0.5000	0.3669	0.2465	0.1487	0.0782	0.0343	0.0117	0.0027	0.0003	0.0000+
	6	1.0000−	1.0000−	1.0000−	0.9997	0.9980	0.9924	0.9784	0.9499	0.9006	0.8262	0.7258	0.6029	0.4672	0.3317	0.2103	0.1146	0.0504	0.0159	0.0028	0.0001
	7	1.0000−	1.0000−	1.0000−	1.0000−	0.9998	0.9988	0.9957	0.9878	0.9707	0.9390	0.8867	0.8089	0.7037	0.5744	0.4304	0.2867	0.1611	0.0694	0.0185	0.0016
	8	1.0000−	1.0000−	1.0000−	1.0000−	1.0000−	0.9999	0.9994	0.9980	0.9941	0.9852	0.9673	0.9348	0.8811	0.7999	0.6873	0.5448	0.3826	0.2212	0.0896	0.0152
	9	1.0000−	1.0000−	1.0000−	1.0000−	1.0000−	1.0000−	1.0000−	0.9998	0.9993	0.9976	0.9941	0.9861	0.9698	0.9394	0.8870	0.8029	0.6779	0.6078	0.3028	0.1019
	10	1.0000−	1.0000−	1.0000−	1.0000−	1.0000−	1.0000−	1.0000−	1.0000−	1.0000−	0.9996	0.9995	0.9986	0.9964	0.9912	0.9802	0.9578	0.9141	0.8327	0.6862	0.4312
	11	1.0000−	1.0000−	1.0000−	1.0000−	1.0000−	1.0000−	1.0000−	1.0000−	1.0000−	1.0000−	1.0000−	1.0000−	1.0000−	1.0000−	1.0000−	1.0000−	1.0000−	1.0000−	1.0000−	1.0000−
12	0	0.8864	0.5404	0.2824	0.1422	0.0687	0.0317	0.0138	0.0057	0.0022	0.0008	0.0002	0.0001	0.0000+	0.0000+	0.0000+	0.0000+	0.0000+	0.0000+	0.0000+	0.0000+
	1	0.9938	0.8816	0.6590	0.4435	0.2749	0.1584	0.0850	0.0424	0.0196	0.0083	0.0032	0.0011	0.0003	0.0001	0.0000+	0.0000+	0.0000+	0.0000+	0.0000+	0.0000+
	2	0.9998	0.9804	0.8891	0.7358	0.5583	0.3907	0.2528	0.1513	0.0834	0.0421	0.0193	0.0079	0.0028	0.0008	0.0002	0.0000+	0.0000+	0.0000+	0.0000+	0.0000+
	3	1.0000−	0.9978	0.9744	0.9078	0.7946	0.6488	0.4925	0.3467	0.2253	0.1345	0.0730	0.0356	0.0153	0.0056	0.0017	0.0004	0.0001	0.0000+	0.0000+	0.0000+
	4	1.0000−	0.9998	0.9957	0.9761	0.9274	0.8424	0.7237	0.5833	0.4382	0.3044	0.1938	0.1117	0.0573	0.0255	0.0095	0.0028	0.0006	0.0001	0.0000+	0.0000+
	5	1.0000−	1.0000−	0.9995	0.9954	0.9806	0.9456	0.8822	0.7873	0.6652	0.5269	0.3872	0.2607	0.1582	0.0846	0.0386	0.0143	0.0039	0.0007	0.0001	0.0000+
	6	1.0000−	1.0000−	0.9999	0.9993	0.9961	0.9857	0.9614	0.9154	0.8418	0.7393	0.6128	0.4731	0.3348	0.2127	0.1178	0.0544	0.0194	0.0046	0.0005	0.0000+
	7	1.0000−	1.0000−	1.0000−	0.9999	0.9994	0.9972	0.9905	0.9745	0.9427	0.8883	0.8062	0.6956	0.5618	0.4167	0.2763	0.1576	0.0726	0.0239	0.0043	0.0002
	8	1.0000−	1.0000−	1.0000−	1.0000−	0.9999	0.9996	0.9983	0.9944	0.9847	0.9644	0.9270	0.8655	0.7747	0.6533	0.5075	0.3512	0.2054	0.0922	0.0256	0.0022
	9	1.0000−	1.0000−	1.0000−	1.0000−	1.0000−	1.0000−	0.9998	0.9992	0.9972	0.9921	0.9807	0.9579	0.9166	0.8487	0.7472	0.6093	0.4417	0.2642	0.1109	0.0196
	10	1.0000−	1.0000−	1.0000−	1.0000−	1.0000−	1.0000−	1.0000−	0.9999	0.9997	0.9989	0.9968	0.9917	0.9804	0.9576	0.9150	0.8416	0.7251	0.5565	0.3410	0.1184
	11	1.0000−	1.0000−	1.0000−	1.0000−	1.0000−	1.0000−	1.0000−	1.0000−	1.0000−	0.9999	0.9998	0.9992	0.9978	0.9943	0.9862	0.9683	0.9313	0.8578	0.7176	0.4596
	12	1.0000−	1.0000−	1.0000−	1.0000−	1.0000−	1.0000−	1.0000−	1.0000−	1.0000−	1.0000−	1.0000−	1.0000−	1.0000−	1.0000−	1.0000−	1.0000−	1.0000−	1.0000−	1.0000−	1.0000−
15	0	0.8601	0.4633	0.2059	0.0874	0.0352	0.0134	0.0047	0.0016	0.0005	0.0001	0.0000+	0.0000+	0.0000+	0.0000+	0.0000+	0.0000+	0.0000+	0.0000+	0.0000+	0.0000+
	1	0.9904	0.8290	0.5490	0.3186	0.1671	0.0802	0.0353	0.0142	0.0052	0.0017	0.0005	0.0001	0.0000+	0.0000+	0.0000+	0.0000+	0.0000+	0.0000+	0.0000+	0.0000+
	2	0.9996	0.9638	0.8159	0.6042	0.3980	0.2361	0.1268	0.0617	0.0271	0.0107	0.0037	0.0011	0.0003	0.0001	0.0000+	0.0000+	0.0000+	0.0000+	0.0000+	0.0000+
	3	1.0000−	0.9945	0.9444	0.8227	0.6482	0.4613	0.2969	0.1727	0.0905	0.0424	0.0176	0.0063	0.0019	0.0005	0.0001	0.0000+	0.0000+	0.0000+	0.0000+	0.0000+
	4	1.0000−	0.9994	0.9873	0.9383	0.8358	0.6865	0.5155	0.3519	0.2173	0.1204	0.0592	0.0255	0.0093	0.0028	0.0007	0.0001	0.0000+	0.0000+	0.0000+	0.0000+
	5	1.0000−	0.9999	0.9978	0.9832	0.9389	0.8518	0.7216	0.5643	0.4032	0.2608	0.1509	0.0769	0.0338	0.0124	0.0037	0.0008	0.0001	0.0000+	0.0000+	0.0000+
	6	1.0000−	1.0000−	0.9997	0.9984	0.9819	0.9434	0.8689	0.7548	0.6098	0.4622	0.3036	0.1818	0.0950	0.0422	0.0152	0.0042	0.0008	0.0001	0.0000+	0.0000+
	7	1.0000−	1.0000−	1.0000−	0.9994	0.9958	0.9827	0.9500	0.8868	0.7869	0.6535	0.5000	0.3465	0.2131	0.1132	0.0500	0.0173	0.0042	0.0006	0.0000+	0.0000+
	8	1.0000−	1.0000−	1.0000−	0.9999	0.9992	0.9958	0.9848	0.9578	0.9050	0.8182	0.6964	0.5478	0.3902	0.2452	0.1311	0.0566	0.0181	0.0036	0.0003	0.0000+
	9	1.0000−	1.0000−	1.0000−	1.0000−	0.9999	0.9992	0.9963	0.9876	0.9662	0.9231	0.8491	0.7392	0.5988	0.4357	0.2784	0.1484	0.0611	0.0168	0.0022	0.0001

	10	1.0000–	1.0000–	1.0000–	1.0000–	1.0000–	0.9999	0.9993	0.9972	0.9907	0.9745	0.9408	0.8796	0.7827	0.6481	0.4845	0.3135	0.1642	0.0617	0.0127	0.0006
	11	1.0000–	1.0000–	1.0000–	1.0000–	1.0000–	1.0000–	0.9999	0.9995	0.9981	0.9937	0.9824	0.9576	0.9095	0.3273	0.7031	0.5387	0.3518	0.1773	0.0556	0.0055
	12	1.0000–	1.0000–	1.0000–	1.0000–	1.0000–	1.0000–	1.0000–	0.9999	0.9997	0.9989	0.9963	0.9883	0.8729	0.9383	0.8732	0.7639	0.6020	0.3958	0.1841	0.0382
	13	1.0000–	1.0000–	1.0000–	1.0000–	1.0000–	1.0000–	1.0000–	1.0000–	1.0000–	0.9999	0.9995	0.9983	0.9948	0.9858	0.9647	0.9198	0.8329	0.6814	0.4510	0.1710
	14	1.0000–	1.0000–	1.0000–	1.0000–	1.0000–	1.0000–	1.0000–	1.0000–	1.0000–	1.0000–	1.0000–	0.9999	0.9995	0.9984	0.9953	0.9866	0.9648	0.9126	0.7941	0.5367
	15	1.0000–	1.0000–	1.0000–	1.0000–	1.0000–	1.0000–	1.0000–	1.0000–	1.0000–	1.0000–	1.0000–	1.0000–	1.0000–	1.0000–	1.0000–	1.0000–	1.0000–	1.0000–	1.0000–	1.0000–
20	0	0.8179	0.3585	0.1216	0.0388	0.0115	0.0032	0.0008	0.0002	0.0000+	0.0000+	0.0000+	0.0000+	0.0000+	0.0000+	0.0000+	0.0000+	0.0000+	0.0000+	0.0000+	0.0000+
	1	0.9831	0.7358	0.3917	0.1756	0.0692	0.0243	0.0076	0.0021	0.0005	0.0001	0.0000+	0.0000+	0.0000+	0.0000+	0.0000+	0.0000+	0.0000+	0.0000+	0.0000+	0.0000+
	2	0.9990	0.9245	0.6769	0.4049	0.2061	0.0913	0.0355	0.0121	0.0036	0.0009	0.0002	0.0000+	0.0000+	0.0000+	0.0000+	0.0000+	0.0000+	0.0000+	0.0000+	0.0000+
	3	1.0000–	0.9841	0.8670	0.6477	0.4114	0.2252	0.1071	0.0444	0.0160	0.0049	0.0013	0.0003	0.0000+	0.0000+	0.0000+	0.0000+	0.0000+	0.0000+	0.0000+	0.0000+
	4	1.0000–	0.9974	0.9568	0.8298	0.6296	0.4148	0.2375	0.1182	0.0510	0.0189	0.0059	0.0015	0.0003	0.0000+	0.0000+	0.0000+	0.0000+	0.0000+	0.0000+	0.0000+
	5	1.0000–	0.9997	0.9887	0.9327	0.8042	0.6172	0.4164	0.2454	0.1256	0.0553	0.0207	0.0064	0.0016	0.0003	0.0000+	0.0000+	0.0000+	0.0000+	0.0000+	0.0000+
	6	1.0000–	1.0000–	0.9976	0.9781	0.9133	0.7858	0.6080	0.4166	0.2500	0.1299	0.0577	0.0214	0.0065	0.0015	0.0003	0.0000+	0.0000+	0.0000+	0.0000+	0.0000+
	7	1.0000–	1.0000–	0.9996	0.9941	0.9679	0.8982	0.7723	0.6010	0.4159	0.2520	0.1316	0.0580	0.0210	0.0060	0.0013	0.0002	0.0000+	0.0000+	0.0000+	0.0000+
	8	1.0000–	1.0000–	0.9999	0.9987	0.9900	0.9591	0.8867	0.7624	0.5956	0.4143	0.2517	0.1308	0.0565	0.0196	0.0051	0.0009	0.0001	0.0000+	0.0000+	0.0000+
	9	1.0000–	1.0000–	1.0000–	0.9998	0.9974	0.9861	0.9520	0.8782	0.7553	0.5914	0.4119	0.2493	0.1275	0.0532	0.0171	0.0039	0.0006	0.0000+	0.0000+	0.0000+
	10	1.0000–	1.0000–	1.0000–	1.0000–	0.9994	0.9961	0.9829	0.9468	0.8725	0.7507	0.5881	0.4086	0.2447	0.1218	0.0480	0.0139	0.0026	0.0002	0.0000+	0.0000+
	11	1.0000–	1.0000–	1.0000–	1.0000–	0.9999	0.9991	0.9949	0.9804	0.9435	0.8692	0.7483	0.5857	0.4044	0.2376	0.1133	0.0409	0.0100	0.0013	0.0001	0.0000+
	12	1.0000–	1.0000–	1.0000–	1.0000–	1.0000–	0.9998	0.9987	0.9940	0.9790	0.9420	0.8684	0.7480	0.5841	0.3990	0.2277	0.1018	0.0321	0.0059	0.0004	0.0000+
	13	1.0000–	1.0000–	1.0000–	1.0000–	1.0000–	1.0000–	0.9997	0.9985	0.9935	0.9786	0.9423	0.8701	0.7500	0.5834	0.3920	0.2142	0.0867	0.0219	0.0024	0.0000+
	14	1.0000–	1.0000–	1.0000–	1.0000–	1.0000–	1.0000–	1.0000–	0.9997	0.9984	0.9936	0.9793	0.9447	0.8744	0.7546	0.5836	0.3828	0.1958	0.0673	0.0113	0.0003
	15	1.0000–	1.0000–	1.0000–	1.0000–	1.0000–	1.0000–	1.0000–	1.0000–	0.9997	0.9985	0.9941	0.9811	0.9490	0.8818	0.7625	0.5852	0.3704	0.1702	0.0432	0.0026
	16	1.0000–	1.0000–	1.0000–	1.0000–	1.0000–	1.0000–	1.0000–	1.0000–	1.0000–	0.9997	0.9987	0.9951	0.9840	0.9556	0.8929	0.7748	0.5886	0.3523	0.1330	0.0159
	17	1.0000–	1.0000–	1.0000–	1.0000–	1.0000–	1.0000–	1.0000–	1.0000–	1.0000–	1.0000–	0.9998	0.9991	0.9964	0.9879	0.9645	0.9087	0.7939	0.5921	0.3231	0.0755
	18	1.0000–	1.0000–	1.0000–	1.0000–	1.0000–	1.0000–	1.0000–	1.0000–	1.0000–	1.0000–	1.0000–	0.9999	0.9995	0.9979	0.9924	0.9757	0.9308	0.8244	0.6083	0.2642
	19	1.0000–	1.0000–	1.0000–	1.0000–	1.0000–	1.0000–	1.0000–	1.0000–	1.0000–	1.0000–	1.0000–	1.0000–	1.0000–	0.9998	0.9992	0.9968	0.9885	0.9612	0.8784	0.6415
	20	1.0000–	1.0000–	1.0000–	1.0000–	1.0000–	1.0000–	1.0000–	1.0000–	1.0000–	1.0000–	1.0000–	1.0000–	1.0000–	1.0000–	1.0000–	1.0000–	1.0000–	1.0000–	1.0000–	1.0000–

Note: 0.0000 + means the probability is 0.0000 rounded to four decimal places. However, the probability is not zero.
1.0000 – means the probability is 1.0000 rounded to four decimal places. However, the probability is not one.

TABLE A.7 Quantiles of Lilliefors' Test for Normality

The entries in the table are the values $d_{n,1-\alpha}$ such that $P\left(D > d_{n,1-\alpha}\right) = \alpha$ when the underlying distribution is a normal distribution.

n	$1-\alpha$: 0.80	0.85	0.90	0.95	0.99	0.999
4	0.303	0.321	0.346	0.376	0.413	0.433
5	0.289	0.303	0.319	0.343	0.397	0.439
6	0.269	0.281	0.297	0.323	0.371	0.424
7	0.252	0.264	0.280	0.304	0.351	0.402
8	0.239	0.250	0.265	0.288	0.333	0.384
9	0.227	0.238	0.252	0.274	0.317	0.365
10	0.217	0.228	0.241	0.262	0.304	0.352
11	0.208	0.218	0.231	0.251	0.291	0.338
12	0.200	0.210	0.222	0.242	0.281	0.325
13	0.193	0.202	0.215	0.234	0.271	0.314
14	0.187	0.196	0.208	0.226	0.262	0.305
15	0.181	0.190	0.201	0.219	0.254	0.296
16	0.176	0.184	0.195	0.213	0.247	0.287
17	0.171	0.179	0.190	0.207	0.240	0.279
18	0.167	0.175	0.185	0.202	0.234	0.273
19	0.163	0.170	0.181	0.197	0.228	0.266
20	0.159	0.166	0.176	0.192	0.223	0.260
25	0.143	0.150	0.159	0.173	0.201	0.236
30	0.131	0.138	0.146	0.159	0.185	0.217
Approximation for $n > 30$	$\frac{0.74}{\sqrt{n}}$	$\frac{0.77}{\sqrt{n}}$	$\frac{0.82}{\sqrt{n}}$	$\frac{0.89}{\sqrt{n}}$	$\frac{1.04}{\sqrt{n}}$	$\frac{1.22}{\sqrt{n}}$

Adapted from Table 1 of G. E. Dallal and L. Wilkinson, An analytical approximation to the distribution of Lilliefors' test statistic. *The American Statistician* 40 (1986): 294–296. Reprinted with permission from *The American Statistician*.

SOLUTIONS TO EXERCISES

1.1

a. Qualitative.
b. Qualitative.
c. Qualitative.
d. Quantitative.
e. Qualitative.
f. Quantitative.
g. Quantitative.
h. Quantitative.
i. Qualitative.
j. Qualitative.
k. Qualitative.
l. Quantitative.

1.2

a. Discrete.
b. Discrete.
c. Continuous.
d. Continuous.
e. Continuous.
f. Discrete.
g. Continuous.
h. Continuous.

1.3

a. Nominal.
b. Nominal.
c. Ordinal.
d. Ordinal.
e. Ordinal.

1.4

a. Yes.
b. Yes.
c. No.
d. Yes.
e. Yes.
e. No.

Statistical Inference: A Short Course, First Edition. Michael J. Panik.

2.1

X	f_j	f_j/n	$100 \times f_j/n$ (%)
5	2	2/25	8
6	3	3/25	12
7	5	5/25	20
8	2	2/25	8
9	4	4/25	16
10	5	5/25	20
11	2	2/25	8
12	1	1/25	4
13	1	1/25	4
	25	1	100

2.2 a. Range $= 59$. b. $59/6 = 9.83 \approx 10$.

c.	d. Classes of X	f_j	f_j/n	$100 \times f_j/n$ (%)
40–49	40–49	2	2/40	5
50–59	50–59	1	1/40	2.5
60–69	60–69	7	7/40	17.5
70–79	70–79	8	8/40	20
80–89	80–89	14	14/40	35
90–99	90–99	8	4/40	20
		40	1	100

e. Class Marks	e. Class Boundaries
44.5	39.5–49.5
54.5	49.5–59.5
64.5	59.5–69.5
74.5	69.5–79.5
84.5	79.5–89.5
94.5	89.5–99.5

2.3

Class Marks	Class Boundaries
26.5	24.5–28.5
30.5	28.5–32.5
34.5	32.5–36.5
38.5	36.5–40.5
42.5	40.5–44.5

Yes.

2.4 a. Range = 137. b. $137/16 = 8.56 \approx 9$.

c. 55–70
71–86
87–102
103–118
119–134
135–150
151–166
167–182
183–198

d.

Classes of X	f_j	f_j/n	$100 \times f_j/n$ (%)
55–70	2	2/48	4.16
71–86	6	6/48	12.50
87–102	3	3/48	6.25
103–118	14	14/48	29.16
119–134	9	9/48	18.75
135–150	5	5/48	10.41
151–166	2	2/48	4.16
167–182	3	3/48	6.25
183–198	4	4/48	8.33
	48	1	100

e.

Class Marks	Class Boundaries
62.5	54.5–70.5
78.5	70.5–86.5
94.5	86.5–102.5
etc.	etc.

2.5 $n = 50, K = 66 \approx 7$ classes.
$n = 100, K = 7.6 \approx 8$ classes.

2.6 $n = 200, K = 8.6 \approx 9$ classes. Suggested classes are:

10–17
18–15
26–33
34–41
42–49
50–57
58–65
66–73
74–81

2.7

Class Marks	Class Boundaries
37	24.5–49.5
62	49.5–74.5
87	74.5–99.5
112	99.5–124.5
137	124.5–149.5

Class Interval = 25

Classes of X	f_j/n	$100 \times f_j/n$ (%)
25–49	16/100	16
50–74	25/100	25
75–99	30/100	30
100–124	20/100	20
125–149	9/100	9
	1	100

2.8

Classes of X	f_j
20–39	5
40–59	20
60–79	15
80–99	10
	50

2.9

Number of Defects	$100 \times f_j/n$ (%)
0	18
1	20
2	18
3	16
4	14
5	10
6	2
7	2
	100

a. 28
b. 86
c. 56
d. 38

3.1
a. $a_1X_1^2 + a_2X_2^2 + a_3X_3^2 + a_4X_4^2$.
b. $aX_1 + a^2X_2 + a^3X_3 + a^4X_4$.
c. $a - X_1 + a^2 - X_2 + a^3 - X_3$.
d. $-X_2^3 + X_3^4 - X_4^5 + X_5^6$.

3.2
a. $\sum_{i=1}^{3} X_iY_i$.
b. $\sum_{i=1}^{3} a_iX_i^2$.
c. $\sum_{i=1}^{3} (X_i - Y_i)$.
d. $\sum_{i=1}^{3} (-1)^i X_i/Y_i$.

3.3 a. –15. b. 9. c. 125. d. 22.

3.4 a. No. b. No.

3.5

Mean	Median	Mode	Range	Variance	Standard Deviation
a. 8	9	—	11	20	4.472
b. 8	8	4, 8, 10	12	16	4
c. 0	0	−4, 0, 2	12	16	4
d. 14	15	—	22	64	8

3.6

$9-8=1$
$11-8=3$
$2-8=-6$
$5-8=-3$
$13-8=5$
0

3.7 $\overline{X}_w = \dfrac{3(3)+3(3)+5(4)+4(2)}{3+3+5+4} = \dfrac{46}{15} = 3.066.$

(The numerical values of the letter grades are weighted by the number of credit hours.)

3.8 $\overline{X}_w = \dfrac{4(3.50)+3(2.75)+2(2.25)}{4+3+2} = \dfrac{26.75}{9} = \$2.97.$

(The cost per pound values are weighted by the number of pounds purchased.)

3.9 a. 68%. b. 95%. c. 99.7%. d. 5%. e. 2.5%.

3.10 a. 68%. b. 32%. c. 2.5%. d. 0.15%.

3.11 a. At least 84%, (1.10, 1.60).
b. At least 88.9%, (1.05, 1.65).
c. $1.35 \pm k(0.10)$. Solve for k from either $1.35 - k(0.10) = 1.15$ or $1.35 + k(0.10) = 1.55$. Thus $k = 2$. At least 75%.

3.12 a. At least 55.6%.
b. $27 \pm 1.5(8)$ or $(15, 39)$.
c. $27 \pm k8$ or $(3, 51)$. Thus $k = 3$. At least 88.9%.

3.13 $\overline{X} = 7, s = 5.70.$
$\frac{X-\overline{X}}{s}: -0.877, -0.702, -0.526, 0.702, 1.404.$

3.14 $X - 1135: 21, 28, -22, -35.\ s^2 = 1{,}020.66.$

3.15 $\mathrm{GM} = (5 \cdot 7 \cdot 10 \cdot 11 \cdot 20)^{\frac{1}{5}} = 9.49 < \overline{X}.$

3.16 a. Distribution B since $s_B > s_A$.
b. Distribution A since $v_A = 21.74\% > v_B = 11.62\%$.
c. Distribution C since $v_C > v_B > v_A$.

3.17 a. $sk = 3(8-9)/4.472 = -0.67$ (skewed left).
b. $sk = 3(8-8)/4 = 0$(approximately symmetric).
c. $sk = 3(0-0)/4 = 0$(approximately symmetric).
d. $sk = 3(14-15)/8 = -0.38$(skewed left).

3.18 a. $Q_1 = 2.56, Q_2 = 3.98, Q_3 = 5.39$.
b. $D_3 = 2.95, D_6 = 4.15, D_9 = 6.27$.
c. q.d. $= 1.42$.
d. $P_{10} = 1.22, P_{65} = 4.52, P_{90} = 6.27$.
e. $X = 2.34$ at the 15th percentile; $X = 5.20$ at the 70th percentile.

3.19 a. $Q_1 = 13.5, Q_2 = 28.5, Q_3 = 42.75$.
b. $D_4 = 21.8, D_7 = 41.7, D_8 = 50.2$.
c. q.d. $= 14.63$.
d. $P_{10} = 7.1, P_{35} = 20.35, P_{90} = 87.2$.
e. $X = 13$ at the 20th percentile; $X = 41$ at the 65th percentile.

3.20 $k = 0.282$ (has a coefficient of kurtosis similar to that of a normal distribution). Are you comfortable with this result?

3.21 $k = 0.182$ (has a coefficient of kurtosis greater than that for a flat curve but less than that for a normal distribution). Are you comfortable with this result?

3.22 $k_4 - 3 = -0.853$ (has a peak that is flatter than that of a normal distribution).

3.23 $l = -1.685, u = 9.635$ [no outliers according to Equation (3.20)]. Using (3.20.1), the lower and upper fences are respectively -2.19 and 10.15 (no outliers).

3.24 $l = -30.38, u = 86.63$ [89 and 92 are outliers according to Equation (3.20)]. Using (3.20.1), the lower and upper fences are respectively -35.27 and 92.27 (no outliers).

3.25 $med = 8, mad = 3$. Using (3.19.1),

$$\frac{|1-18|}{4.45} = 3.82, \frac{|28-18|}{4.45} = 2.25, \frac{|40-18|}{4.45} = 4.94.$$

Hence 1, 28, and 40 are deemed outliers.

3.26 $\alpha = 0.20, \alpha n = 1.8 \approx 2$. We thus delete 1, 14, 28, and 40. The resulting α-trimmed mean is 17.8. The untrimmed mean is $\overline{X} = 19.11$.

3.27 a. $\min X_i = 0.99, Q_1 = 2.56, Q_2 = 3.98, Q_3 = 5.39, \max X_i = 7.75,$
$l = -1.685, u = 9.635.$

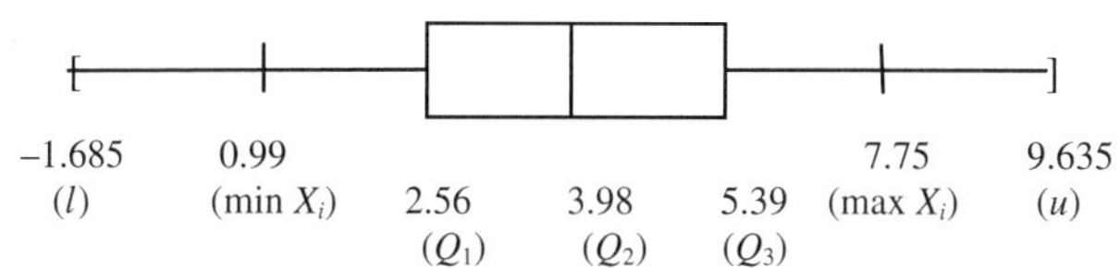

b. $\min X_i = 1, Q_1 = 13.5, Q_2 = 28.5, Q_3 = 42.75, \max X_i = 92,$
$l = -30.38, u = 86.63.$

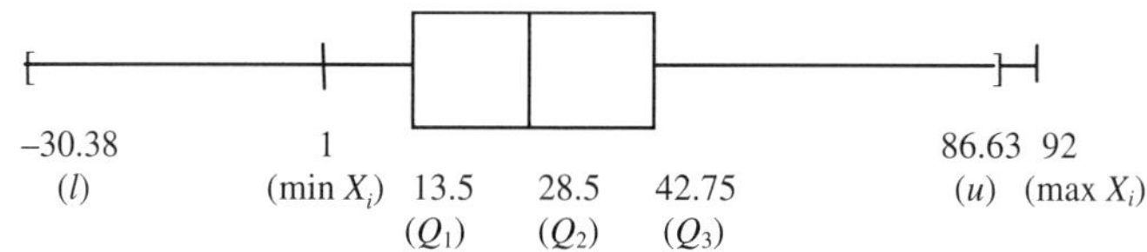

3.28 $c = 6, \overline{X} = 40.38$, median $= 40.97$, mode $= 42.37, s = 10.35, Q_1 = 33.25,$
$Q_3 = 41.29, D_3 = 34.93, D_6 = 43.5, P_{10} = 26.10, P_{33} = 35.89.$

4.1 1.2, −0.7, −1.4.

4.2 a. $A \cap B = \varnothing$. b. $C \cup D = S, C \cap D = \varnothing$.

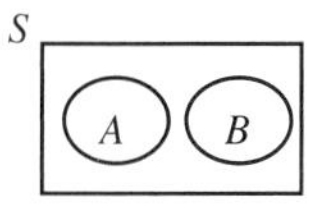

S

C | D

$\bar{C} = D, \bar{D} = C$

4.3 a. $P(A \cup B) = P(A) + P(B) = 0.70$.
b. $P(A \cup B) = P(A) + P(B) - P(A \cap B) = 0.55$.
c. $1 - P(B) = 0.55$.
d. $P(A \cap B) = P(A) + P(B) - P(A \cup B) = 0.05$.

4.4 $P(A) = P(A \cup B) - P(B) + P(A \cap B) = 0.47$.

4.5 a. $P(A) = 3/8$. b. $P(B) = 1/2$. c. $P(C) = 1/2$.
d. $P(D) = 1/2$. e. $P(A \cap B) = 1/4$.

f. $P(A \cup B) = P(A) + P(B) - P(A \cap B) = 5/8$. g. 1.
h. $P(C \cap D) = 0$. i. $1 - P(A) = 5/8$. j. $P(C \cup D) = P(C) + P(D) = 1$.
k. $1 - P(C) = 1/2$. l. $P(\overline{A} \cap \overline{B}) = P(\overline{A \cup B}) = 1 - P(A \cup B) = 3/8$.
m. $P(\overline{A} \cup \overline{B}) = P(\overline{A \cap B}) = 1 - P(A \cap B) = 3/4$.

4.6 a. $13/20$. b. $1/2$. c. $13/20$. d. $13/20$.

4.7 $25/52$.

4.8 $8/52$.

4.9 $(1/2)^5 = 1/32$.

4.10 $P(A \cap B) = 1 - P(5\text{ heads}) - P(5\text{ tails}) = 1 - \frac{1}{32} - \frac{1}{32} = \frac{30}{32}$.

4.11 $0.4/0.6 = 0.66$.

4.12 0.28.

4.13 No.

4.14 No. A and B are mutually exclusive.

4.15 $P(M_1 \cap M_2) = P(M_1) \cdot P(M_2|M_1) = 2/7$.

4.16 a. 1/2. b. 1/2. c. 1/4. d. 1/4.

4.17 a. 0.06. b. $P(\overline{A} \cap \overline{B}) = (1 - P(A))(1 - P(B)) = 0.56$. c. 0.44.

4.18 a. 0.09. b. $P(W_1 \cap W_2) = P(W_1) \cdot P(W_2|W_1) = 0.066$.

4.19 a. 0.3. e. 0.14.
b. 0.7. f. 0.66.
c. 0.1. g. 0.2.
d. 0.9. h. 0.6.

4.20 $P(A) = 3/9, P(B) = 2/9, P(C) = 4/9$.

4.21 a. $P(B|D) = 110/405 = 0.27$.
b. $P(D|B) = 110/1510 = 0.07$.
c. $P(A|D) = 45/405 = 0.11$.
d. $P(B|\overline{D}) = 1400/9595 = 0.15$.

4.22

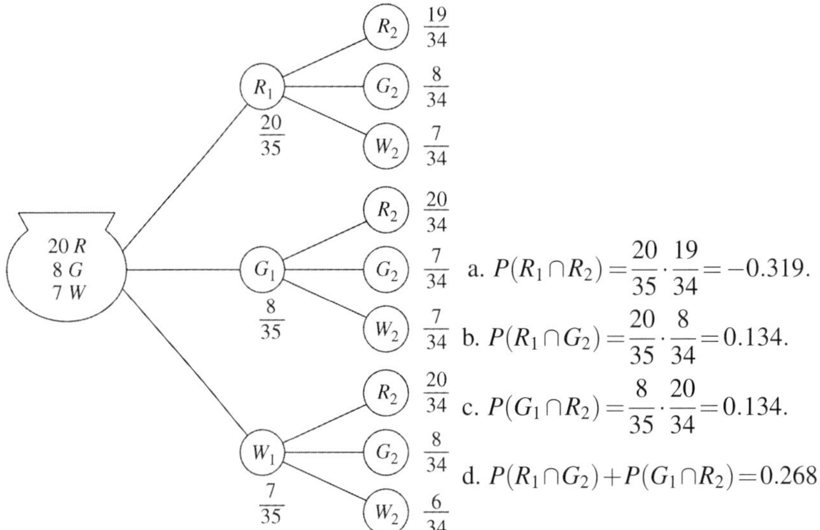

a. $P(R_1 \cap R_2) = \dfrac{20}{35} \cdot \dfrac{19}{34} = -0.319.$

b. $P(R_1 \cap G_2) = \dfrac{20}{35} \cdot \dfrac{8}{34} = 0.134.$

c. $P(G_1 \cap R_2) = \dfrac{8}{35} \cdot \dfrac{20}{34} = 0.134.$

d. $P(R_1 \cap G_2) + P(G_1 \cap R_2) = 0.268.$

4.23 a. 0.4. b. 0.8. c. 0.70. d. 4/7. e. 0.25.

No $\left(\text{e.g., } P(Y \cap M) \neq P(Y) \cdot P(M)\right)$.

4.24 $P(B_1 \cap B_2 \cap B_3) = 1/30.$
$P(B_1 \cap R_2 \cap R_3) = 1/6.$

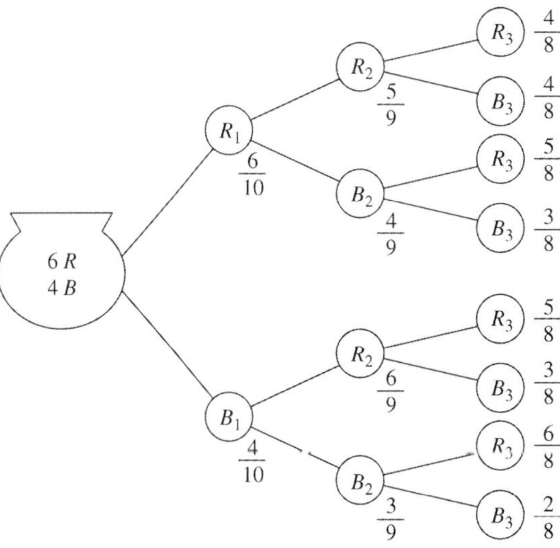

4.25 a. Yes. e. Yes.
b. No. f. No.
c. No. g. No.
d. Yes. No. No. h. Independent.

4.26 a. 0. e. 191/295.
b. 175/295. f. 22/130.
c. 10/295. g. 57/165.
d. 43/295. h. No (e.g., $P(S \cap Y) \neq P(S) \cdot P(Y)$)

4.27 a. 3/5. b. 2/5. c. 4:1. d. 1:4.

5.1 All are discrete.

5.2

Candidate	Probability
Mary Smith	100/300
William Jones	50/300
Charles Pierce	40/300
Harold Simms	110/300
	1

Random Sampling.

5.3 a. Distributions (a) and (b).

5.4 a. 0.4. b. 0.5. c. 0.9. d. 0.6.

5.5

X	1	2	3	4	5
Probability	0.07	0.13	0.05	0.51	0.24

a. 0.75. b. 0.20. c. 3.72.

5.6 $E(X) = 2.00.$

$$V(X) = (0-2)^2(0.15) + (1-2)^2(0.15) + (2-2)^2(0.35) + (3-2)^2(0.25) + (4-2)^2 0.10 = 1.4.$$

$$V(X) = (0)^2(0.15) + (1)^2(0.15) + (2)^2(0.35) + (3)^2(0.25) + (4)^2(0.10) - (2)^2 = 1.4.$$

$S(X) = 1.183.$

5.7 $E(X) = 1.$
$V(X) = 1.$
$S(X) = 1.$

5.8 You made a mistake. The random variable X is constant at a single value having probability 1.

5.9 15, 500.

5.10 Outcome points are: (H, H, H), (H, H, T), (H, T, T) (T, T, T), (H, T, H), (T, H, T), (T, H, H), (T, T, H).

X	$f(X)$
0	1/8
1	3/8
2	3/8
3	1/8
	1

$E(X) = 1.5$, $V(X) = 0.75$, $S(X) = 0.866$.

5.11 $E(X) = 2.53$. The number of children per 100 married couples after ten years of marriage is 253.

5.12

X ($ amount)	$f(X)$ (probability)
10,000	1/100,000 (=0.00001)
5,000	3/100,000 (=0.00003)
100	1000/100,000 (=0.01)
3	10,000/100,000 (=0.10)
0	88,966/100,000 (=0.88966)
	1

$E(X) = \$1.50$. If you play this lottery game again and again (week after week) for a large number of times, then, in the long run, your average winnings will be $1.50. Expected long-run loss for a single lottery ticket is thus $1.50.

5.13 a. 210. b. 720. c. 56. d. 2184 e. 35 f. 270,725. g. 6. h. 15.

5.14 ${}_7P_3 = 210$.

5.15 ${}_{10}C_5 = 252$.

5.16 a. Binomial. b. Binomial. c. Binomial. d. No (height is continuous). e. No (mileage is continuous). f. No (must sample with replacement). g. No (number of trials must be fixed).

5.17 a. 0.2786. c. 0.2184.
b. 0.2150. d. 0.0123.

5.18 See Table A.5.

5.19 a. 0.0000. d. 0.9993.
b. 0.0030. e. 1 − P(none) = 1 − 0.0000 = 1.
c. 0.9382.

5.20 a. $E(X) = 2.80, V(X) = 1.82$. c. $E(X) = 12.75, V(X) = 1.91$.
b. $E(X) = 4.00, V(X) = 2.40$. d. $E(X) = 12.00, V(X) = 1.80$.

5.21 a. 0.2225. c. 0.8883.
b. 0.8655. d. 0.5186.

5.22 a. 0.0596. c. 0.6389.
b. 0.9841. d. 0.0000.

6.1 All continuous.

6.2 a. 0. b. 1/3. c. 1/3.

6.3 a. 1/4. b. 3.4. c. 1. d. 1/2.

6.4 a. 2.5. b. 0.6. c. 0.2.

6.5 $Z = (X - 50)/10$. Z is $N(0, 1)$.

6.6 a. −0.67. b. 0.67. c. 0.495.

6.7 a. 0.0075. d. 0.5.
b. 0.3409. e. 0.9995.
c. 0.9082.

6.8 a. 0.9997. d. 0.5.
b. 0.7088. e. 0.0004.
c. 0.0135.

6.9 a. 0.9782. d. 0.0639.
b. 0.1950. e. 0.5.
c. 0.2116.

6.10 a. −0.84. d. −0.39.
b. 0.68. e. 1.04.
c. 0.

6.11 a. −1.04, 1.04. b. −2.58, 2.58. c. −1.28, 1.28.

6.12 68%, 95%, and 99% respectively.

6.13 a. 2.50. d. −0.25.
b. 0.15. e. 1.53.
c. −1.04.

6.14 a. 0.0161. d. 0.9074.
b. 0.2375. e. 0.2354.
c. 0.

6.15 0.5328.

6.16 a. 1.96. b. 1.645. c. 2.58.

6.17 71.

6.18 10.

6.19 a. 0.1056. b. 0.0301. c. 0.2342. d. 0.0003.

6.20 a. 19.76 b. 40.24. c. 36.72 d. 30.

6.21 a. 0.8413. b. 0.5789. c. 344 hr. d. 285 hr to 363 hr.

6.22 a. 0.1112. b. 0.3889. c. 40.50. d. $23.12 to $36.88.

6.23 56.

6.24 Your Z-score is 2.07. Thus only about 1.9% did better. (Wow!)

6.25 $np(1-p) = 16 > 10$.

a. $P(X \leq 15) = P(0 \leq X \leq 15)$. The binomial variable does not have an infinite range (it has an upper bound of n and a lower bound of 0). Then $P(-0.5 \leq X \leq 15.5) = 0.1293$.

b. $P(18 \leq X \leq 22) - P(1.75 \leq X \leq 22.5) = 0.4713$.

c. $P(X \geq 10) = P(10 \leq X \leq 100) = P(9.5 \leq X \leq 100.5) = 0.9959$.

d. $P(X = 20) = P(19.5 \leq X \leq 20.5) = 0.1034$.

6.26 a. 6. b. 2.39. c. 0.1174.
d. $P(X \geq 4) = P(4 \leq X \leq 100) = P(3.5 \leq X \leq 100.5) = 0.8531$.
e. $P(6 \leq X \leq 10) = P(5.5 \leq X \leq 10.5) = 0.5538$.

7.1 ${}_5C_2 = 10$ possible samples.

$\overline{X}$	$f(\overline{X})$
3	1/10
4	1/10
5	2/10
6	2/10
7	2/10
8	1/10
9	1/10
	1

a. $\mu = 6, \sigma = \sqrt{8}$.

b. $E(\overline{X}) = 6, V(\overline{X}) = \frac{8}{3}\frac{3}{4} = 3$

c. $\sqrt{V(\overline{X})} = \sigma_{\overline{X}} = \sqrt{3}$.

d. $(-3)\frac{1}{10} + (-2)\frac{1}{10} + \cdots + (3)\frac{1}{10} = 0$.

7.2 $n = 25, P\left(-4 < \overline{X} - \mu < 4\right) = P\left(\frac{-4\sqrt{n}}{\sigma} < \overline{Z} < \frac{4\sqrt{n}}{\sigma}\right) = P\left(-2 < \overline{Z} < 2\right)$
$= 0.9545$.

$n = 50, P\left(-2.82 < \overline{Z} < 2.82\right) = 0.9952$.

$n = 100, P\left(-4 < \overline{Z} < 4\right) = 0.9999$.

7.3 $\sigma = 10, P\left(-2.5 < \overline{Z} < 2.5\right) = 0.9876$.

$\sigma = 13, P\left(-1.92 < \overline{Z} < 1.92\right) = 0.9452$.

$\sigma = 16, P\left(-1.56 < \overline{Z} < 1.56\right) = 0.8788$.

7.4 ${}_6C_3 = 20$ possible samples.

$\overline{X}$	$f(\overline{X})$
2	1/20
7/3	1/20
8/3	2/20
3	3/20
10/3	3/20
11/3	3/20
4	3/20
13/3	2/20
14/3	1/20
5	1/20
	1

$E(\overline{X}) = \mu = 3.5$.

$V(\overline{X}) = \frac{\sigma^2}{n}\frac{N-n}{N-1} = \frac{2.916}{3}\frac{6-3}{6-1} = 0.583$.

$\sigma_{\overline{X}} = \sqrt{V(\overline{X})} = 0.764$.

7.5 $E(\overline{X}) = 60, V(\overline{X}) = \sigma^2/n = 9.77.$

7.6 $\overline{X}$ is $N(\mu, \sigma/\sqrt{n}) = N(25, 8/\sqrt{10}).$

7.7 No, provided n is "large enough" (via the Central Limit Theorem). Since $n = 64$ is "large enough,"$\overline{X}$ is $N(\mu, \sigma/\sqrt{n}) = N(60, 8/\sqrt{40}).$

7.8 The population must be normal. $\overline{X}$ is $N(64, 16/\sqrt{15}).$

a. 0.3156. b. 0.0003. c. 0.7580. d. 0.9545.

7.9 $\overline{X}$ is $N(32, 4/\sqrt{n}).$

a. 0.3086. b. 0.2148. c. 0.0268.

7.10 $P\left(-a < \overline{X} - \mu < a\right) = P\left(\frac{-a}{0.04} < \overline{Z} < \frac{a}{0.04}\right) = 0.95.$-
$a/0.4 = 1.96, a = 0.08.$ Interval length is 0.16.

7.11 $P\left(-5 < \overline{X} - \mu < 5\right) = P\left(\frac{-5}{13/\sqrt{n}} < \overline{Z} < \frac{5}{13/\sqrt{n}}\right) = 0.95.$
$5\sqrt{n}/13 = 1.96, n = 26.$

7.12

Index j	X_j	P_j	Expected X_{Pj}
1	33.29	0.10	−1.28
2	33.35	0.26	−0.64
3	33.90	0.42	−0.20
4	34.16	0.58	0.20
5	34.41	0.74	0.64
6	34.55	0.90	1.28

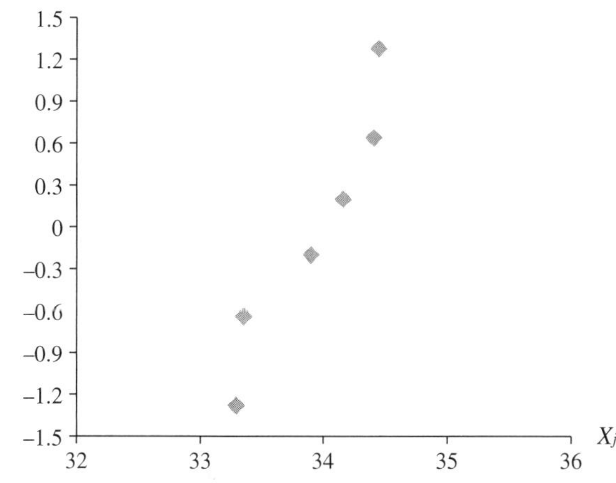

We get a good approximation to a straight line—we can assume that the population is close to normal.

8.1 a. (11.69, 18.31). b. (12.22, 17.78). c. (10.64, 19.36).

8.2 a. 0.59. b. 62. c. 2,032.5.

8.3 a. (18.04, 21.96). b. (46.84, 57.15). c. (113.42, 126.58).

8.4 912.

8.5 458.

8.6 a. 1.316. b. 1.697. c. −2.552. d. −1.943. e. 2.205. f. 1.309.

8.7 a. 2.771. b. 2.145. c. 1.684.

8.8 a. (−0.612, 8.612). b. (2.154, 5.846). c. (17.869, 22.131). d. (15.872, 24.128).

8.9 a. (3.567, 12.433). b. (0.708, 7.292).

8.10 (98.416, 121.584).

8.11 (37.8, 42.2).

8.12 (1.936, 6.064). Yes, a mean difference of 0 (no effect) is not included in the interval.

8.13 (0.543, 6.823) in hundreds of \$.

8.14 $\overline{X} = 23.88$. med $= X_{(26/2)} = X_{(13)} = 13$. $k = 7.1$. $X_{(k)} = X_{(7)} = 4$ and $X_{(n-k+1)} = X_{(19)} = 20$.

9.1 a. $\hat{p} = 0.40, S_{\hat{p}} = 0.098$. b. $\hat{p} = 0.50, S_{\hat{p}} = 0.063$. c. $\hat{p} = 0.64, S_{\hat{p}} = 0.048$. d. $\hat{p} = 0.25, S_{\hat{p}} = 0.021$.

9.2 $\hat{p} = 0.672, \quad S_{\hat{p}} = 0.014$.

9.3 For large n, if $np(1-p) > 10$, then $U = \dfrac{\hat{p} - p}{\sigma_{\hat{p}}}$ is approximatel $N(0,1)$.

a. $U = (\hat{p} - 0.07)/0.015$ is approximately $N(0,1)$.

b. $U = (\hat{p} - 0.04)/0.024$ is approximately $N(0,1)$.

c. $U = (\hat{p} - 0.10)/0.009$ is approximately $N(0,1)$.
d. $U = (\hat{p} - 0.75)/0.043$ is approximately $N(0,1)$.

9.4 If $np(1-p) > 10$, $U = (\hat{p} - p)/\sqrt{p(1-p)/n} = (\hat{p} - 0.8)/0.046$ is approximately $N(0,1)$.

a. $P(X \geq 65) = P(\hat{p} \geq 0.866) = P(U \geq 1.43) = 0.0764$.
b. $P(X < 25) = P(\hat{p} < 0.333) = P(U < -10.152) = 0$.
c. $P(35 \leq X \leq 60) = P(0.466 \leq \hat{p} \leq 0.80) = P(-7.26 \leq U \leq 0) = 0.5$.
d. $P(\hat{p} < 0.24) = P(U < -12.17) = 0$.
e. $P(\hat{p} > 0.63) = P(U > -3.69) = 0.9999$.

9.5 If $np(1-p) > 10, U = (\hat{p} - 0.07)/0.0114$ is approximately $N(0,1)$.
a. $P(\hat{p} > 0.10) = P(U > 2.63) = 0.0043$.
b. $P(X < 50) = P(\hat{p} < 0.10) = P(U < 2.63) = 0.9957$.

9.6 Since $n/N = 0.10$, we use $\sigma_{\hat{p}} = 0.0114\sqrt{(N-n)/(N-1)} = 0.0114$ $(0.9487) = 0.0108$.

a. $P(\hat{p} > 0.10) = P(U > 2.77) = 0.0028$.
b. $P(X < 50) = P(\hat{p} < 0.10) = P(U < 2.77) = 0.9972$.

9.7 (0.262, 0.318).

9.8 (0.097, 0.302).

9.9 (0.041, 0.071).

9.10 a. 971. b. 1,067.

9.11 3,959.

10.1 a. Right-tailed. b. Two-tailed. c. Left-tailed.
d. Invalid—equality must be a part of H_0.

10.2 a. $H_0 : p = 0.15, H_1 : p > 0.15$.
b. $H_0 : \mu = \$200, H_1 : \mu \neq \$200$.
c. $H_0 : \mu = \$2.50, H_1 : \mu > \$2.50$.
d. $H_0 : \mu = 85, H_1 : \mu < 85$.
e. $H_0 : p = 0.35, H_1 : p > 0.35$.

10.3 $R = \{\overline{Z} | \overline{Z} < -\overline{Z}_{0.05} = -1.645\}$(left-tail). $\overline{Z}_0 = -1.02$. Do not reject H_0. p-value $= 0.1539$.

10.4 $R = \{\bar{Z} | \bar{Z} > \bar{Z}_{0.01} = 2.33\}$(right-tail). $\bar{Z}_0 = 3.33$. Reject H_0. p-value $= 0.0004$.

10.5 $R = \{\bar{Z} | |\bar{Z}| > \bar{Z}_{0.05} = 2.58\}$(two-tail). $|\bar{Z}_0| = 3.423$. Reject H_0. p-value $= 0.0006$.

10.6 $H_0 : \mu = 565, H_1 : \mu \neq 565$. $R = \{\bar{Z} | |\bar{Z}| > Z_{0.025} = 1.96\}$. $|\bar{Z}_0| = 1.126$. Do not reject H_0. p-value $= 0.2076$.

10.7 a. $t_o = -2.5, -t_{0.01,24} = -2.492$. Reject H_0. $0.005 < p\text{-value} < 0.01$.

b. $|t_o| = 13.33, t_{0.25,24} = 2.064$. Reject H_0. $p\text{-value} < 0.0005$.

c. $t_o = 5.4, t_{0.10,8} = 1.397$. Reject H_0. $p\text{-value} < 0.001$.

d. $|t_o| = 55, t_{0.025,15} = 2.131$. Reject H_0. $p\text{-value} < 0.0005$.

e. $|t_o| = 1.5, t_{0.05,8} = 1.86$. Do not reject H_0. $0.05 < p\text{-value} < 0.10$.

f. $t_o = -1.782, -t_{0.05,24} = -1.711$. Reject H_0. $0.025 < p\text{-value} < 0.05$.

10.8 $H_0 : \mu = 180, H_1 : \mu \neq 180$. $t_o = 1.597, t_{0.005,11} = 3.106$. Do not reject H_0. $0.05 < p\text{-value} < 0.10$.

10.9 $H_0 : \mu \leq 6, H_1 : \mu > 6$. $t_o = 4.203, t_{0.05,9} = 1.833$. Reject H_0. $p\text{-value} < 0.001$.

10.10 $H_0 : \mu = 33.5, H_1 : \mu > 33.5$. $t_o = 8.52, t_{0.05,34} = 1.691$. Reject H_0. $p\text{-value} < 0.0005$.

10.11 a. $U_o = 0.375, U_{0.05} = 1.645$. Do not reject H_0. p-value $= 0.233$.
b. $U_o = -0.833, -U_{0.10} = -1.28$. Do not reject H_0. p-value $= 0.203$.
c. $|U_o| = 6.84, U_{0.025} = 1.96$. Reject H_0. $p\text{-value} < 0.005$.
d. $U_o = -4.32, -U_{0.01} = -2.33$. Reject H_0. $p\text{-value} < 0.005$.
e. $|U_o| = 1.27, U_{0.025} = 1.96$. Do not reject H_0. p-value $= 0.204$.

10.12 $H_0 : p \geqslant 0.90, H_1 : p < 0.90$. $U_o = -2.0, -U_{0.05} = -1.645$. Reject H_0. p-value $= 0.023$.

10.13 $H_0 : p = 0.10, H_1 : p > 0.10$. $U_o = 0.6053, U_{0.05} = 1.645$. Do not reject H_0. p-value $= 0.2709$.

10.14 $H_0 : p = 0.50, H_1 : p \neq 0.50$. $|U_o| = 2.997, U_{0.025} = 1.96$. Reject H_0. p-value $= 0.0014$.

10.15 $H_0 : p = 0.94, H_1 : p. > 0.94$. $U_o = 0.63, U_{0.01} = 2.33$. Do not reject H_0. p-value $= 0.2643$.

10.16 $R = 19, n_1 = 19, n_2 = 19. |\overline{Z}_R| = 0.233 < 1.96 = Z_{0.025}$. We cannot reject H_0: the order of the sample data is random.

10.17 $W^+ = 65, m = 24, |Z_{W^+}| = 3.156 > Z_{0.025} = 1.96$. Reject H_0: $MED = 15$.

10.18 $\overline{X} = 1522, s = 250.95, D = 0.2090, R = \{D|D > 0.262\}$. We cannot reject H_0: the random sample is from a normal CDF. p-value > 0.20.

11.1 $5 \pm 2.58(2.07)$ or $(-0.34, 10.34)$.

11.2 $0.32 \pm 1.96(0.0208)$ or $(0.28, 0.36)$.

11.3 $2,250 \pm 2.024(249.978)$ or $(1,744.04, 2,755.96)$.

11.4 $\overline{X} = 10.4, \overline{Y} = 9.0, s_p = 1.549$. $1.40 \pm 2.074(0.6324)$ or $(0.089, 2.711)$

11.5 $t_{\alpha/2,\phi} = t_{0.025,28} = 2.048, -0.64 \pm 2.048(0.3817)$ or $(-1.422, 0.142)$.

11.6 $t_{\alpha/2,\phi} = t_{0.025,22} = 2.074, 1.40 \pm 2.074(0.5003)$ or $(0.3624, 2.4376)$.

11.7 $t_{\alpha/2,\phi} = t_{0.025,112} = 1.984, 6.72 \pm 1.984(2.3514)$ or $(2.06, 11.39)$.

11.8
a. Independent. b. Independent. c. Dependent.
d. Dependent. e. Dependent.

11.9 $0.258 \pm 2.201(0.1024)$ or $(0.0326, 0.4834)$.

11.10 $0.30 \pm 2.571(0.1367)$ or $(-0.0516, 0.6516)$.

11.11 $0.625 \pm 3.499(0.4596)$ or $(-0.983, 2.233)$.

11.12 $0.04 \pm 1.96(0.05)$ or $(-0.058, 0.138)$.

11.13 $-0.10 \pm 2.58(0.0375)$ or $(-0.1968, -0.0032)$.

11.14 $0.0148 \pm 1.645(0.0071)$ or $(0.0032, 0.0264)$.

11.15 $Z'_{\delta_0} = 2.42 > 1.645 = Z_{0.05}$. Reject H_0. p-value $= 0.0078$.

11.16 $H_0 : \mu_X - \mu_Y = 0, H_1 : \mu_X - \mu_Y > 0$. $Z'_{\delta_0} = 15.38 > 1.28 = Z_{0.10}$. Reject H_0. p-value < 0.0001.

11.17 $Z''_{\delta_0} = 1.0001 < 1.686 = t_{0.05,38}$. Do not reject H_0. $0.15 < p$-value < 0.20.

11.18 $Z''_{\delta_0} = 2.214 > 2.074 = t_{0.025,22}$. Reject H_0. $0.02 < p$-value < 0.04.

11.19 $T'_{\delta_0} = -1.677 > -1.701 = t_{0.05,28}$. Do not reject H_0. $0.05 < p$-value < 0.10.

11.20 $T'_{\delta_0} = -0.119 > -1.660 = t_{0.05,112}$. Do not reject H_0. p-value > 0.50.

11.21 $T = 2.517 > 1.796 = t_{0.05,11}$. Reject H_0. $0.01 < p$-value < 0.025.

11.22 $T = 2.19 < 4.032 = t_{0.005,5}$. Do not reject H_0. $0.05 < p$-value < 0.10.

11.23 $H_0 : \mu_D < 0, H_1 : \mu_D > 0$. $T = 1.36 < 1.895 = t_{0.05,7}$. Do not reject H_0. p-value > 0.10.

11.24 $Z_{\delta_0} = 0.80 < 1.645 = Z_{0.05}$. Do not reject H_0. p-value > 0.22.

11.25 $Z_{\delta_0} = -2.66 < -2.33 = Z_{0.01}$. Reject H_0. p-value < 0.004.

11.26 $Z_{\delta_0} = 2.085 > 1.96 = Z_{0.025}$. Reject H_0. p-value < 0.018.

11.27 $R = 19, E(R) = 21, \sqrt{V(R)} = 3.1215, Z_R = -0.64$. $-Z_{0.05} = -1.645$. Do not reject H_0: population distributions are identical.

11.28 $R_1 = 337.5, U = 272.5, E(U) = 200, \sqrt{V(U)} = 36.96, |Z_U| = 1.962 > 1.96 = Z_{0.025}$. Too close for comfort—reserve judgement.

11.29 $m = 25, W^+ = 129, Z_{W^+} = -0.37$. $|Z_{W^+}| = 0.37 < 1.96 = Z_{0.025}$. Do not reject H_0 : population distributions are identical.

12.1 a. $\hat{\beta}_0 = -0.80, \hat{\beta}_1 = 0.90$.
b. $\hat{Y}_0 = 3.7$.

12.2 a. $\hat{\beta}_0 = -2.464, \hat{\beta}_1 = 1.892$.
b. $\hat{Y}_0 = 22.132$.

12.3 a. $\hat{\beta}_0 = -32.892, \hat{\beta}_1 = 2.107$.
b. $\hat{Y}_0 = 9.248$.

12.4

SS	d.f.	MS
SSR = 32.40	1	MSR = 32.40
SSE = 8.80	3	MSE = 2.93
SST = 41.20	4	$r^2 = 0.7864$

12.5

SS	d.f.	MS
SSR = 100.32	1	MSR = 100.32
SSE = 1.678	5	MSE = 0.336
SST= 102.00	6	$r^2 = 0.9835$

12.6

SS	d.f.	MS
SSR = 124.32	1	MSR = 124.32
SSE = 9.107	5	MSE = 1.82
SST= 133.429	6	$r^2 = 0.9317$

12.7 a. $\hat{\beta}_0 = 2.5, \hat{\beta}_1 = -3$.

b.

SS	d.f.	MS
SSR = 90.00	1	MSR = 90.00
SSE = 70.00	28	MSE = 2.50
SST= 160.00	29	$r^2 = 0.5625$

12.8 a. $\sum x_i^2 = 400, \sum y_i^2 = 128, \sum x_i y_i = -120, \hat{\beta}_0 = 2.90, \hat{\beta}_1 = -0.30$.

b.

SS	d.f.	MS
SSR = 36	1	MSR = 36.00
SSE = 92	23	MSE = 4.00
SST= 128	24	$r^2 = 0.2813$

12.9 a. β_0 lies within (1.7176, 3.2823); β_1 lies within (−4.0231, −1.9769).
b. $|t| = 6.54, R = \{t \| t| > 2.048\}$. Reject H_0 (p-value < 0.01).
c. $|t| = 6.00, R = \{t \| t| > 2.048\}$. Reject H_0 (p-value < 0.01).
d. $|t| = 1.40, R = \{t \| t| > 2.048\}$. Do not reject H_0 ($0.10 < p$-value < 0.20).

12.10 a. β_0 lies within (1.4965, 4.3035); β_1 lies within (−0.5807, −0.0193).

b. $t = 5.80, R = \{t | t > 2.50\}$. Reject H_0 (p-value < 0.005).

c. $t = -3, R = \{t | t > -2.50\}$. Reject H_0 (p-value < 0.005).

d. $|t| = 0.5, R = \{t \| t| > 2.807\}$. Do not reject H_0 (p-value < 0.20).

12.11 a. (3.4146, 9.3854).
b. (0.1854, 12.6146).

12.12 a. (8.2635, 9.5223).
b. (7.2759, 10.5098).

12.13 a. $\hat{\beta}_0 = 4.00, \hat{\beta}_1 = 4.00$.
b. $\hat{Y}_o = 28$.

c. (20.9131, 35.0869).

d. (8.5692, 47.4305).

12.14 a. $r = 0.8868$. X and Y follow a joint bivariate distribution.

b. $T = 3.32$. $R = \{T||T| > 3.182\}$. Reject H_0 ($0.02 < p$-value < 0.05). X and Y follow a joint bivariate normal distribution.

12.15 a. $r = 0.9917$. X and Y follow a joint bivariate distribution.

b. $T = 17.25$. $R = \{T|T > 3.365\}$. Reject H_0 (p-value < 0.005). X and Y follow a joint bivariate normal distribution.

12.16 a. $r = 0.80$. X and Y follow a joint bivariate distribution.

b. $T = 4.00$. $R = \{T|T > 2.821\}$. Reject H_0 (p-value < 0.005). X and Y follow a joint bivariate normal distribution.

13.1 a. $\chi^2_{0.95,12} = 21.03$.

b. $\chi^2_{0.99,5} = 15.09$.

c. $\chi^2_{0.99,20} = 37.51$.

d. $\chi^2_{0.95,8} = 15.51$.

13.2 If the ratio is 8:1, one out of every nine students is an athlete. If the ratio has not changed, the proportion of athletes is $p = \frac{1}{9}$. Test $H_1 : p = p_o = \frac{1}{9}$ versus $H_1 : p \neq \frac{1}{9}$. $X = 68, n - X = 384, np_o = 50, n(1 - p_o) = 400$. Then from (13.2.1), $U_o = 7.29$. Since $R = \{U|U > \chi^2_{0.95,1} = 3.84\}$, we reject H_0.

13.3 $H_0 : p_1 = p_1^o = \dfrac{9}{16}, p_2 = p_2^o = \dfrac{3}{16}, p_3 = p_3^o = \dfrac{3}{16}, p_4 = p_4^o = \dfrac{1}{16}; H_1$: not H_0.

13.4 $H_0 : p_1 = \dfrac{1}{2}, p_2 = \dfrac{1}{4}, p_3 = \dfrac{1}{4}; H_1$: at least one of these equalities does not hold. $R = \{U|U > \chi^2_{0.99,2} = 9.21\}$. $o_1 = 101, o_2 = 60, o_3 = 89; e_1 = np_1 = 125,$

$e_2 = np_2 = 62.5, e_3 = np_3 = 62.5$. From (13.5), $U_o = 15.94$. Reject H_0.

13.5 H_0: number of calls is uniformly distributed; H_1: not H_0.

$R = \left\{U|U > \chi^2_{0.95,3} = 7.81\right\}$. $o_1 = 57, o_2 = 65, o_3 = 60, o_4 = 66; n = 248,$

$p_i = \dfrac{1}{4}$ for all i so that $e_1 = np_i = 62$ for all i. $U_o = 0.871$. Do not reject H_0.

13.6 $\overline{X} = 65.1, s = 16.96$. H_0: X is normally distributed; H_1: not H_0.

$$\begin{aligned}
\hat{p}_1 &= P(-\infty < X \leq 39.5) = P(-\infty < Z < -1.509) = 0.0668,\\
\hat{p}_2 &= P(39.5 \leq X \leq 49.5) = P(-1.509 \leq Z \leq -0.919) = 0.1132,\\
\hat{p}_3 &= P(49.5 \leq X \leq 59.5) = P(-0.919 \leq Z \leq -0.330) = 0.1919,\\
\hat{p}_4 &= P(59.5 \leq X \leq 69.5) = P(-0.330 \leq Z \leq 0.259) = 0.2319,\\
\hat{p}_5 &= P(69.5 \leq X \leq 79.5) = P(0.259 \leq Z \leq 0.849) = 0.1997,\\
\hat{p}_6 &= P(79.5 \leq X \leq 89.5) = P(0.849 \leq Z \leq 1.438) = 0.1213,\\
\hat{p}_7 &= P(89.5 \leq X < +\infty) = P(1.438 \leq Z < +\infty) = 0.0764.
\end{aligned}$$

The expected frequencies for the first and last classes are $\hat{e}_1 = n\hat{p}_1 = 3.34$ and $\hat{e}_7 = n\hat{p}_7 = 3.82$ respectively. Since the expected frequency of each class must be at least 5, let us combine the first and second classes and the sixth and seventh classes to obtain the following revised distribution. Now

Classes of X	Observed Frequency o_i	$\hat{e}_i = n\hat{p}_i$
30–49	10	8.94
50–59	8	9.60
60–69	12	11.60
70–79	9	9.99
80–99	11	9.89
	50	50.00

$\hat{p}_1 = P(-\infty < X \le 49.5) = P(-\infty < Z \le -0.919) = 0.1788$,
$\hat{p}_2 = 0.1919, \hat{p}_3 = 0.2319$,
$\hat{p}_4 = 0.1997,\ \hat{p}_5 = P(79.5 \le X < +\infty) = P(0.849 \le Z < +\infty) = 0.1977$.
Then from (13.51),
$U_o = 0.629$. $R = \{U|U > \chi^2_{0.95,2} = 5.99)$, we cannot reject H_0.

13.7

		Degree		
Homecoming	B.A.	M.A.	Ph.D	Row Totals
Attended	47 (47.92)	42 (43.93)	46 (43.13)	135
Did not attend	13 (12.07)	13 (11.07)	8 (10.86)	34
Column totals	60	55	54	169

$H_0 : p_{ij} = (p_{i.})(p_{.j}); H_1 :$ not H_0. $R = \{U|U > \chi^2_{0.95,2} = 5.99\}$.

$$e_{11} = \frac{(60)(135)}{169} = 47.92, e_{12} = \frac{(55)(135)}{169} = 43.93,$$

$$e_{13} = \frac{(54)(135)}{169} = 43.13, \text{and so on}$$

From (13.14), $U_o = 1.45$. Do not reject H_0.

13.8

Fertilizer	Germinated	Did not germinate	
A	120	30	150
B	110	15	125
	230	45	275

$H_0 : p_{ij} = (p_{i.})(p_{.j}); H_1 :$ not H_0. $R = \left\{U \middle| U > \chi^2_{0.95,1} = 3.84\right\}$.
From (13.14.1),

$$U'_o = \frac{275\left(|(120)(15)-(110)(30)| - \frac{275}{2}\right)^2}{(230)(45)(150)(125)} = 2.63.$$

Do not reject H_0.

13.9

Job Classification	Degree Plan 1	Plan 2	Plan 3	Row Totals
Clerical	39 (36)	50 (43.2)	19 (28.8)	108
Labor	52 (50)	57 (60)	41 (40)	150
Manager	9 (14)	13 (16.8)	20 (11.2)	42
Column totals	100	120	80	300

$H_0: p_{ij} = (p_{i.})(p_{.j}); H_1:$ not H_0. $R = \{U | U > \chi^2_{0.99,4} = 13.28\}$.

$$e_{11} = \frac{(100)(108)}{300} = 36, e_{12} = \frac{(120)(108)}{300} = 43.2,$$

$$e_{13} = \frac{(80)(108)}{300} = 28.8, \text{ and so on}$$

From (13.14), $U_o = 14.47$. Reject H_0.

13.10

		Customer Satisfaction Poor	Good	Excellent	Row Totals
Days of Training	1	10 (5.92)	15 (14.21)	2 (6.86)	27
	2	7 (5.70)	10 (13.68)	9 (6.61)	26
	3	5 (6.14)	15 (14.73)	8 (7.12)	28
	4	3 (7.23)	20 (17.36)	10 (8.39)	33
Column totals		25	60	29	114

$H_0: p_{ij} = (p_{i.})(p_{.j}); H_1:$ not H_0. $R = \{U | U > \chi^2_{0.95,6} = 12.59\}$. $U_o = 11.93$. Do not reject H_0.

13.11

Ward	Candidate A	Candidate B	
1	65	40	$n_1 = 105$
2	80	80	$n_2 = 160$
3	40	40	$n_3 = 80$
4	50	90	$n_4 = 105$
	235	250	485

The wards have been polled independently and randomly.

$H_0 : p_1 = p_2 = p_3 = p_4$; H_1: not $H_{0.}$ $R = \{U|U > \chi^2_{0.95,3} = 7.81\}$.
$\hat{p} = 235/485 = 0.485, 1 - \hat{p} = 0.515$.

$\hat{e}_{11} = 105(0.485) = 50.93$ $\quad$ $\hat{e}_{12} = 105(0.515) = 54.08$
$\hat{e}_{21} = 160(0.485) = 77.60$ $\quad$ $\hat{e}_{22} = 160(0.515) = 82.40$
$\hat{e}_{31} = 80(0.485) = 38.80$ $\quad$ $\hat{e}_{32} = 80(0.515) = 41.20$
$\hat{e}_{41} = 140(0.485) = 67.90$ $\quad$ $\hat{e}_{42} = 140(0.515) = 72.10$

From (13.18), $U_{\hat{p}} = 16.92$. Reject H_0.

13.12

	In favor	Opposed	
Investment bankers	37	13	$n_1 = 50$
Environmentalists	10	40	$n_2 = 50$
High school teachers	25	25	$n_3 = 50$
Construction industry	40	10	$n_4 = 50$
Small business assn.	30	20	$n_5 = 50$
	142	108	250

$H_0 : p_1 = \cdots = p_5; H_1 :$ not H_0. $R = \{U|U > \chi^2_{0.49,3} = 13.28\}$.

$\hat{p} = 142/250 = 0.568, 1 - \hat{p} = 0.432$.

$\hat{e}_{i1} = 28.4, \hat{e}_{i2} = 21.6, i = 1, \ldots, 5$.

From (13.18), $U_{\hat{p}} = 57.1$. Reject H_0.

13.13 a. (Exercise 13.8) $\Phi^2 = 2.63/(275)(2) = 0.005$.
Strength of association is weak.
b. (Exercise 13.9) $\Phi^2 = 14.47/(300)(3) = 0.016$.
Strength of association is weak.
c. (Exercise 13.10) $\Phi^2 = 11.93/(114)(3) = 0.034$.
Strength of association is weak.

13.14 $s^2 = 9.39; \chi^2_{0.025,14} = 5.63, \chi^2_{0.975,14} = 26.12$. From (13.23), a 95% confidence interval is (5.03, 23.35).

13.15 $s^2 = 5.36; \chi^2_{0.005,19} = 6.84, \chi^2_{0.995,19} = 38.58$. From (13.23), a 99% confidence interval is (1.62, 3.85).

13.16 a. 5.06 $\quad$ b. 3.37 $\quad$ c. $F_{0.99,7,5} = \dfrac{1}{F_{0.01,5,7}} = \dfrac{1}{7.46} = 0.13$

d. $F_{0.99,4,12} = \dfrac{1}{F_{0.01,12,4}} = \dfrac{1}{14.37} = 0.07$.

13.17 $H_0 : \beta_1 = 0; H_1 : \beta_1 \neq 0. R = \{F | F > F_{\alpha,1,n-2} = F_{0.05,1,3} = 10.13\}.$
$F = \text{MSR/MSE} = 11.05$. Reject H_0.

13.18 $H_0 : \beta_1 = 0; H_1 : \beta_1 \neq 0. \quad R = \{F | F > F_{\alpha,1,n-2} = F_{0.01,1,5} = 16.26\}.$
$F = \text{MSR/MSE} = 298.57$. Reject H_0.

13.19 $H_0 : \beta_0 = \beta_1 = 0; H_1 : \text{not } H_0. R = \{F | F > F_{\alpha,2,n-2} = F_{0.05,2,28} = 3.34\}.$

$$F = \frac{\left[(30)(1)^2 + (-3)(-30)\right]/2}{1.58} = 37.97. \text{ Reject } H_0.$$

REFERENCES

Anderson, J., Brown, R., *Risk and Insurance P-21-05*, 2nd printing: Society of Actuaries, (2005).

Aristotle, Metaphysics, trans. by W. D. Ross, in *The Basic Works of Aristotle* (Richard M., ed.). New York: Random House, (1941).

Aristotle, *Posterior Analytics*, trans. by E. S. Bouchier in *Aristotle's Posterior Analytics*. Oxford: Blackwell, (1901).

Austin, P. C., A Critical Appraisal of Propensity—Score Matching in the Medical Literature from 1996 to 2003. *Statistics in Medicine* **27**: 2037–2049, (2008).

Beltrami, E., *What is Random?* New York: Springer-Verlag, Inc., (1999).

Blom, G., *Statistical Estimates and Transformed Beta-Variables*. New York: John Wiley and Sons, (1958).

Cole, P., Causality in Epidemiology, Health Policy, and Law. *Journal of Marketing Research* **27**: 10279–10285, (1997).

Collier, D., Sekhon, J., Stark, J., *Statistical Models and Causal Inference*. New York: Cambridge University Press, (2010).

Croucher, John S., *Cause and Effect in Law, Medicine, Business, and Other Fields*. Review of Business Research **8**(6): 67–71, (2008).

David, Philip, Seeing and Doing: The Pearlian Synthesis. In: *Heuristics, Probability, and Causality* (Dechter, R., Geffner, H. Halpern, J., eds.). New York: Cambridge University Press, (2010).

Dechter, R., Geffner, H., Halpern, J., eds., *Heuristics, Probability, and Causality*. London: College Publications, (2010).

Finkelstein, M. O., Levin, B., *Statistics for Lawyers*. New York: Springer-Verlag, (1990).

Statistical Inference: A Short Course, First Edition. Michael J. Panik.

Freedman, D., Statistical Models and Shoe Leather. *Sociological Methodology* **21**: 291–313 (Marsden, P. V., ed.). Washington, D.C.: American Sociological Association, (1991).

Freedman, D., From Association to Causation: Some Remarks on the History of Statistics. *Statistical Science* **14**: 243–258, (1999).

Freedman, D., *Statistical Models and Causal Inference*. (Collier, D., Sekhon, J., Stark, J., eds.). New York: Cambridge University Press, (2010).

Freedman, D., Stark, P. B., The Swine Flu Vaccine and Guillian-Barre Syndrome: A Case Study in Relative Risk and Specific Causation. *Evaluation Review* **23**: 619–647, (1999).

Glymour, M. M., Greenland, S., Causal Diagrams. *In*: *Modern Epidemiology* (Rothman, K. J., Greenland, S. Lash, T. L., eds.). Philadelphia: Lippincott, Williams & Wilkins, (2008).

Godfrey-Smith, P., *Theory and Reality: An Introduction to the Philosophy of Science*. Chicago: University of Chicago Press, (2003).

Greenland, S., *Confounding*. In: *Encyclopedia of Biostatistics* (Armitage, P., Colton, T., eds.). New York: John Wiley & Sons, (1998).

Greenland, S., Brumback, B., An Overview of Relations Among Causal Modelling Methods. *International Journal of Epidemiology* **31**: 1030–1037, (2002).

Holland, P. M., Statistics and Causal Inference (with discussion). *Journal of the American Statistical Association* **81**: 945–971, (1986).

Hull, T. E., Dobell, A. R., Random Number Generators. *SIAM Review* **4**: 230–254, (1962).

Humphreys, P., *The Chances of Explanation: Causal Explanations in the Social, Medical, and Physical Sciences*. Princeton: Princeton University Press, (1997).

Humphreys, P., *Extending Ourselves: Computational Science, Empiricism, and Scientific Method*. New York: Oxford University Press, (2004).

Knight, F., *Risk, Uncertainty, and Profit*. New York: Harper, (1921).

Lorenz, E., *Predictability: Does the Flap of a Butterfly's Wings in Brazil Set Off a Tornado in Texas?* Address given at the American Assn. for the Advancement of Science, (1972).

Morgan, S., Winship, C., *Counterfactuals and Causal Inference*, 7th printing. New York: Cambridge University Press, (2010).

Pearl, J., Embracing Causality in Formal Reasoning. *Artificial Intelligence* **35**(2): 259–271, (1988).

Pearl, J., *Causality*, 2nd ed., New York: Cambridge University Press, (2009).

Psillos, S., *Scientific Realism: How Science Tracks Truth*. London: Routledge, (1999).

Rao, C. R., *Statistics and Truth*, 2nd ed., London: World Scientific Publishing Co. (1997).

Rubin, D. B., Estimating Causal Effects of Treatments in Randomized and Non-randomized Studies. *Journal of Education Psychology* **66**: 688–701, (1974).

Rubin, D. B., Assignment of Treatment Groups on the Basis of a Covariate. *Journal of Educations Statistics* **2**: 1–26, (1977).

Rubin, D. B., Bayesian Inference for Causal Effects: The Role of Randomization. *The Annals of Statistics* **6**: 34–58, (1978).

Rubin, D. B., *Discussion of Randomization Analysis of Experimental Data: The Fisher Randomization Test*, by D. Basu, *Journal of the American Statistical Association* **75**: 591–593, (1980).

Rubin, D. B., The Design Versus the Analysis of Observational Studies for Causal Effects: Parallels with the Design of Randomized Trials. *Statistics in Medicine* **26**: 20–36, (2007).

Snow, J. (1855), *On the Mode of Communication of Cholera*, Reprinted ed. New York: Hafner, (1965).

Tsonis, A., *Randomnicity*. London: Imperial College Press, (2008).

INDEX

Statistical Inference: A Short Course, First Edition. Michael J. Panik.